T0134990

Mechatronics by Bond Graphs

Vjekoslav Damić · John Montgomery

Mechatronics by Bond Graphs

An Object-Oriented Approach to Modelling and Simulation

Second Edition

 Springer

Vjekoslav Damić
University of Dubrovnik
Dubrovnik
Croatia

John Montgomery
formerly The Nottingham Trent University
Nottingham
Great Britain

and

Pulheim-Brauweiler
Germany

ISBN 978-3-662-56967-2   ISBN 978-3-662-49004-4   (eBook)
DOI 10.1007/978-3-662-49004-4

Printed on acid-free paper

This Springer imprint is published by SpringerNature
The registered company is Springer-Verlag GmbH Berlin Heidelberg

I will have my bond.

The Merchant of Venice (Wm. Shakespeare)

To Mira, Renata, and Dražen

—Vjekoslav Damić

To Helga, Lorna, and Stuart

—John Montgomery

Preface to the Second Edition

The present edition continues with the approach to bond graph modelling used in the first edition. There are also some improvements that are mostly the result of developments of BondSim program taking place over the last more than ten years. Its appearance is now quite different, and we hope more user-friendly. There were some other important changes as well. One is the use of .NET technology for efficient model solving during the simulation phase.

Also there is a range of new modelling components, in particular for digital signal processing, which enable modelling and simulation of mechatronics systems in its entity including the embedded digital signal processing. This was illustrated on an example of Coriolis mass flowmeter in Chap. 10. There is also support for 3D visualization and inter-process communications. It is now possible to visualize motion of complex mechanical system in space and their interactions with its surrounding.

The version of the modelling and simulation program environment that was used in the book is BondSim 2014. It is freely available to the readers. We recommend the readers to download it and use while reading the book. We also encourage the reads to try to solve their own problems using the approaches described in this book.

We would also like to thank Dr. Christoph Baumann of Springer-Verlag, Heidelberg, for his help, kindness, and patience during the preparation of the manuscript. We are also grateful to Ms. Petra Jantzen and Ms. Carmen Wolf for their help.

Dubrovnik   Vjekoslav Damić
June 2015   John Montgomery

Preface to the First Edition

A short history of this book

This book had its origins in the authors' common interest in modelling and simulating dynamic engineering systems, especially those related to *mechatronics*. These interests date from the early 1970s.

We well remember—even somewhat nostalgically—our experiences with one of the first digital computer simulation tools that became available: IBM's Continuous System Simulation Program (CMSP) for IBM 1130 computers. C.W. Gear's famous DIFSUB code for solving stiff differential equations—the forerunner of modern differential-algebraic equation solvers—also appeared around the same time. Then, in 1975, Karnopp and Rosenberg's classic book, *System Dynamics: A Unified Approach*, was published. It introduced a system analysis methodology based on bond graphs. We loved it from the start for it laid a solid foundation for the development of a systematic approach to modelling complex mechatronics systems. We were also aware of developments in the field of electronic circuit modelling that led to the famous Berkley's SPICE program.

The difficulties posed by solving real-world design problems motivated the first author to begin development of a methodology for computer-aided modelling and simulation of engineering—particularly mechatronics—systems. It was targeted to developing a methodology that supports systematic model development by decomposition. Bond graphs were taken as modelling formalism, because they are well suited to modelling different physical processes taking place in a typical mechatronic system. It was expanded, however, by developing the concept of bond graph word models into complete component models. More attention was given to component ports as interfaces of the components. The ports are treated as objects in themselves that enable representation of the complex interconnections inside the components. This way, a model of a system can be built as a complex multilevel structure, in a form that mimics how a real system is built. The component can be reused as well to build the models.

Another departure from classical bond graphs and the Continuous System Simulation Language (CSSL) philosophy was made by putting aside the causality issues. Strict input–output relationships in the models are not supported. Thus, instead of mathematical models in state-space equation form, differential-algebraic equation models are used. This enables separation of modelling and model solving tasks. We believe that, taken together, this extends the applicability of methods to solving real engineering problems.

The first implementation of this methodology was made in the beginning of the 1980s with the release of *Simulex*. This program was implemented using FORTRAN and run on Digital VAX-750 computers. *Simulex* models were described with SPICE-like scripts. The resulting equations were solved with a version of Gear's DIFSUB. *Simulex* was applied successfully to a range of practical problems in servo-systems and robotics.

The revolutionary appearance of PCs in the mid-1980s, followed by development of operating systems that supported user-friendly visual interfaces in the 1990s, spurred the next phase of development. This was also influenced by the paradigm shift in programming languages: The truly object-oriented languages were replacing the procedural languages—such as FORTRAN and C—that we had all been using. Another important technological development became available around the same time—symbolic computational algebra.

In the beginning of the 1990s, the shift to object-oriented modelling paradigm was made. Class hierarchies were developed that enabled representing component models as objects. Also, computational algebra methods were developed that, as explained in the present book, simplified some important user interface problems and the solution of model equations. Methods for solving differential-algebraic equations were further developed to support model solving during simulation. These all were implemented in a visual modelling and simulation program, *BondSim*, the first version of which appeared in the mid-1990s. It fully automated many important operations. Thus, there was no need for the developer to use any traditional programming; rather, models were developed and solved simply by mouse clicks.

In 1995, the authors met at The Nottingham Trent University and started collaborative work on *Dynamic System Simulations Using Bond Graphs*, a project funded partially through an ALIS (Academic Links and Interchange Scheme) award (1995–1998), sponsored jointly by the Croatian Ministry of Science and Technology and the British Council. This cooperation continued *via* e-mail and reciprocal visits to Nottingham (England) and Dubrovnik (Croatia). One result of this fruitful joint work is this book that we here offer to the reader.

What is this book about?

The title suggests that this book is about mechatronics; this is, indeed, one of its central themes. It is not, however, another book on what mechatronics *is*; rather, it is about how mechatronic problems can be solved by a systematic approach employing bond graphs. *Why bond graphs?* Because they offer an efficient means of modelling interdisciplinary problems, such as those commonly found in mechatronics. (The book, by the way, assumes no previous experience with bond graphs, though it certainly would be useful.)

The book shows, in step-by-step fashion, how models are developed systematically and then simulated in a way that permits thorough analysis of the problem under study. Every chapter that deals with an engineering application starts with the exposition and solution of a simple problem relevant to that chapter. Then, the solution of related—though much more difficult—problems is explained.

The book is divided into two parts: *Fundamentals* and *Applications*.

Part I, Fundamentals, consists of five chapters on bond graph modelling. It starts with an introduction to the subject and then proceeds with describing a systematic object-oriented approach to modelling; implementation of object-oriented modelling in a visual environment; and the numerical and symbolic solution of the underlying model equations.

Part II, Applications, consists of five chapters that apply bond graphs and component model techniques to mechanical systems, electrical systems, control systems, multibody dynamics, and continuous systems. Great attention is given to modelling electrical components and systems, including semiconductors. The same holds for multibody systems, both rigid and deformable, such as found in various mechanisms and robots.

What readers can gain from the book?

There are several ways in which this book can be used, depending mainly upon the background and interests of the reader.

Researchers in mechatronics and micro-mechanics design, for example, can use it to find out how difficult problems in their disciplines can be solved using a combination of bond graphs and component model techniques.

For the reader interested in simulation technology, the book provides an introductory description of the object-oriented visual approach to modelling and simulation.

The reader whose background is in one of the applied disciplines covered herein can gain valuable insight into how bond graphs may be used to solve problems particular to his area of interest.

We also think that the book can be useful as a textbook, or as a supplementary text, in courses on physical modelling of engineering systems in general. We believe that it can help students learn the system way of solving a problem in electrical and mechanical engineering, as well as coupled problems that span disciplines.

Finally, it is our sincere wish that the text and software will aid the reader in his work. We invite, and will appreciate, all constructive feedback.

BondSim Research Pack

A special version of BondSim—*BondSim Research Pack* (beta version)—is bundled with this book. It provides a visual development environment for the modelling and simulation of engineering and mechatronics systems based on bond graphs. The problems presented in the book are solved using the *BondSim Research Pack*.

It runs on the Windows 2000 Professional operating system, but can be used on other Windows platforms, too. The reader can use this version of *BondSim* to analyse all of the problems presented in the book. (These are found in *BondSim*'s program library.) The projects that a reader might develop on his or her own are somewhat more restricted. The interested reader can order the complete version of *BondSim* from the first author. (*See* Appendix for details.)

Acknowledgements

A number of people have reviewed the initial outline (and the drafts) of this book. We are most grateful to them for their time and expertise.

Our special thanks go to the following people and institutions:

The Polytechnic of Dubrovnik (*Veleučilište u Dubrovniku—Collegium Ragusinum,* now University of Dubrovnik) for facilities provided to both authors. We are especially grateful to the rector, Professor Dr. Mateo Milković, for his encouragement and support.

Vlado Jaram, Mr. Sc., for initiating the whole publishing project and his help on getting the book published, as well as on his suggestions during writing the book.

Professor Barry Hull of the Department of Mechanical Engineering of the Nottingham Trent University for his support.

The Croatian Ministry of Science and Technology and the British Council for the funds provided through the ALIS award.

Dr. Nick Staresinic, EcoMar Mariculture, for his careful reading of the manuscript and his helpful editing suggestions.

We would also like to thank Dr. Dieter Merkel of Springer-Verlag, Heidelberg, for his help, kindness, and patience during the preparation of the manuscript. We are also grateful to Ms. Petra Jantzen and Ms. Gaby Mass for their help.

And last, but in no way least, we wish to express our deep appreciation and love to our wives—Mira and Helga—for their love, support, patience, and sacrifice during the long period over which this book was produced.

Dubrovnik   Vjekoslav Damić
June 2002   John Montgomery

Contents

Part I
Fundamentals

Chapter 1
Basic Forms of Model Representation

1.1 Objectives

The solution of complex, real-world problems is based on modelling. A model simplifies the system of interest by abstracting some subset of its observable attributes. This focuses attention on those features of the system relevant to the problem of interest, and excludes others deemed not to be of direct relevance to the problem. The level of detail included in a model thus depends on the problem to be solved—as well as on the problem solver. Based on such an idealised picture, the system is described in a suitable form that is used as a basis for deriving a solution (Fig. 1.1).

After obtaining a solution, results are interpreted with respect to the real-world context of the original system. Thus, as well as being able to create a valid model of the system and to solve it, it is of great importance to represent the solution in a form that can be understood readily and communicated.

The traditional modelling approach used in engineering is mathematical. That is, real-world physical processes are described by mathematical relationships that are solved using suitable analytical or numerical techniques. As real engineering systems are very complex, it is not an easy task to create a valid model and solve it. An added practical consideration is that problems must be solvable efficiently in terms of resources and time. Advances in computer technology have dramatically improved the solvability; it is now possible to solve problems that formerly were intractable.

Solving problems with a computer means that a problem posed in one physical domain is solved in another physical domain, the computer domain. This naturally leads to the topic of simulation modelling.

Simulation models mimic the behaviour of engineering processes in their environment. By experimenting on models of equipment instead of on real equipment, the system's behaviour can be studied even before the hardware is built. Simulation

© Springer-Verlag Berlin Heidelberg 2015
V. Damić and J. Montgomery, *Mechatronics by Bond Graphs*,
DOI 10.1007/978-3-662-49004-4_1

Fig. 1.1 Model approach to problem solution

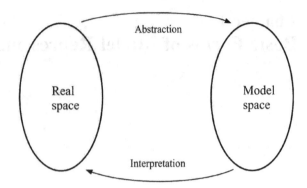

models can be used at various stages of design, from the early stages of conceptual design to final prototype testing. There are many fields in which this technique has been applied profitably.

The fundamental question that naturally arises is: *How are such models best constructed*? A well-known adage suggests that modelling is more art than science. It is, in fact, a bit of both. On the "scientific" side, there are a number of approaches, methods, and tools that can be mastered, and then applied, to develop effective models; the "art", perhaps, is the insight that a modeller accumulates through practice and familiarity with the system being studied.

This chapter reviews some of the more promising approaches and methods at the foundation of modelling. The perspective adopted is motivated mainly by problems in mechatronics. There are many definitions of mechatronics.[1] One that we choose states that *mechatronics is a synergistic combination of precision mechanics, electronic control, and system thinking*.

Perhaps the best-known examples of mechatronic design are found in robotics. There are also other, no-less-important applications of this design philosophy.

In today's highly competitive and demanding development environment, classical solutions without embedded microprocessors have little chance of success. Modelling and simulation play an even greater role in product design. To promote efficient solutions, the computer modelling and simulation environments should relieve designers of many routine, low-level tasks [1], as well as support collaborative work.

---

[1]In www.engr.colostste.edu/∼dga/mechatronics/definitions.html definitions of Mechatronics are collected from different resources.

1.2   The General Modelling Approach

The concept of system plays a central role in model building. An efficient model need not embrace the entire universe to design just a part of it. This is not only an impossible task, but also an unnecessary one. We thus pay attention only to that part of the problem in which we are interested. This is termed the system for the given problem. Everything not included in the system constitutes its environment (Fig. 1.2).

The system might consist of the engineering equipment that is the subject of the problem, but it can include other parts, as well. In this system-centred approach it is tacitly assumed that the environment determines the behaviour of the system. Thus, the environment influences the system and can change its behaviour.

It is, of course, also of interest to study the influence of the system on its environment, e.g. the current drawn by a system from an external source. In the case in which the system can change its environment in a way that there is a 'backwards' influence on its own behaviour, the system should, in most cases, be enlarged to include this part of its environment.

It is often useful to decompose a system into components. For example, the simple actuator illustrated in Fig. 1.3 consists of an electric motor driven by a controller. The motor shaft is connected to a nut. The shaft rotation is transformed by the nut into a translation of an actuator shaft, which moves a load. The position information of the load is fed back via a sensor to the controller.

Every part of such a system, i.e. the electronic drive unit, motor, shaft, etc., may be modelled as a separate component. The complete model of the drive thus may be depicted as a system of interconnected components.

Decomposition of the complete system into its components generally simplifies the modelling task and gives a sharper insight into the system's structure. Such a representation is a great help in interpreting model behaviour in terms of the real engineering system.

System decomposition can proceed to ever-lower levels—essentially treating each component as another system that, in turn, consists of even simpler components. At some point a level of detail will be reached at which the components may be considered as elementary, i.e. not admitting any further useful decomposition.

Fig. 1.2 System and its environment

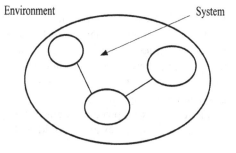

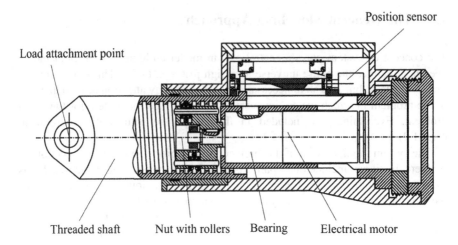

Fig. 1.3 A simple electro-mechanical actuator

Such elementary components are modelled as entities and define the limit of detail of the model in question.

It should perhaps be pointed out again that a model is an abstraction of the real world: It is not necessary—or even possible—that the structure of the model represents the original physical system in all of its complexity. Model development, however, generally is an iterative process: Additional details may be added as the model matures, or expands to address additional problems.

Decomposing an engineering system into components also suggests a natural decomposition of tasks among members of a modelling team. Each group might be assigned development of one component. The overall model can then be built up by combining the separate sub-models.

Yet another advantage of this approach is that the component models can be reused, e.g. components developed for one particular application might serve as building blocks for another, unrelated, application.

1.3 Physical Modelling, Analogies, and Bond Graphs

Both "top-down" decomposition and "bottom-up" composition are powerful modelling techniques. To use their full power requires describing those components treated as entities (elementary components), and modelling the interactions between them. In engineering, these considerations are based on physical reasoning derived from the various branches of physics. Such an approach is sometimes termed physical modelling [2].

Processes taking place in engineering systems thus may be classified generally as belonging to, for example, rigid and solid body mechanics, fluid mechanics,

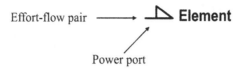

Fig. 1.4 Generic bond graph element

electricity and magnetism, semiconductor physics, thermodynamics, and so forth.
Each of these branches has its particular methodology for solving problems. Thus,
if the problem in question deals with a single physical domain, it is natural to apply
the methodology of the field in question, including any specialised computational
methods that may be available. This same approach can be applied even in
multi-domain problems if the interactions between domains are weak; but this is
rarely the case in engineering in general and in the field of mechatronics in particular. We thus must cope with interacting, multi-domain physical processes.

One well-known approach designed to deal with multi-domain engineering
problems is the bond graph method elucidated by Henry Paynter.[2] He presented this
methodology for the first time in the lecture "Ports, Energy, and Thermodynamic
Systems" delivered on April 24, 1959, at the Massachusetts Institute of
Technology. This work later was published [3].

The application of Paynter's bond graph method began with the works of
Karnopp, Rosenberg, Thoma, and others [4–13]. Over the last 40 years there have
been many publications dealing with the theory and application of bond graphs in
different branches of engineering.[3]

The method uses the *effort-flow* analogy to describe physical processes [7, 10].
These processes are represented graphically in the form of elementary components
(bond graph elements) with one or more ports (Fig. 1.4). The ports represent places
where interactions with other processes take place.

The process "seen" at a port is described by a pair of variables, *effort* and *flow*.
These are termed power variables, and their product is power. Through every port
there is flow of power, either in or out of the component. The direction of power
flow is depicted by a *half-arrow*.

In addition to the power variables, there also are internal variables that represent
the accumulations of effort and flow over time. These variables are called *generalized momenta* and *generalized displacements*, respectively.

The typical association of Bond Graph variables in various domains is given in
Table 1.1. It should be noted that thermal effort and flow variables, as defined in the
last row of the table, are not power variables because their product is not power.

---

[2]Interested readers can review web page www.me.utexas.edu/~longoria/paynter/hmp/index.html
for more information.

[3]More information on the Bond graph method can be found in The Bond Graph Compendium held
at http://www.ece.arizona.edu/~cellier/bg.html.

Table 1.1 Bond graph variables

Domain	Effort	Flow	Momentum	Displacement
Mechanical translation	Force	Velocity	Momentum	Displacement
Mechanical rotation	Torque	Angular velocity	Angular momentum	Angle
Electrical	Voltage	Current	Flux linkage	Charge
Hydraulic	Pressure	Volume flow rate	Pressure momentum	Volume
Thermal	Temperature	Heat flow	–	Heat energy

Bond graphs corresponding to variables having this property are usually termed *pseudo-bond graphs* [10, 13].

All physical processes are described using several elementary components, or elements:

- Sources of effort and of flow (denoted as SE and SF respectively),
- Accumulation of effort and of flow (I and C respectively),
- Dissipation of power (R),
- Transformers of power (Transformers and Gyrators) (TF and GY), and
- Branches of efforts and flows (denoted as 1 and 0 respectively)

The processes that these components represent are described by constitutive relations expressed in terms of port and internal variables (generalized variables).

The ports of components are joined with a line. These lines are termed bond lines, or bonds, for short. They imply that the power variables at connected ports are equal. The graph that results is a bond graph model of the component.

A simple mechanical system consisting of a body, a spring, and a damper provides a useful illustration (Fig. 1.5a). The corresponding bond graph model is shown in Fig. 1.5b.

The source effort SE represents an applied external force, component I represents the inertia of the body, component C describes the elasticity of the spring, and R represents friction in the damper. Branching element 1 denotes the summation of all

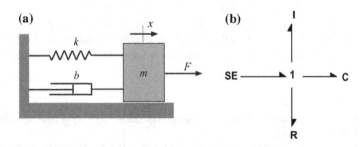

Fig. 1.5 A simple mechanical system. **a** Scheme. **b** Bond graph model

forces acting on the body. This diagram, taken together with the corresponding component constitutive relations, completely defines the mathematical model of the system; it can thus be used as a basis for simulating the system. The bond graph also shows the structure of the model in a way that resembles the structure of the real system. This proves useful in efficiently communicating details of the model to interested parties outside of the modelling team.

There is also another analogy, introduced by Firestone [7, 14] and based on *across*- and *through*-variables, that can be used. A variable defined at a point in space with respect to another point in space is termed an *across* variable. For example, velocity, voltage, pressure, and temperature all may be interpreted as across-variables. On the other hand, a variable defined at a single point without respect to any other point is termed a *through* variable. Examples of through-variables include force, current, and fluid flow. The across- and through-variable analogy naturally leads to representation of a model in terms of linear graphs.

There is no general agreement on which analogy is preferable. The effort-flow analogy corresponds to the force-voltage electromechanical analogy; and the across-through analogy corresponds to the force-current electromechanical analogy. We use the effort-flow analogy, as it perhaps better explains *efforts* as *intensities* and *flows* as *extensities*.

It should be stressed that system decomposition combined with the bond graph modelling method readily leads to a lumped-parameter model. For the case in which variables inside a component change continuously over some region of space, it is necessary to apply discretization; that is, to represent the model by a finite number of components. This can be done in various ways, such as using the well-known finite-element discretization method.

In spite of the attention this approach has attracted, the bond graph method has not received the widespread acceptance expected by its proponents. In the opinion of the authors, one of the drawbacks of the classical bond graph modelling technique is its "flat" structure. That is, the model is constrained to be represented as a single-level structure. This leads to quite complicated diagrams even for relatively simple systems, and these can be difficult to interpret. One remedy for this is to pay more attention to the modelling of components in general, as well as to their ports. This enables more systematic model development.

There is another important concept embedded in bond graph theory. This is the concept of *causality*. This refers to cause (input) and effect (output) relationships [10].

Thus, as part of the bond graph modelling process, a causality assignment is implicitly introduced. This leads to the description of bond graphs in the form of state-space equations. The problem that arises is that such a model is restrictive. Furthermore, as pointed out in [15], there is no true notion of causality in physical laws. For example, there are no purely physical reasons to interpret a force on a body as the cause of its motion; or to interpret a voltage on electrical terminals of a component as the cause of the current flowing through it. Thus, causality will not be part of our focus. The point of view taken is that modelling can—and should—be treated as separate from the mathematical model developed and the solution derived thereof. Used in this way, and together with the general modelling technique mentioned in the immediately previous section, bond graphs can be used as a powerful *visual modelling language*.

1.4 Block Diagrams

Block diagrams are often used to denote input-output relations (Fig. 1.6). They have been used traditionally in control engineering, but also have found popular application in other fields, such as computer science, economics, and ecology.

Block diagrams depict operations on signals (information). The symbol inside the block represents a process applied to an input signal to generate an output. By connecting the output of one block to the input of another block, we can illustrate a procedure for the calculation of some quantity in which we are interested.

This approach can be used for modelling and simulating systems, too [2]. As an illustration, Fig. 1.7 shows a block diagram model of the simple mechanical system introduced earlier (Fig. 1.5a). The block diagram shows how to evaluate model variables given the time-history of the applied force and values of position and velocity of the body at the start of the motion.

Many simulation programs are based on a modelling approach of this type (e.g. Matlab-SIMULINK). Unfortunately, many practical systems cannot be treated this way, except at some simplified conceptual level. In addition, modelling the system by block diagrams can be complicated and error prone, and also more difficult to

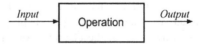

Fig. 1.6 Block diagram notion

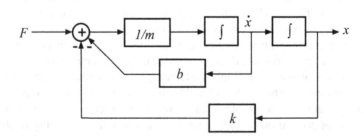

Fig. 1.7 Block diagram model of simple mechanical system

interpret. This technique, however, can be useful as a complement to a more general modelling method, such as bond graphs, and is used at certain points of the present work, too.

Signals can be used for monitoring processes, describing control actions, and processing outputs. In the bond graph approach, signals are termed activated bonds, and there is no power associated with the transfer of signals (information only). We allow bond graph components to have ports for the input or output of the signals as well. Such ports are depicted by a *full-arrow* and are called the *control ports*.

1.5 Symbolic Model Solving

Solution of a mathematical equation representing a system is usually accomplished by applying a suitable analytical procedure or, much more commonly, through some numerical routine. Developments in software engineering, however, have opened up the possibility of obtaining certain solutions symbolically. This approach relies on the tools of computational algebra.

Computational algebra software manipulates mathematical expressions symbolically [16]. There is currently a number of general-purpose symbol-manipulation applications available that can be applied to a wide range of practical mathematical problems. Some of the better-known packages are REDUCE, MAPLE, MATHEMATICA, and AXIOM. In the field of simulation, the best-known systems are MATHCAD and MATLAB.[4]

Solving problems symbolically requires processing power and memory resources usually well beyond the capability of most PCs and workstations. It thus becomes preferable to combine symbolical and numerical approaches to the practical problems that arise in engineering systems [1, 17, 18]. Thus, symbolic manipulation can be used to generate mathematical expressions that subsequently may be solved with numerical or analytical techniques. In this way, modelling of the system (the symbolic description) can be separated from its solution (the numeric result).

Once a model based solely on the physical considerations and requirements of the problem has been developed, the equations can be generated automatically in symbolic form. These equations can be simplified, either during the model's generation or after it. For example, constant expressions may be evaluated, trivial equations eliminated, or other actions taken to simplify the solution.

As the next step, the equations are prepared for numerical solution. This can entail a variety of operations, such as generation of functions that must be evaluated at run time, generation of the symbolic Jacobian matrix, and generation of any supplementary relationships necessary at start-up or for simulating across

---

[4]More information on symbol manipulations can be found on the Symbolic Mathematical Computation Information Center page at http://www.symbolicnet.mcs.kent.edu.

discontinuities. Symbolic manipulation also can be used for post-processing and the control of the complete problem-solving task.

This procedure can be accomplished without using the complete machinery of computational algebra. Simple arithmetical and logical operations plus symbolic differentiation is usually sufficient. What is really important is that the computational algebra should be seamlessly integrated with the other parts of the modelling and simulation environment.

It should also be pointed out that another technique for the evaluation of differentials has attracted much attention: automatic differentiation [19]. Automatic differentiation is a numerical technique for the evaluation of a differential based on an algorithmic approach using the rules of differentiation. This technique is often confused with symbol manipulation. It is usually argued that this technique is superior in terms of efficiency and memory usage; but symbolic manipulation is constantly being improved and can be applied to problems other than those that require the evaluation of a differential.

The approach taken here is based partly on the symbolic manipulation technique described in [20]. The constitutive relations for modelling the elements described in the last section are held in symbolic form, together with parameter expressions and model structure data. Before the start of a simulation, the model is assembled in the form of byte codes that can be evaluated efficiently during simulation. The implementation of the method may be based on a combination of the interpretative and compiled approaches. A very powerful technique that can be also used is based on .NET assembly technology. Thus, it is not necessary to recompile the model and relink it.

1.6   The Object-Oriented Approach

The object-oriented paradigm represents a major achievement in software engineering that facilitates modelling complex real-world problems [17, 21]. When properly applied, it yields robust models consisting of reusable, easy-to-maintain components. Only a very brief introduction is provided here. More information on object-oriented programming (OOP) can be found in [22, 23].

The focus of this approach is an object. An object is an abstraction of reality described by *attributes* and *methods* (Fig. 1.8).

Attributes define an object's state and can be represented as variables of the fundamental types (such as integers, Booleans, characters, or real numbers), other objects, or collections of various types.

Methods describe an object's behaviour. From a related perspective, methods represent the services that an object provides.

Both attributes and methods are members of an object. Attributes sometimes are referred to as data-members, and methods as member-functions. An object is said to encapsulate its members.

Fig. 1.8 Object attributes and
methods

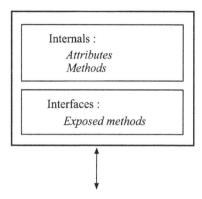

One of the key concepts of the OO approach is information hiding. This means that an object exposes only those of its members that are required for use by other objects; all other attributes and methods are "hidden" within the object, inaccessible to the outside world.

To enable interaction with other objects in its environment, an object must provide *an interface*. An object interacts with its environment only via its interfaces. The interfaces expose the access methods.

Most real-world objects cannot be modelled adequately with only a single fundamental data type. The OO paradigm thus introduces the concept of a *class*. A class is a generalised data type that defines the attributes and methods shared by all objects of that class. The terms *object* and *class* are sometimes (improperly) used interchangeably; strictly, an object is an instance of a class. In practice, an object is created in the computer memory (constructed) at run time using the class definition as a template; and removed from memory (destructed) when no longer required.

Yet another fundamental principle of the object-oriented approach *is inheritance*. The inheritance mechanism permits classes to be organised in a logical hierarchy that describes their interrelationships. Thus, a class derived from an existing class inherits the members of its parent. The derived class may define additional attributes and functionality, as well as modify those inherited from its parent.

Relative to the derived (child) class, the parent class is a generalization, a super class, or, in the terminology of C++, a base class.

Viewed from the complementary perspective, the derived class is a specialization of the parent class; it is a kind of the base class. At the head of the hierarchy there is usually a class that does nothing more than define the interfaces of all lower level classes in a consistent way. Such a class is an *abstract base class*. We treat the elementary component of Fig. 1.4 as a kind of general component, which likewise is a kind of more abstract object class (Fig. 1.9).

Another OO principle is *polymorphism*. Polymorphism enables methods with the same name to exhibit different functionalities. For example, methods in the lineage of a class hierarchy can have the same function header—that is, the same name, argument list, and return type—but different behaviour. Such methods are

Fig. 1.9 Component class
hierarchy

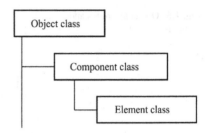

declared virtual in the base class. The particular method invoked by a function call
depends on the object's class and is implemented through a mechanism termed
dynamic binding. The decision on which method to call thus is made at run time.

A consequence of polymorphism is also that two methods may share the same
name, but have different argument (parameter) types and return types. The specific
method executed depends on the type of the arguments passed by the calling
function. This is an example of the so-called function overloading. In contrast to the
previous example, these functions are not virtual: They have a unique signature
(parameter list), so the decision on which version of an over-loaded function to call
is decided at compile time.

The general modelling approach set out in Sect. 1.2 cannot be supported using
class hierarchies alone, as components generally contain other components; this is
how real devices are built. To represent a model of a component we use two
objects: a *component* and its accompanying *document*.

A component is represented by a component object. Such an object is compounded and contains port objects that serve as interfaces to other components
(Fig. 1.10, component `Platform`). If the component is simple and hence doesn't
contain other components, then it is just an elementary component. But, if it does
contain other components, we use another object to define its internal structure.
Such an object is not a kind of any other component, but a separate entity—a
document (Fig. 1.10, bottom-right box).

The document object has external ports that correspond exactly to the component object's ports and provide access to the component internals. In this way, the
component model representation is based on two associated classes—the component class and the document class. These classes are designed in a way that supports
access to the document through its accompanying component object, or by its ports.

This is very close to the way in which we deal with real components. That is, to
see what is inside a component, we first have to find a component by its name or
manufacturer's designation, or maybe how it is connected to other components;
only then we can "open" it to look at what is inside. The component is usually
contained in another component, this again in yet another component, etc. The set
of all components constitutes the system. In this way a model of a system can be
represented as a tree of components.

Elementary components are treated somewhat differently, as they do not contain
other components. Ports of such components provide access to the component

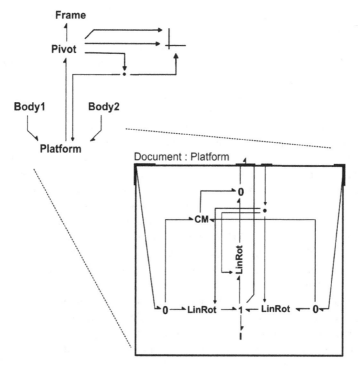

Fig. 1.10 Component model

constitutive relations that describe the mathematical model of the particular elementary component (Sect. 1.3). Thus, the elementary components can be looked at as leaves of the model tree.

1.7   Computer Aided Modelling

One of the first computer programs developed for modelling and simulation of practical engineering systems was SPICE [24], which was developed at the University of California, Berkeley at the beginning of the 1970s for integrated electronic circuits. The program was accepted quickly by leading semiconductor manufacturers. Since that time, SPICE has undergone continuous development to follow technological advances in semiconductors. Today, it is the de facto standard for electrical simulations. All of the main semiconductor manufacturers offer SPICE models of their components.

The success of SPICE—and of similar products that emerged shortly afterwards—was owed partly to the fact that, in electrical engineering, and particularly in electronics, systems and devices normally are modelled using electrical schemes. SPICE uses an input file that contains a description of electrical schemes using a simple textual language.

Another successful modelling and simulation system in electrical engineering is SABER.[5] It uses the MAST modelling language. Recently, under the umbrella of the IEEE, a new modelling language has been developed: VHDL-AMS [25]. This offers a more uniform approach to modelling mixed analogue-digital systems. VHDL-AMS does not assume any causality, generates models in differential-algebraic form, and permits discontinuities.

From the start, bond graphs were looked at as a unified approach to the modelling of general engineering systems, in some ways similar to electrical schematics in electrical engineering. Many programs based on bond graphs have been developed over the last two decades. The first, ENPORT [26], uses a textual description of the bond graph model topology (similar to SPICE) as input. It supports macro capabilities to simplify bond graph model creation. The program was developed for workstations. There is also a PC version, but with somewhat reduced capabilities.

Historically, the next program to appear was TUTSIM, developed at Twente University, Netherlands. This application was originally developed for block diagram models, but was modified to accept causality-augmented bond graphs. The program translates models into state-space form and hence cannot treat models with dependent storage (masses, capacitors etc.) and algebraic loops.

The same research group later developed the CAMAS program [27], which is now known as 20-SIM [28, 29]. This program uses bond graphs for model input and icons for component representation. The icons serve as placeholders for the component models. The models are expressed as equations using the SIDOPS language and are organised as model fragments. For real reusability of the fragments adequate caution must be exercised. Sub-model storage is not yet implemented as a complete database facility. System model processing, from model equations to assignment statements, is performed using computational causality analysis. The program has its own simulator based on routines from NETLIB.[6]

The MS1 program is designed for the modelling and simulation of dynamic systems with continuous elements.[7] Models are developed in the form of causality augmented bond graphs and solved using ACSL, Matlab-SIMULINK, and other programs. Similarly CAMP-G [30] was developed as a pre-processor for ACSL, Matlab-SIMULINK, and other programs, and is suitable for modelling systems that can be represented in state-space form only. There are other bond graph based programs that implement a similar philosophy.

Finally, the Dynamic Modelling Language (Dymola) [17] should be mentioned. Dymola was developed using another philosophy: the through/across physical analogy and the theory of graphs. It supports bond graphs, thus bond graph

---

[5]Originally the SABER was product of Analogy, Inc., Beaverton, USA. Currently it is distributed by Synopsys, Inc., http://www.synopys.com.

[6]The NETLIB is web based public library repository that can be accessed at http://netlib.org.

[7]Lorenz Simulation SA, http://www.lorsim.be.

elements and structures can be coded directly. There is some incompatibility between these two approaches, but this can be circumvented. Dymola helps the user edit and compose programs, manipulate models from sub-models, and generate code for continuous system simulation languages such as ACSL, SIMULINK, and some others. It can also generate code in C or Fortran. This can be executed in a module for the simulation of time-continuous systems, which supports both ordinary differential equations (ODE) and differential-algebraic equations (DAE). Recently, under the auspices of EUROSIM, a new modelling language MODELICA has been developed [31, 32].

This book describes an approach to the development of an integrated automated computer environment for visual model development and experimentation by simulation that is implemented in BondSim.[8] To help the designer during the tedious task of model development, the modelling environment supports the fundamental modelling approach—systematic problem decomposition and model creation at every level of decomposition. The number of levels of decomposition depends on the complexity of the engineering system being modelled and the depth of abstraction. For simple problems, a single level suffices, i.e. a flat model. But real systems, such as are usually found in mechatronics, consist of many components that are themselves constructed of many other components. Thus, two or more levels of model decomposition may be necessary.

The model can be created as a multi-level structure. This helps in understanding the model and in providing for its maintenance. Thus, the model for a problem can be treated as a tree with component models as its branches. The modelling environment also supports building the model from the "pieces", i.e. the component models. A combination of these two approaches is also possible. The environment also has facilities for the creation of suitable libraries in which the system and/or the component models are stored.

The simulation tool, BondSim, supports the collaboration that has become essential in the development of complex models. Model development usually is teamwork, hence the distribution of modelling tasks and the integration of results is welcome, if not an absolute necessity. Also, the exchange of models with colleagues, or use of component models from manufacturers (similar to the SPICE models), is very useful. This aspect of the modelling environment also is implemented in BondSim.

BondSim's general concept of modelling and simulation is illustrated in Fig. 1.11. Development is done entirely visually—without coding—using the support of the Windows system [1]. In this visual modelling approach, the modelling process consists of creating objects in the computer memory that are depicted on the screen as bond graphs or block diagrams. It is also possible to represent them by common electrical and mechanical schemes. The model of a component discussed in the previous section plays a central role in this process.

---

[8]Program BondSim© is available from the author's web page www.bondsimulation.com.

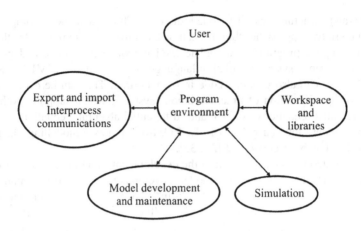

Fig. 1.11 Visual modelling and simulation environment

This approach should not to be confused with Microsoft's Component Object Model (COM) technology. Components used in BondSim are not placeholders for component models, but real objects that serve as interfaces to the document that contains the model. Thus, we can open a component to inspect it in more detail, edit it, and continue with multi-level model development. Such a component can be stored in a component library, or may be sent by e-mail to somebody else. Likewise, a component received by e-mail can be imported into the modelling environment.

The BondSim modelling environment automates many common utility operations, such as saving, loading, copying, deleting, inserting, models export and import, interprocess communications, etc. The constitutive relations specifying the characteristics of the elementary components or block diagram operations are described as simple linear or non-linear algebraic expressions.

After a bond graph model has been developed, its mathematical representation is built by examining the component tree. This results in a model in the form of differential-algebraic equations (DAEs), which then are solved by appropriate numerical routines. Symbolic processing of the model equations plays an important role in the solution.

BondSim also supports generating reports: Practically all of the bond graph diagrams contained in this book were created using BondSim's print-to-file support.[9] The application was also used to solve all of the mechatronics problems presented in this book.

---

[9]The BondSim supports, for documentation purposes, print of screen images into a file in emf (Enhanced Windows Metafile) format, which is supported by main word or graphic programs including MS Word, Corel Draw, etc.

1.8 The Book Summary

This book consists of two parts. The first part—Fundamentals—describes the basics of the design of the visual object oriented environment for the modelling and simulation of general engineering systems, with emphasis on mechatronics systems. This part contains five chapters.

This chapter gives a short overview of the approaches and methods used for model representation. After a short discussion of the objectives, the general modelling approach is described which lies at the root of the modelling philosophy—model development by systematic decomposition. The concepts of system, environment and of component are discussed. Then physical modelling, analogies and bond graphs are described as well as an alternative approach to modelling based on the through/across variables approach. The block diagram approach is described next as a model representation method on its own or in combination with bond graphs. A short introduction to symbolic model solving is also given and it's potential for use in combination with numerical methods. The next important technique described is object-oriented programming with an emphasis on a component-based approach to the modelling of engineering systems. Finally computer aided modelling is described together with the concept of modelling and simulation in a visual environment.

Chapter 2 gives an overview of the bond graph modelling technique. The perspective taken is an extension of conventional bond graphs to multilevel modelling. Starting from the concept of word models, ports, bonds, and power variables, the component model development approach is described. The elementary bond graph components used for modelling of the basic physical processes are defined. Also the components corresponding to basic block diagram operations are introduced, both for continuous-time and discrete-times models. The systematic decomposition approach to bond graph model development is illustrated on examples of mechanical and electrical systems. Finally the notion of causality in bond graphs is discussed.

Chapter 3 deals with the systematic object-oriented approach to modelling. The basic idea is to do the modelling visually, i.e. without any coding, but solely by interacting with the modelling system in a suitably designed visual environment. The concept of the component model is introduced as the basic mechanism for systematic simulation models development. The special class hierarchies are designed to support creation of models for given problems as trees of linked objects in the computer memory. The models are represented visually as bond graphs and are stored as a set of linked files. Underlying physical processes are represented in terms of elementary bond graph components, which constitutive relations are described symbolically using a simple specially designed language. The design of a suitable modelling environment to support the modeller is outlined.

Chapter 4 describes an implementation of the object-oriented approach of Chap. 3. A program BondSim is described that offers a visual environment for the modelling and simulation of engineering and mechatronics systems. The program implements several services that are accessible to a user through a window system.

The two basic services are Modelling and Simulation. The program also supports model database maintenance, library support, as well as collaborative support for models exports and imports.

Chapter 5 closes the first part of the book and describes the methods used for automatically generation of the mathematical model equations and their solution. This is divided into two distinctive phases—the model building and the execution of simulation runs. During the first the mathematical model equations are generated based on the model object tree. The methods for generating the mathematical models implied by the component's structure are described. The model equations are machine generated in the form of differential-algebraic equations (DAEs). During the simulation phase such equations have to be solved. The well-known backward differentiation formula (BDF) is used. A special implementation of the method is described based on the variable coefficient formula. Also the problem of starting values and of discontinuities in the model equations are also discussed. The methods developed depend to a great extent on computational algebra support implemented in the program. This also adds to flexibility in the modelling.

The second part of the book deals with Applications to mechatronics of the bond graph modelling technique. It is divided into Chaps. 6–10.

Chapter 6 deals with simple mechanical problems. The intention is to familiarize the reader with using the BondSim program on relatively simple problems. These problems are also of interest on their own right. Thus, the well-known Body Spring Damper problem is used as the introduction to the bond graph modelling and simulation of mechanical systems. It is also shown that it is possible to visually represent mechanical components by mechanical schemes. After that the effects of dry friction are studied. Next another class of discontinuous problems is studied—impact. Using a simple model of impact the classical problem of a ball bouncing on a vibrating table is studied. The problem is better known for its chaotic behaviour. The chapter ends with a description of a see-saw problem. It is a pendulum, but it can be looked at as a multibody problem. It is shown that such problems—of interest in Mechatronics—can be readily solved by bond graphs.

Chapter 7 is devoted to the modelling of electrical systems. It is shown that the component models approach can be readily used for the modelling of electrical and electromechanical components and systems. This is important for this enables both the mechanical and electrical part of a system can be modelled and analysed on the same basis, i.e. from the bond graph point of view. Models of the most important electrical components are developed in terms of bond graphs such as resistors, inductors, capacitors etc., and also fundamental semiconductor components such as the diodes and the transistors, using SPICE like models. An important feature of the approach is the visual representation of bond graph electrical component models as electrical schemes. The chapter ends with analysing an electro-magnetic system.

Chapter 8 describes modelling of control systems in terms of block diagrams. The approach is very popular in other fields as well. It is shown how block diagram models can be developed and is illustrated on a simple control system problem. Some details of modelling that are specific to block diagram components are given. Then a short overview of the modelling approach to control systems is given concentrated mostly on the modelling of PID controllers in servo loops. Finally the modelling and simulation of a DC motor servo is given.

Chapter 9 is devoted to multibody systems. The models of planar rigid bodies and the basic joints are developed and applied to some practical multibody problems. The approach is then extended to multibody systems in space. The components developed are used for solving control and 3D visualization of robots. This demonstrates that the component model approach developed in this book can be used for solving complex problems in mechatronics.

Chapter 10 is the last chapter of the book and deals with the modelling and simulation of continuous systems. Continuous systems are important in many engineering disciplines, mechatronics included. The approach used here is based on the method of lines. The system is discretized and represented as an assemblage of finite elements in the form of bond graph component models. The model equations are then generated and solved using the DAEs solver. The approach is applied first to a problem of the modelling of electric transmission lines. Then a bond graph component model of a beam element based on the classical Euler-Lagrange and Timoshenko theory is developed and applied to the solution of two practical problems—a package vibration testing system and the analysis of a Coriolis mass flow meter.

References

1. Damic V, Montgomery J (1998) Bond graph based automated modelling approach to functional design of engineering systems. In: Gentle GR, Hull JB (eds) Mechanics in design international conference. The Nottingham Trent University, Nottingham, pp 377–386
2. Ljung L, Glad T (1994) Modelling of dynamic systems. PTR Prentice Hall, Englewood Cliffs
3. Paynter HM (1961) Analysis and design of engineering systems. MIT Press, Boston
4. Blundell AJ (1982) Bond graphs for modelling engineering systems. Ellis Horwood Limited, Chichester
5. Breedveld PC (1984) Physical systems theory in terms of bond graphs, PhD thesis, Technische Hochschool Twente, Entschede
6. Gawthrop P, Smith L (1996) Metamodelling: bond graphs and dynamic systems. Prentice Hall, Hemel
7. Hezemans PMAL, van Geffen LCMM (1991) Analogy theory for a systems approach to physical and technical systems. In: Fishwick PA, Luker PA (eds) Qualitative simulation modeling and analysis. Springer, New York, pp 170–216
8. Karnopp DC, Rosenberg RC (1975) System dynamics: a unified approach. Wiley, New York
9. Karnopp DC, Margolis DL, Rosenberg RC (1990) System dynamics: a unified approach, 2nd edn. Wiley, New York
10. Karnopp DC, Margolis DL, Rosenberg RC (2000) System dynamics: modeling and simulation of mechatronic systems, 3rd edn. Wiley, New York
11. Thoma JU (1975) Introduction to bond graphs and their applications. Pergamon Press, Oxford
12. Thoma JU (1990) Simulation by bondgraphs. Springer, Berlin Heidelberg
13. Thoma J, Bousmsma BO (2000) Modelling and simulation in thermal and chemical engineering. A bond graph approach. Springer, Berlin Heidelberg
14. Firestone FA (1933) A new analogy between mechanical and electrical systems. J Accoustical Soc 4:249–267

15. Celier FE, Elmqvist H, Otter M (1995) Modelling from physical principles. In: Levine WS (ed) The control handbook. CRC Press, Boca Raton, pp 99–108
16. Shei TK, Steeb WH (1998) Symbolic C++: an introduction to computer algebra using object-oriented programming. Springer, Singapore
17. Cellier FE (1996) Object-oriented modeling: means for dealing with system complexity. In: Proceedings of the 15th Benelux meeting on systems and control, Mierlo, pp 53–64
18. Kreuter EJ (1994) Generation of symbolic equations of motion of multibody systems. In: Kreuzer E (ed) Computerized symbolic manipulation in mechanics. Springer, Wien, New York, pp 1–66
19. Iri M (1991) History of automatic differentiation and rounding error estimation. In: Griewank A, Corliss GF (eds) Automatic differentiation of algorithms: theory, implementation and applications, proceedings. of first SIAM workshop on automatic differentiation, pp 3–16
20. Reverchon A, Ducamp M (1993) Mathematical software tools in C++. Wiley, Chichester
21. Schroeder W, Martin K, Lorensen B (1998) The visualization toolkit, 2nd edn. Prentice-Hall PTR, Upper Saddle River
22. Rumbaugh J, Blacha M, Premerlani W, Eddy F, Lorenson W (1991) Object-oriented modeling and design. Prentice Hall
23. Stroustrup B (1998) C++ programming language, 3rd edn. Addison-Wesley, Reading
24. Vladimirescu A (1994) The spice book. Wiley, New York
25. Christen E, Bakalar K, Dewey AM, Moser E (1999) Analog and mixed signal modeling using VHDL-AMS Language (tutorial). The 36th Design Automation Conference, New Orleans
26. Rosenberg RC (1974) A user's guide to ENPORT-4. Wiley, New York
27. Broenink JF (1990) Computer-aided physical systems modelling and simulation: a bond graph approach. PhD thesis, University of Twente, Enschede, Netherlands
28. Breunese APJ, Broenink JF (1997) Modeling mechatronic systems using the SIDOPS+ language. In: Cellier FE, Granda JJ (eds) 1997 International conference on bond graph modeling and simulation, Phoenix, Arizona, pp 301–306
29. Broenink JF, Kleijn C (1999) Computer-aided design of mechatronic systems using 20-SIM 3.0. In: Roberts GN, Tubb CAJ (eds) Proceedings of 2nd workshop on European scientific and industrial cooperation, Newport, UK, pp 27–34
30. Granda JJ (1982) Computer-Aided Modelling Program (CAMP): a bond graph processor for computer aided design and simulation of physical systems using digital simulation languages. Ph.D. Dissertation, Department of Mechanical Engineering, University of California, Davis
31. Fritzon P, Engelson V (1997) Modelica—a unified object-oriented language for system modeling and simulation. In: Modelica home page http://Dynasim.se/Modelica
32. Frizon P (2014) Principles of object-oriented modeling and simulation with Modelica 3.3: a cyber-physical approach, 2nd edn. Wiley-IEEE Press

Chapter 2
Bond Graph Modelling Overview

2.1 Introduction

The bond graph physical modelling analogy provides a powerful approach to modelling engineering systems in which the power exchange mechanism is important, as is the case in mechatronics. In this chapter we give an overview of the bond graph modelling technique. The intention is not to cover bond graph theory in detail, for there are many good references that do this well, e.g. [1–3]. The purpose is to introduce the reader to the basic concepts and methods that will be used to develop a general, systematic, object-oriented modelling approach in Chap. 3.

2.2 Word Models

Many engineering systems consist of components, e.g. electric motors, gears, shafts, transistors etc. (Fig. 1.3). Simulation models of such components can be represented as objects in the computer memory and depicted on the screen by their *word model*, i.e. a word description chosen to describe the component (Fig. 2.1).

The component name is useful for reference to the model. But, what is more important, the word model represents also how the component is connected to other components. When we look at a component, the internals of its design are usually hidden (e.g. by its housing). What are seen are the locations where it is connected to other components. These places—such as are electrical terminals, output shafts, fixing places, hydraulic ports, and boundary surfaces across which heat transfer takes place—are termed ports. In Fig. 2.1 the ports are shown by short lines.

When the component is connected and the system is energized from a suitable power source, there is a flow of power through these ports. Also, some ports serve to monitor or control the component. Thus, the ports serve as places where power or information exchange takes place. This is explained in Sect. 2.3.

© Springer-Verlag Berlin Heidelberg 2015
V. Damić and J. Montgomery, *Mechatronics by Bond Graphs*,
DOI 10.1007/978-3-662-49004-4_2

Fig. 2.1 Component word
model

A component represented by its word description (its "name") and its ports is taken as the most fundamental representation of a component model and is termed the *word model*. The word model is used as the starting point of component model development.

2.3   Ports, Bonds, and Power Variables

Ports, as noted in Sect. 2.2, are places where interactions between components take place. These interactions can be looked on as power or information transfer. Thus, two types of ports are defined.

Ports characterised by power flow into or out of a component are termed *power ports*. Such ports are depicted by a *half arrow* (Fig. 2.2). The half arrow pointing to the component describes *power inflow*. It is assumed that at such a port there is positive power transfer into the component. Similarly, a half arrow pointing away from the component depicts *power outflow* from the component and the corresponding power transfer is then taken as negative.

Another type of port is characterised by negligible power transfer, but high information content. These are termed *control ports* and are depicted by a *full arrow*. The arrow pointing to the component denotes transfer of information into the component (*control input*). Similarly the port arrow pointing away from the component denotes information extracted from the component (*control output*).

The word model, i.e. the component represented by the name and the ports, is taken as the lowest level of component abstraction (Fig. 2.2). Components interact with other components through their ports. These interactions are looked on as power or information transfer between components and are depicted by lines connecting corresponding component ports (Fig. 2.3a). The lines that connect power ports are termed *bond lines*, or *bonds* for short. A bond line joins a *power*

Fig. 2.2 Component word
model with the ports defined

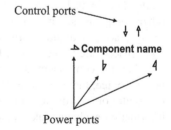

(a)

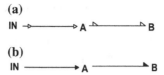

(b)

IN ⟶ A ⟶ B

Fig. 2.3 Connecting components: **a** connecting ports by bond lines, **b** line and connected ports represented by bond line only

outflow port of one component and a *power inflow* port of the other and clearly shows the assumed direction of power transfer between components. Similarly, lines connecting control *output ports* and control *input ports* are termed *active* or *control* bonds. These lines show the direction of information transfer between components. When a bond line is drawn, ports and connecting lines appear as a single line with a half or full arrow at one of its ends (Fig. 2.3b). In the bond graph literature emphasis is put on the bonds, with ports playing a minor role. In our approach just the opposite point of view is taken: Ports, the places where inter-component actions take place, receive the emphasis.

Power or information exchange between component ports can be quite complex. It generally depends on the processes taking place in the components. In the simplest case the process in the component as seen at a power port can be described by a pair of *power variables*, the *effort* and *flow* variables. Their product is the power through the port (Sect. 1.3). Connecting such ports by a bond simply implies that effort and flow variables of interconnected ports are equal. Similarly, information at component control ports can be described by a single *control variable* (signal). Connecting an output port of a component to an input port of the other just means that these two control variables are equal.

In general, the situation is not this simple. Thus, the revolute joint illustrated schematically in Fig. 2.4a may be used to connect robot links or, a door in a door-frame. The joint can be represented by a word model (Fig. 2.4b), with ports representing the parts of the joint provided for the connection.

The function of the joint is to enable rotation of the connected bodies about the joint axis. To describe the interactions at the joint connection properly, pairs of effort and flow *vectors* are used. The effort vector can be represented by three rectangular components of the forces and torques, and likewise the flow vector by the rectangular components of linear and angular velocities. The meaning of these

Fig. 2.4 Revolute joint: **a** scheme of joint, **b** word model representation

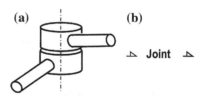

variables can be explained by defining a detailed model of the joint and the bodies in question. Hence the connection of a body to the joint can be represented by a bond, which denotes again that the efforts and flows of connected parts are equal. This time the power variables are not simple one-dimensional variables, but vector quantities.

Complex interactions at the ports can also be represented using multidimensional bond notation known as *multibonds* [4, 5]. We do not use this approach here, but instead treat the component ports as *compounded*. This means that the component ports are not simply objects, but define the structure of the mathematical quantities that describe the processes taking place inside the component. The bond lines simply define which port is connected to which, and hence which mathematical quantities should be equal. To define the structure of the ports the component model is developed in more detail.

2.4 Component Model Development

The detailed model of a component represented by the word model (Fig. 2.5, top-left) can be described in a document framed by a rectangle (Fig. 2.5, on the right). We call it a document because it will be represented on the computer screen in a document window and saved in a file (Chap. 3). The document title uses the name of the component that it models. The document contains ports represented by short strips placed just outside of the frame rectangle. These document ports correspond to the ports of the component: Every component port has a document port. The component Comp A in Fig. 2.5 has three ports: a power-in port, a power-out port, and a control-out port. Thus, there are exactly three document ports of the same type. The document ports are depicted in the positions around frame rectangle that corresponds to the position of the component ports around the component text (name). This way it is easy to see which port corresponds to which.

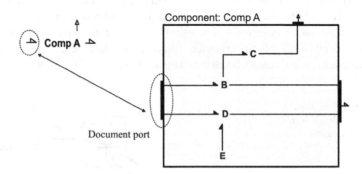

Fig. 2.5 Concept of component model

To develop the model, the component is analysed to identify the components of which it consists. Each is represented by its word model. Thus, in Fig. 2.5 there are four such components, named B, C, D, and E. Next we determine how these components are interconnected. Some are connected only internally, and this is represented by bonds connecting their respective ports, e.g. component B to component C. Some of the components are connected to the outside. In this case the respective component ports should be connected by the bonds to the document port strips, e.g. component B ports should be connected to the left and the right document ports.

Thus, document ports serve for the internal connection of the contained components. The document strip allows more than one bond to be connected to the port. We also assume that these bond connections are ordered, e.g. from top to bottom and from left to right. Hence, a component port (Fig. 2.5, left top) is represented by an array of internal component connections (Fig. 2.5 on the right).

When all the word models of the contained components are connected, we obtain a diagrammatical representation of the component model structure termed the *bond graph*. To complete the model it is necessary to continue developing the models of all contained components, which are represented by their word models, e.g. components B, C, etc. (Fig. 2.5).

The important question is how and when we end this process of systematic model decomposition. Formally, this happens when we get to a component that is fundamental, i.e. it doesn't contain simpler components. This is the problem of the level of abstraction we use when developing a model.

Normally we start model development from the *system level* (Sect. 1.2). At that level we define the model as a bond graph of the components. This is the lowest level of problem abstraction and component word models at this level usually correspond to the real-world components. In the next step we describe a model of the component by identifying the basic physical effects in the component, ignoring other, less important effects. Thus, the electrical resistor or mechanical spring can be described by an elementary model, such as Ohms law or the linear spring force-extension relationship. But we can also include the inductivity effect of the resistor or inertial effects in the spring. Hence, even in such simple cases we can use either simple or compounded component models. In other more complex devices such as robot arms, we identify real components that constitute such an arm, e.g. links, joints, base, etc. But even then we reach a stage at which we decide on the level of detail to be included in the underlying model. Physical processes are usually distributed over the component space, not restricted to small regions only. In such cases distributed models are usually discretized and can be represented by bond graphs of the components.

The bond graph modelling analogy enables the representation of models of basic physical processes taking place in engineering systems in the form of elementary components (Sect. 1.3). These components are described in more detail in Sect. 2.5. In addition, signal processing can also be described by several elementary operations (Sect. 2.6). Thus, starting from the system level, it is possible to develop the model gradually by applying the component decomposition technique. At every

level of decomposition the components can be represented as elementary, or by a
word model that is developed further. The resulting model thus can have one or
more levels of decomposition. This depends on the system under study and the
accepted level of abstraction of the problem being solved.

2.5 Modelling Basic Physical Processes

2.5.1 Elementary Components

The notion of elementary components has already been introduced in Sects. 1.2 and
1.3. These have a simple structure and serve as the building blocks of complex
component models. In the bond graph method such components represent basic
physical processes. Sometimes such components can be used as simplified representations of real components, such as bodies, springs, resistors, coils, or
transformers.

There are, altogether, nine such components that represent underlying physical
processes in a unique way. These are

- Inertial (I), Capacitive (C) and Resistive (R) components
- Sources of efforts (SE) and of flows (SF)
- Transformers (TF) and Gyrators (GY)
- Effort (1) and flow (0) junctions

The standard symbols used for the components are given in the parentheses.

In this way multi-domain physical processes, typical of mechatronics and other
engineering systems, can be modelled in a unified and consistent way. A review of
all the elementary components is given in Fig. 2.6. Components are described by
their constitutive relations in terms of variables and physical parameters.

The components can have one or more power ports. The processes seen at these
ports are described by pairs of power variables: *effort* e and *flow* f. In addition,

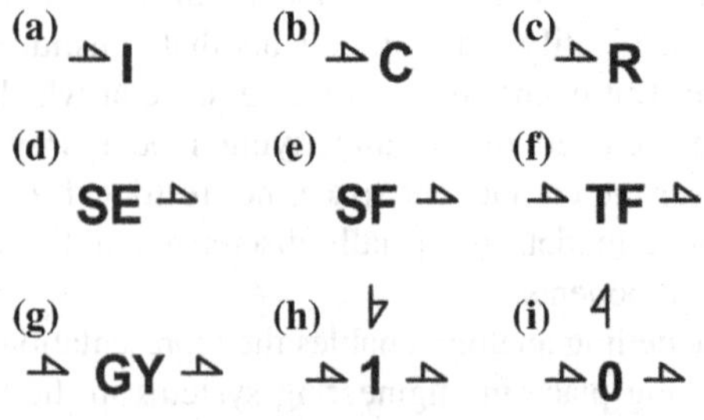

Fig. 2.6 Elementary components: **a** inertial, **b** capacitive, **c** resistive, **d** source effort, **e** source
flow, **f** transformer, **g** gyrator, **h** effort junction, **i** flow junction

certain components have internal state variables. The next sub-sections give a detailed description of each component (Sects. 2.5.2–2.5.7). In Sect. 2.5.8 the controlled elementary components are described, i.e. common components with added control ports. At the control ports a control variable c is defined that is used for supplying information to, or extracting information from, the component.

2.5.2 The Inertial Components

The *inertial component* is identified by the symbol I and has at least one power port (Fig. 2.6a). This component is used to describe the inertia of a body in translation or rotation, or the inductivity of an electrical coil.

The port variables are effort e and flow f. In addition, there is an energy variable, *generalised momentum p*, defined by the relationship

$$e = \dot{p} \tag{2.1}$$

The generalised momentum can be viewed as the accumulation of effort in the component,

$$p = p_0 + \int_0^t e \, dt \tag{2.2}$$

The constitutive relation of the process reads

$$p = I \cdot f \tag{2.3}$$

where I is a parameter. The constitutive relation also can be non-linear, of the form

$$p = \Phi(f, par) \tag{2.4}$$

or, alternatively,

$$f = \Phi^{-1}(p, par) \tag{2.5}$$

where Φ is a suitable non-linear function and *par* denotes the parameters.

If the component has n ports, the constitutive relation at the ith port generally has the form

$$p_i = \Phi_i(f_j, par), (i, j = 1, \ldots, n) \tag{2.6}$$

or, alternatively,

$$f_i = \Phi_i^{-1}(p_j, par), (i, j = 1, \ldots, n) \tag{2.7}$$

where Φ_i are suitable multivariate functions.

A process represented by an inertial component is characterised by the accumulation of power flow into the component in form of energy

$$E = E_0 + \int_0^t e \cdot f \, dt$$

Using (2.1) we get

$$E(p) = E(p_0) + \int_0^t f \, dp \tag{2.8}$$

2.5.3 The Capacitive Components

The *capacitive* component is identified by the symbol C and has at least one power port (Fig. 2.6b). This component is used to model mechanical springs, electrical capacitors, and similar processes.

The port variables are effort e and flow f. In addition, there is an energy variable, *generalised displacement q*, defined by relation

$$f = \dot{q} \tag{2.9}$$

Thus, generalised displacement can be viewed as the accumulation of the *flow* in the component,

$$q = q_0 + \int_0^t f \, dt \tag{2.10}$$

The constitutive relation of the process reads

$$q = C \cdot e \tag{2.11}$$

where C is a parameter. The constitutive relation also can be nonlinear, i.e.

$$q = \Phi(e, par) \tag{2.12}$$

or, alternatively,

$$e = \Phi^{-1}(q, par) \tag{2.13}$$

where Φ is a suitable non-linear function and *par* denotes parameters.

If the component has n ports, the constitutive relation at the ith port generally is of the form

$$q_i = \Phi_i(e_j, par), \quad (i, j = 1, \ldots, n) \tag{2.14}$$

or, alternatively,

$$e_i = \Phi_i(q_j, par), \quad (i, j = 1, \ldots, n) \tag{2.15}$$

and Φ_i are suitable multivariate functions.

A process represented by a capacitive component is characterised by the accumulation of power flow into the component in form of energy

$$E = E_0 + \int_0^t e \cdot f \, dt$$

or by (2.9)

$$E(q) = E(q_0) + \int_0^t e \, dq \tag{2.16}$$

2.5.4 The Resistive Components

The *resistive component* is identified by the symbol R and, like the inertial and capacitive components, has at least one port (Fig. 2.6c). This component models friction in mechanical systems, or electrical resistors.

The port variables are effort e and flow f. The component constitutive relation is given by

$$e = R \cdot f \tag{2.17}$$

where R is a parameter. The constitutive relation can also be non-linear

$$e = \Phi(f, par) \tag{2.18}$$

or, alternatively,

$$f = \Phi^{-1}(e, par) \tag{2.19}$$

where Φ is a suitable non-linear function and *par* denotes parameters.

If the component has n ports, the constitutive relation at the ith port generally has the form

$$e_i = \Phi_i(f_j, par), \quad (i, j = 1, \ldots, n) \tag{2.20}$$

or,

$$f_i = \Phi_i^{-1}(e_j, par), \quad (i, j = 1, \ldots, n) \tag{2.21}$$

and Φ_i are suitable multivariate functions.

2.5.5 The Sources

Sources are components that represent power generation (or power sinks) such as voltage and current sources, certain types of forces (e.g. gravity), volume flow sources (such as pumps) etc. In these sources efforts or flows are almost independent of the other power variable. It is possible to define two types of source components: *source efforts*, designated by SE; and *source flows*, designated by SF (Fig. 2.6d, e). These are, basically, single port components. Denoting the port effort by e and port flow by f, the corresponding constitutive relations are given by the following relationships depending on the source type.

2.5.5.1 Source Efforts SE

$$e = E_0 \tag{2.22}$$

or, more generally,

$$e = \Phi(t, par) \tag{2.23}$$

2.5.5.2 Source Flows SF

$$f = F_0 \tag{2.24}$$

or, more generally,

$$f = \Phi(t, par) \tag{2.25}$$

In the relationships above, E_0, F_0, and par are suitable parameters, and Φ is a function of time t.

2.5.6 The Transformers and Gyrators

The *transformer* TF and the *gyrator* GY are two important components that represent transformations of the power variables between their ports (Fig. 2.6f, g). Both have two ports; power is directed into the component at one port, and out of the component at the other. Thus, power is assumed to flow through the component.

An important characteristic of these components is the conservation of power flow, i.e. power inflow is equal to power outflow. If we denote the corresponding power ports effort-flow variables by e_i and f_i ($i = 0,1$), this fact can be expressed by the relationship

$$e_0 f_0 = e_1 f_1 \tag{2.26}$$

2.5.6.1 Transformer TF

The transformer models the levers, gears, electrical transformers, and similar devices. In robotics and multi-body mechanics, transformers are extensively used for the transformation of power variables between body frames.

In the transformer there is a linear relationship between the same types of port variables, i.e. effort to effort and flow to flow. Denoting the transformation ratio by m, we have

$$\left.\begin{aligned} e_1 &= m \cdot e_0 \\ f_0 &= m \cdot f_1 \end{aligned}\right\} \tag{2.27}$$

These relationships satisfy the power conservation relationship given by (2.26). It is sufficient to define one of these relationships; the other follows due to the power conservation requirement. There is some ambiguity in how to define the transformation ratio because the power conservation relation is also satisfied by the inverse equations

$$\left.\begin{aligned} e_0 &= k \cdot e_1 \\ f_1 &= k \cdot f_0 \end{aligned}\right\} \tag{2.28}$$

The transformation ratio k in the last pair of the equations is just the reciprocal of ratio m in the former equations, i.e. $k = 1/m$. The form to use is left to the discretion of the modeller.

2.5.6.2 Gyrators GY

The gyrator is similar to the transformer, but relates the different types of ports variables, i.e. the efforts to flows. Denoting the gyrator ratios by m and k, the corresponding equations are

$$
\left.\begin{array}{l}
e_0 = m \cdot f_1 \\
e_1 = m \cdot f_0
\end{array}\right\} \tag{2.29}
$$

and alternatively,

$$
\left.\begin{array}{l}
f_0 = k \cdot e_1 \\
f_1 = k \cdot e_0
\end{array}\right\} \tag{2.30}
$$

The gyrators have their roots in the gyration effects well known from mechanics. Their use is essential in rigid-body dynamics. The gyrator is a more fundamental component than the transformer [1]. Two connected gyrators are equivalent to a transformer. A gyrator and an inertial component are equivalent to a capacitive element. Similarly, a source effort connected to a gyrator is equivalent to a source flow. Using such combinations makes it possible to reduce the set of elementary components necessary for physical modelling. We do not follow this approach here; there is little to be gained by using a smaller number of elementary components, as the resulting model would be more complicated and more abstract than necessary.

2.5.7 The Effort and Flow Junctions

Physical processes interact in such a way that there are restrictions on the possible values that efforts and flows can attain. Many physical laws express such constraints. In mechanics, forces and moments—including inertial effects—are governed by the momentum and the moment-of-momentum laws. In electricity, there is the Kirchhoff voltage law, and there are similar laws in other fields. Similar constraints on flows in rigid body mechanics are governed by the kinematical relative velocity laws, by the law of continuity of fluid flow in fluid mechanics, the Kirchhoff current law in electricity, etc. To satisfy such laws elementary components defined previously are connected to the junctions that impose constraints on efforts or flows. Such junctions are known as *effort* and *flow junctions* (Fig. 2.6h, i).

2.5.7.1 Effort Junctions

The *effort junction* is a multi-port component into which power flows in or out. The traditional symbol for this junction is 1. This junction also is called a *common flow* junction because the flows at all junction ports are the same, i.e.

$$f_0 = f_1 = \cdots = f_{n-1} \tag{2.31}$$

where n is the number of ports at the junction. There is no power accumulation within the junction; thus the sum of the power inflows and outflows equals zero,

$$\pm e_0 f_0 \pm e_1 f_1 \cdots \pm e_{n-1} f_{n-1} = 0 \tag{2.32}$$

In this equation the plus sign is used for the ports pointing towards the junction (positive power) and the minus sign for ports pointing away from the junction (negative power). Using (2.31) we get equation of effort balance at the junction

$$\pm e_0 \pm e_1 \cdots \pm e_{n-1} = 0 \tag{2.33}$$

2.5.7.2 Flow Junctions

The *flow junction* is similar to the effort junction, with the roles of efforts and flows exchanged. The flow junction is a multi-port component traditionally denoted by the symbol 0. This junction is also known as a *common effort* junction, as the efforts at all ports are the same, i.e.

$$e_0 = e_1 = \cdots = e_{n-1} \tag{2.34}$$

There also holds the conservation of power of flows through the junction (2.32). Thus, by (2.34) we get an equation of balance of flows at the junction

$$\pm f_0 \pm f_1 \cdots \pm f_{n-1} = 0 \tag{2.35}$$

2.5.8 Controlled Components

The component constitutive relations introduced so far depend on port and internal variables only (and *time* which is the global variable). In many instances it is also necessary to permit dependence on some external variables. This is the case when modelling controlled hydraulic restrictions in valves, variable resistors; capacitors, sources and other controlled components in electronics; and coordinate transformations in multi-body mechanics. For this purpose bond graphs use so called modulated components—modulated source efforts MSE and sources flows MSF, modulated transformers MTF and gyrators MGY. Some authors introduce other modulated components as well. We do not introduce such special components, but allow components to have control ports in addition to power ports.

Figure 2.7 shows components with added control ports. The most elementary components in Fig. 2.6 can have can have an input port. The only components that cannot have control input ports are effort and flow junctions. The components with control input ports are called *controlled* and their constitutive relations (see Sects. 2.5.2–2.5.6) depend also on the corresponding control variables. The transformers and gyrators must satisfy also the power conservation requirement. But this is not a problem because it is satisfied not only by constant transformer and gyrator ratios, but also by the ratios that dependent on a control variable c. Thus, e.g. the corresponding constitutive relations for controlled transformer and gyrators can have the same forms as given by (2.27)–(2.30), but with variable transformer and gyration ratios, e.g.

$$\left. \begin{array}{l} e_1 = m(c) \cdot e_0 \\ f_0 = m(c) \cdot f_1 \end{array} \right\} \tag{2.36}$$

and,

$$\left. \begin{array}{l} e_0 = m(c) \cdot f_1 \\ e_1 = m(c) \cdot f_0 \end{array} \right\} \tag{2.37}$$

respectively.

In addition we may define one specific component called the *switch*, denoted by Sw (Fig. 2.7b). This component has one power port and one control input port. The constitutive relation for the component is

$$\left. \begin{array}{l} e = 0, \quad c > 0 \\ f = 0, \quad c \leq 0 \end{array} \right\} \tag{2.38}$$

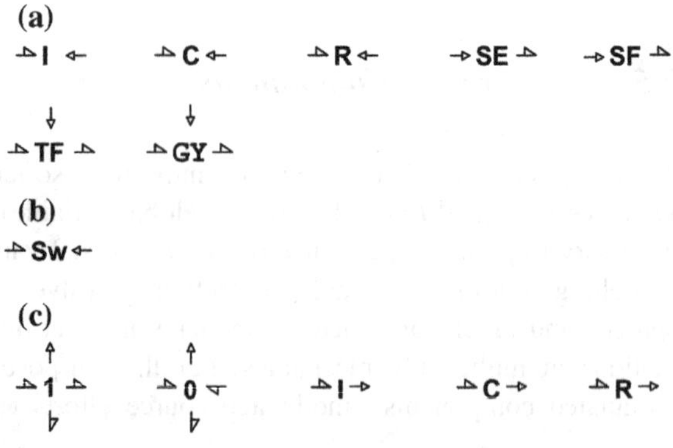

Fig. 2.7 Components with control ports: **a** inputs, **b** switch component, **c** outputs

(a)
⊳I ⟵   ⊳C⟵   ⊳R⟵   ⟶SE⊳   ⟶SF⊳

⊳TF⊳   ⊳GY⊳

(b)
⊳Sw⟵

(c)
⊳1⊳   ⊳0⟵   ⊳I⟶   ⊳C⟶   ⊳R⟶

where e and f are the power port effort and flow variables of, and c is the control variable. This component can be viewed as a controlled source that imposes zero effort or zero flow condition, depending on the sign of the control variable. This component models hard stops and clearances in machines, switches and relays in electronics, and possibly other discontinuous processes. The component can be generalised to allow effort or flow expressions, such as in sources (2.22)–(2.25) and in systems with more complex switching logic than in (2.38).

Finally, control output ports are used to access the component variables that cannot be accessed other way (Fig. 2.7c). Control output ports are commonly used for extraction of information on junction variables (efforts or flows). We also use such ports for access to the internal variables of inertial and capacitive components (momenta and displacements) and for extraction of information from other components, too.

2.6 Block Diagram Components

2.6.1 Introduction

Processes inside a system can be represented, either partially or completely, by signals (see Sect. 1.4). Thus, to complete the arsenal of components for modelling mechatronic systems, we define components that describe the input-output operations (Fig. 2.8). These are in effect the word model components of Fig. 2.2, which has only control ports. They may serve to define the basic block diagram operations in the system, e.g. they can be used to define control laws of mechatronic devices, the processing inside the system, or post processing of the simulation results.

The signals in a system can be broadly classified as *continuous-time* and *discrete-times* ones. The first type describes processes that are defined at every instant of time over some interval. These are commonly termed *analog* processes and usually represent different physical quantities, e.g. voltages, velocities, etc. The basic input-output components used to model continuous-time processes are described in Sect. 2.6.2.

In discrete-time processes the relevant quantities are defined only at discrete points in time. Such processes can be generated by sampling the continuous-time signal at the discrete times (Fig. 2.9). If the sampling frequency f_s is constant then the signal is *uniformly sampled* with sampled time index k given by $t_k = kT_s, (k = 1, 2, \ldots)$, where $T_s = 1/f_s$ is the *sampling interval*. This is typically the case in microprocessor-controlled systems where sampling is achieved by *analog to digital converters* (ADC). The basic components used to model discrete-time processes are discussed in Sect. 2.6.3.

Fig. 2.8 Block-diagram of a component

Fig. 2.9 Ideal sampling of a
continuous-time signal

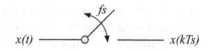

$x(t)$ ——————o ↓ —————— $x(kTs)$

2.6.2   Continuous-Time Components

Some of fundamental input-output components are shown in Fig. 2.10. These
components are well-known from control theory and will be reviewed briefly.

2.6.2.1   Input Components

The input components (Fig. 2.10a) generates control input action. These components can have only single control *output* port. This component generates the output
in the general form

$$c_{out} = \Phi(t, par) \tag{2.39}$$

where Φ is a suitable function of time and the parameters.

These components typically are used to generate step inputs, sinusoidal inputs,
pulse trains, and other input functions.

2.6.2.2   Output Components

The *output components* (Fig. 2.10b) display output signals. They can have one or
more *input* ports. Typically such components are used for collecting signal for
displaying as x-t and x-y plots. These components can be also represented by
graphical symbols which resemble x-y plotters. Such a component is called
`Display`.

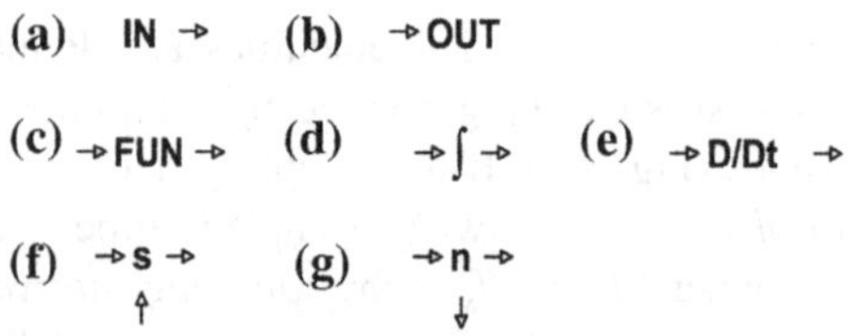

(a)   IN →   (b)   → OUT

(c) → FUN →   (d)   → ∫ →   (e) → D/Dt →

(f) → s →   (g)   → n →
        ↑   ↓

Fig. 2.10 Basic block diagram components: **a** input, **b** output, **c** function, **d** integrator,
e differentiator, **f** summator, **g** node

2.6.2.3 Function Component

The *function components* (Fig. 2.10c) generates output as linear or non-linear functions of its inputs. The component can have one or more control input ports and single output port. The output generated by the function generally can have the form

$$c_{out} = \Phi(c_0, c_1, \ldots, c_{n-1}, par) \tag{2.40}$$

where c_0, c_1,..., c_{n-1} are the inputs, and *par* are the parameters.

Such a function can be used to represent linear gains, multiplications of the inputs, or other non-linear operation on the inputs. Often instead of generic symbol FUN more specific words can be used, which better describe the function, such as k for *k*-gains, Limiter for functions that limits the output, etc.

2.6.2.4 Integrator

As its name implies, this component evaluates the time integral of its input (Fig. 2.10d), i.e.

$$c_{out} = c_{out}(0) + \int_0^t c_{in}(t)dt \tag{2.41}$$

Obviously, this is a single input–single output component. Important parameter of the function is the initial value of the output.

2.6.2.5 Differentiator

In parallel with the integrator we introduce the *differentiator* (Fig. 2.10e), as component which generates the time derivative of its input, i.e.

$$c_{out} = \frac{dc_{in}}{dt} \tag{2.42}$$

However, it is not a proper input-output component, because time derivative of a time function is not well defined operation. This is reason why such an operation is not usually met among block diagram components. Here it is included because, as will be show later, the system model that we use are in the form of semi-implicit differential-algebraic equations (DAEs), which are solved using a method based on differentiation (so called *backward differentiation formula* BDF). Therefore, such a function is legitimate, but will be solved differently than the other input-output functions.

The differentiator can be used to model D component of PID controllers, and also if time-derivatives are explicitly needed.

2.6.2.6 Summator

The *summator* (Fig. 2.10f) gives the sum of its inputs, with optional positive or negative signs, i.e.

$$c_{out} = \pm c_0 \pm c_1 \cdots \pm c_{n-1} \tag{2.43}$$

At every input port there is associated a *plus* or *minus* sign, which indicates whether the corresponding input is added or subtracted when evaluating the output. Often instead of s (for summation), more common symbols such as Σ, $\oplus$, or $\otimes$ are used.

2.6.2.7 Node

The *node* (Fig. 2.10g) serves for branching signals. This component has a single input and one or more outputs. Usually, instead of symbol n (for node), a large dot • is used.

2.6.3   Discrete-Time Components

Many continuous-time input-output components have their discrete-time counterparts. This is the case with the *Input, Output, Function, Summator* and *Node* components. However, there are components that are specific to discrete-time processes, in particular digital ones. These are *Analog to Digital Converters* (ADC), *Digital to Analog Converters* (DAC), *Clocks* and *Memory* (*delay*) components (Fig. 2.11).

The A/D component in Fig. 2.11a describes the quantization of the input signal, which can be described by the relationship.[1]

$$c_{out} = q \cdot round(c_{in}/q) \tag{2.44}$$

---

[1]This component corresponds to Quantizer block in Matlab-Simulink.

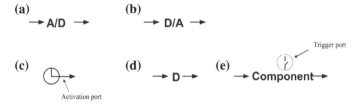

(a) →A/D→ **(b)** →D/A→

(c) Activation port **(d)** →D→ **(e)** Trigger port →Component→

Fig. 2.11 Basic discrete components and ports: **a** analog to digital converter, **b** digital to analog converter, **c** clock with activation port, **d** delay, **e** triggered component

where c_{in} is the *analog input* signal and c_{out} is the *quantized output* signal, and q is the *quantization interval* (Fig. 2.12). Function *round* converts the ratio of input signal and the quantization interval to the lower integer value.

A *clock* component (Fig. 2.11c) can be used to synchronize the discrete operations in the system. It can be created in the form of a unit integrator, i.e.

$$\int_0^t 1\,dt$$

which generates the outputs every sampling interval T_s (Sect. 2.6.3). Note that the clock has a special port called the *activation port* which activates the part of the system following the point of connection of the clock.

As an illustration of the application of the last two components consider the model of the A/D conversion in Fig. 2.13. The input to the A/D conversion is the analog signal, and the output is a digital number. If the range of the input is $c_{inH} - c_{inL}$ than the linear gain of M-bit converter is given by [6]

$$K = \frac{2^M - 1}{c_{inH} - c_{inL}} \tag{2.45}$$

Fig. 2.12 Quantization of the input signal

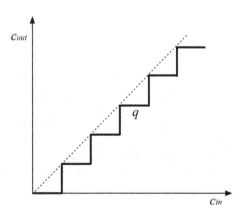

Fig. 2.13 Model of A/D
converter

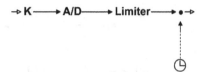

Thus, if the input range of 24 bit converter is ±2.5 V the linear gain is 3355443 1/V. This gain is represented in Fig. 2.13 by continuous-time function K.

The A/D component converts the scaled output to a discrete value using a suitable quantization interval. Ideally, the quantization interval should be based on a number of possible binary values, i.e. $q = 1$ bit. However, due to noise such low resolution cannot be achieved. The minimum change in input voltage required to guarantee a change in the discrete output value is called the *least significant bit* (LSB). Thus in the above example if the LSB is 5 μV (ideally it is $1/K \approx 0.3$ μV) the corresponding quantization interval q = 5 μV • 3355443 $V^{-1} \approx 17$ bits. The Limiter limits the converter output to a maximum and minimum binary values that the converter can generate. Finally, there is a clock connected to the corresponding node. The clock defines the sampling rate of the converter and the starting and ending activation times.

The D/A component in Fig. 2.11b converts a discrete-time signal to the corresponding continuous-time one. There are various possible ways how to do this. The most popular way to generate the signal values over the next sampling interval is to hold the value of the signal at the constant value. Because the constant is zero-order polynomial this method is known as *zero-order hold*. Sometimes a linear interpolation is used (*first-order hold*), but it is generally accepted that zero-order hold is satisfying and is widely used. Note that there is a scaling involved between the digital input and analog output, and a similar gain function should be applied as in Fig. 2.13.

An important component that is often used in digital systems is the *delay* or D component (Fig. 2.11d). This component stores the current input value into a corresponding memory location. Because this value is available at the next sampling interval it is described by the relationship

$$\left. \begin{array}{l} c_{out}[k] = c_{in}[k-1], \quad k > 1 \\ c_{out}[1] = c_{in}[0] = c_0 \end{array} \right\} \tag{2.46}$$

Therefore, the output of the component at the current sampling instance is equal to the value of the input at the previous instance. This function is, therefore, a *unit delay function*. In order to be defined at the first sample instance, i.e. when $k = 1$, it is necessary to define the initial value of the input, $c_{in}[0]$. The D function is analogous to the integrator in continuous-time processes (2.41).

The discrete components discussed so far are fundamental to modelling discrete (digital) processes. It is possible of course to define additional more specific components as well. We will conclude the discussion of discrete component by discussion of one special port introduced and shown in Fig. 2.11e. This is the *trigger port* and serves to activate or deactivate processing inside a discrete block to which it is added if an external signal connected to the port satisfies suitable conditions.[2] Such a component is termed a *triggered component*. Using an external signal it is possible to control when processes in a component are starting and ending. The conditions under which it happens are defined by the trigger port. Thus, it is possible to define that processes starts when a rising input signal goes through zero, or when a dropping signal crosses some value, or when some other condition is satisfied.

2.7   Modelling Simple Engineering Systems

The approach outlined in the previous sections can be used for the systematic computer-aided model development of engineering problems. We apply this approach to two simple, well-known problems, one from mechanical engineering and the other from electrical engineering. The technique is compared with the common bond graph modelling technique as given e.g. in [1]. We also consider a more complicated practical example from mechanical engineering (the See-saw problem).

2.7.1   Simple Body Spring Damper System

The first example models a single-degree-of-freedom mechanical vibration system (Fig. 2.14a). It consists of a body of mass m that translates along a floor, and is connected to a wall by a spring of stiffness k, and by a damper with a linear friction velocity constant b. An external force F acts on the body. We neglect the Coulomb friction between the block and the floor for simplicity, as well as the weight of the body. In Fig. 2.14b the system is decomposed into its basic components. This is the *free-body diagram* well known from engineering mechanics. This decomposition clarifies the power flow direction assignment of the component ports.

The bond graph model of the system is shown at the top of Fig. 2.15. The system consists of three components: Spring, Damper, and Body represented by the corresponding word models. The Wall (and floor) constitutes a component belonging to the system environment and is represented by the word model. The power ports of Spring and Damper and corresponding ports in Wall and Body are connected by the bond lines. The body is acted upon also by an external force, represented by a SE (Source Effort) elementary component, and is connected to the

---

[2]Its function is similar to Triggered Subsystems in Matlab-Simulink.

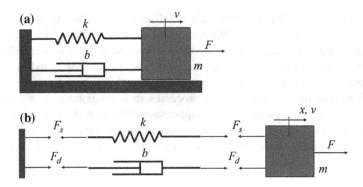

Fig. 2.14 Body spring damper system: **a** schematic representation, **b** free body diagram

Fig. 2.15 Bond graph model of problem in Fig. 2.14

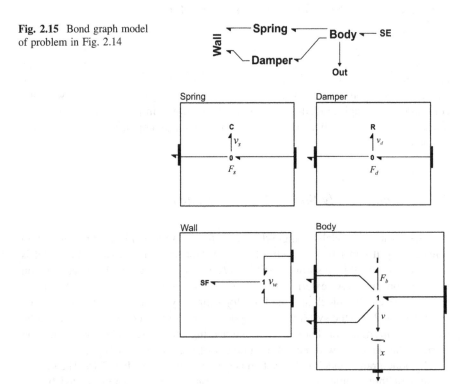

corresponding Body port by a bond. There is also a control-out port on the Body which is used to extract information on the body position. This port is connected to the Out(put) port for display of body position during the simulation.

The model at this level of abstraction has a structure that closely corresponds to the scheme of the system in Fig. 2.14a. The direction of power flow in the model is taken from the SE through the body, then through the spring and damper, and finally to the wall.

This corresponds to the physical situation in Fig. 2.14b. If the sense of the external force and the body velocity are as shown, the power at the external force port is positive; i.e., it is directed *into* the body. Assuming that the spring and damper resist the movement of the body—i.e., the sense of their forces is opposite to that of the body velocity—the powers at the corresponding ports are negative, and the power port arrows are directed *out* of the body. At the spring and damper ports, power again is positive, flowing into these components. Because of the direction of the body's force, according to *Newton's Third Law* their forces act in the opposite sense. A similar conclusion can be drawn regarding the wall side ports. Hence, by joining mechanical ports Newton's Third law is satisfied. Thus, to construct the bond graph model it is not necessary to draw the free-body diagram at all.

Next we develop the component models (Fig. 2.15). The force generated by the spring depends on the relative displacement (extension) of the spring. Thus, the model of the spring can be represented by a flow junction with three ports, two for connecting internally to the spring end ports and the third for the connection of the capacitive element that models the elasticity of the spring (Fig. 2.15 Spring). The junction variable is the force F_s in the spring; and the extension of the spring x_s is the generalized displacement of the capacitive element, with spring stiffness k taken as the element parameter. Thus, the constitutive relations for the capacitive element are (*see* (2.9) and (2.11))

$$v_s = \dot{x}_s \tag{2.47}$$

and

$$F_s = k \cdot x_s \tag{2.48}$$

where v_s is the relative velocity of the spring ends.

The damper has a similar model, with the resistive element used to model mechanical dissipation in the damper (Fig. 2.15 Damper). The junction variable F_d represents the force developed by the damper, v_d is the relative velocity of the damper ends and the velocity constant b is a parameter of the element. Assuming a linear constitutive relation for the resistive element we have (see (2.17))

$$F_d = b \cdot v_d \tag{2.49}$$

The third component of the system is the body of mass m (Fig. 2.15 Body). This component uses an effort junction to describe the balance of forces applied to the body including the inertial force of the body. This junction has four power ports: three for internal connections to the body ports and fourth for the connection of an inertial element. Denoting the body velocity taken as the junction variable by v (see Fig. 2.14a) and the inertial force of the body of mass m taken as parameter by F_b, the constitutive relations of the inertial element are (see (2.1) and (2.3))

$$F_b = \dot{p}_b \tag{2.50}$$

and

$$p_b = m \cdot v \tag{2.51}$$

To calculate the body position, a control output port is added to the junction, and the junction variable is fed to an integrator that outputs the body position x. The corresponding equation can be written as

$$\dot{x} = v \tag{2.52}$$

The next component is simply the source effort element SE, which generates the driving force on the body

$$F = \Phi(t) \tag{2.53}$$

The spring and damper are connected to the fixed wall. The model of the wall is given in Fig. 2.15 Wall. The component uses an effort junction with three ports, which describes the force balance at the wall. Two ports serve for the internal connection to the wall ports, where the spring and damper are connected, and the third is for connecting to the source flow, which imposes a zero wall velocity condition. The junction velocity is v_w. Thus, the relation for the source flow is

$$v_w = 0 \tag{2.54}$$

To complete the mathematical model of the system the equations of the effort and flow junctions are added. Corresponding variables can be found by following the bonds connected to junction ports until some elementary component is found that completes the bond. Thus, for the body effort junction in Fig. 2.15 Body, the port effort variables are the spring force F_s, damper force F_d, inertia force F_b and driving force F, respectively. The equation of the effort balance at the junction thus reads

$$-F_s - F_d - F_b + F = 0 \tag{2.55}$$

If we denote by F_w the total force at the wall, the equation of effort balance at the wall junction reads (Fig. 2.15 Wall)

$$F_s + F_d - F_w = 0 \tag{2.56}$$

A similar equation can be written for the flow junctions. This time the summation is on the flows. Thus, we have (Fig. 2.15 Spring)

$$-v_w - v_s + v = 0 \tag{2.57}$$

and (Fig. 2.15 Damper)

$$-v_w - v_d + v = 0 \tag{2.58}$$

We, therefore, may describe the motion of the system by eight equations of elements that describe the physical processes in the system, i.e. (2.47)–(2.53), and four equations which involve the junctions (2.54)–(2.58). There are, altogether, twelve differential and algebraic equations that have to be satisfied by twelve variables: F_s, v_s, x_s, F_d, v_d, F_b, v, p_b, F, F_w, v_w and x. Although we have arrived at a relatively large number of equations of motion for this simple problem, the equations are very simple, having on average only 27/12 = 2.25 variables per equation.

The structure of the matrix of these equations is very sparse; this simplifies the solution process. We can simplify these equations further. Direct processing can be used to eliminate some, or all, of the algebraic variables (i.e., variables that are not differentiated). We can also simplify the bond graph first, and then write the corresponding equations. We consider the second approach in more detail, as it leads to the sort of bond graphs usually found in the literature.

We can simplify the model by substituting every component at the top of Fig. 2.15 by its corresponding model, given at the bottom part of the same figure. The resulting bond graph is shown in Fig. 2.16a.

The source flow on the left imposes zero wall velocity; thus, we can remove the effort junction and the source flow, as well as the two bonds connecting to the flow junctions. We also remove the corresponding ports at the junctions. This yields a bond graph represented by Fig. 2.16b. We should also eliminate these flow junctions, for they are trivial, having only one power input port and one power output port. Thus, the C and R element ports can be connected directly to the effort junction ports on the right. This results in the bond graph of Fig. 2.16c.

The model in Fig. 2.16c is much simpler than that in Fig. 2.15. The resulting equations now consist of

$$\left. \begin{aligned} v &= \dot{x}_s \\ F_s &= k \cdot x_s \\ F_d &= b \cdot v \\ F_b &= \dot{p}_b \\ p_b &= m \cdot v \\ F &= \Phi(t) \\ -F_s - F_d &- F_b + F = 0 \\ \dot{x} &= v \end{aligned} \right\} \tag{2.59}$$

We have reduced the system to eight equations with eight variables F_s, x_s, F_d, F_b, v, p_b, F, and x. This was achieved, however, by eliminating some variables that can be of interest, e.g. total force transmitted to the wall. This bond graph can be

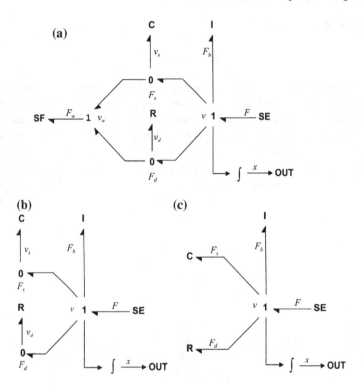

Fig. 2.16 Simplification of the bond graph of Fig. 2.15

developed directly from Fig. 2.16a by the application of classical methods of bond graph modelling, as explained in [1].

Using this form of bond graph model, the equations of motion of the system can be developed in an even simpler form than that given above. It should be noted, however, that this is not true in general for engineering systems of practical interest. We address this matter in more detail in Sects. 2.9 and 2.10. The reduced model is, on the other hand, much more abstract. This makes it more difficult to understand and interpret: Unlike the component model of Fig. 2.15, there is no topological similarity to the system represented in Fig. 2.14a. A change in any part of the model affects the complete model. On the other hand, in the model of Fig. 2.15 we can change some of the components, leaving the others unchanged. Such a model can be refined much more easily, thereby retaining the overall topological similarity to the physical model.

2.7.2   The Simple Electrical Circuit

The second example considers the electrical Resistor Inductor Capacitor circuit (RLC circuit) shown in Fig. 2.17. The circuit consists of a series connection of a

Fig. 2.17 Simple electrical circuit

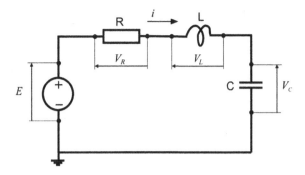

voltage source generating an electromotive force (e.m.f.) VS, a resistor R, an inductor L, and a capacitor C. The polarities of the voltage drop across the electrical components are also shown, as well as the assumed direction of the current flow. We can develop a bond graph model using an approach similar to the mechanical system analysed previously.

We can represent the electrical components by suitable word models. But instead of the component names, we use the common electrical symbols. Standard component names—such as resistor, capacitor, and the like—may be retained internally for compatibility with the usual word model representation. The resulting bond graph is shown at the top of Fig. 2.18.

The source voltage supplies electrical power to other parts of the circuit. The power port corresponding to the positive terminal is taken to be directed outward, and the other port inward. Power from the source flows through the resistor, inductor, and capacitor until the node component is reached where a part of the power flow branches to the ground component. (Later it is shown that there is no power flow to the ground.) The other part returns back to the voltage source. The model has a very similar appearance to the electrical scheme in Fig. 2.17. What is different is the presence of power ports showing the assumed direction of power flow in the circuit. Thus, the correspondence of the bond graph model and the electrical scheme is really very close. Note that for the output component a component graphically resembling x-y plotter (Display) is used.

Component models for the voltage source VS, resistor R, inductor L, and the capacitor C are shown in the lower part of the Fig. 2.18. The components have two ports used for connecting to other components. Their models are represented by three ports effort junctions 1. Two of these are used for internal connection to the component's ports, and the third is used for connection of the elementary components that describe the physical processes in the components. Power flow is chosen to flow into these elementary components. Thus, the effort of the elementary component port represents the voltage difference across the component. We model the components by the idealized linear elements.

Fig. 2.18 Bond graph model
of circuit of Fig. 2.17

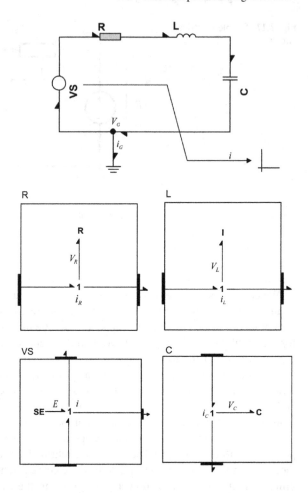

In the case of the resistor, the junction variable is the current i_R flowing through the component, and the voltage drop is V_R. Thus, the constitutive relation for the resistor is given by (2.17)

$$V_R = R \cdot i_R \qquad (2.60)$$

with R the resistance parameter.

Similarly for the inductor, the junction variable is the current i_L through the inductor and the voltage drop is V_L. If we denote the flux linkage of the coil by p_L, the constitutive relations read (see (2.1) and (2.3))

$$V_L = \dot{p}_L \qquad (2.61)$$

and

$$p_L = L \cdot i_L \tag{2.62}$$

where L is the inductance parameter of the inductor.

For the voltage source the joint variable is the current i through the source terminals (ports). The voltage E generated is described by the source effort element. The constitutive relation reads

$$E = \Phi(t) \tag{2.63}$$

Finally, for the capacitor, the junction variable is the current i_C through the component and the voltage drop is V_C. If we denote the capacitor charge by q_C, the constitutive relations can be written as (see (2.9) and (2.11))

$$i_C = \dot{q}_C \tag{2.64}$$

and

$$V_C = q_C/C \tag{2.65}$$

where C is the capacitance.

The node component is simply another representation of the flow junction, and the ground is just the ground potential source effort. The node variable is the ground potential v_G. We take the ground potential as zero, hence the ground component constitutive relation reads

$$V_G = 0 \tag{2.66}$$

If we start from any of the component effort junctions and follow the bonds connected internally to the port, then out of the component to the next component port, and again into the component, we find that all effort junctions are interconnected. We, thus, can treat all these junctions as a single junction, the result being that all junction variables are, in essence, the same variable, i.e.

$$i \equiv i_R \equiv i_L \equiv i_C \tag{2.67}$$

Counting only ports connected to other components, the balance of efforts reads as follows

$$V_G + E - V_R - V_L - V_C - V_G = 0 \tag{2.68}$$

After cancellation of the ground potential v_G, we get

$$E - V_R - V_L - V_C = 0 \tag{2.69}$$

This is the Kirchhoff voltage law for the circuit.

Finally, if we denote the current drawn by the ground by i_G, the balance of flows at the node reads

$$-i - i_G + i = 0 \tag{2.70}$$

Again, due to the cancellation of current i, we obtain

$$i_G = 0 \tag{2.71}$$

This lengthy derivation shows that we arrive at the circuit equation by using the constitutive equations for elementary components and symbolically simplifying the junction equations. Thus, the mathematical model of the circuit consists of nine differential-algebraic equations (2.60)–(2.66), (2.69), and (2.71) with nine variables E, V_R, i, V_L, p_L, V_C, q_C, V_G and i_G. Equations (2.66) and (2.71) are trivial and could be eliminated from the system.

As in the previous example, we can simplify the bond graph instead of the equations. We first substitute the bond graph model of the components from the lower part of Fig. 2.18 into the system bond graph at the top. Also the node is changed to the flow junction and the ground to source effort. We thus obtain bond graph shown in Fig. 2.19a.

Fig. 2.19 Simplification of model of Fig. 2.18

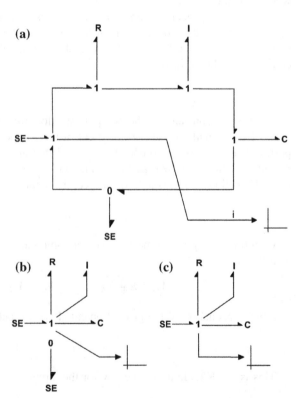

We again see that the four effort junctions are interconnected and can be condensed into a single junction, retaining only the ports connected to other components. In addition, we see that this junction is connected to the same flow junction by two ports of opposite power-flow sense. Hence, such ports can be disconnected, and then removed. The corresponding ports of the flow junction must be removed, too. This yields the bond graph shown in Fig. 2.19b. We now have a flow junction connected only to a ground source effort. These can be removed, too. This results in the final simplified system bond graph (Fig. 2.19c). The last bond graph can be described by the same equations as before, but without (2.66) and (2.71).

The above procedure shows that, instead of the simplification of junction (2.68) and (2.70), we could directly arrive at (2.69) and (2.71) by noting that interconnected effort junctions are connected to the *same* flow junction. Corresponding junction ports then could be treated as internal, and not taken into account when writing the junction equations.

Comparing bond graphs in Figs. 2.18 and 2.19c, we draw similar conclusions as in the previous example: The model in Fig. 2.18 is much easier to interpret and upgrade. Even people who are not too familiar with bond graphs could understand such a model. On the other hand, it retains the advantages that bond graphs enjoy over other modelling methods.

2.7.3 A See-Saw Problem

As a third problem, we develop a model of a simple see-saw often found in children's playgrounds (Fig. 2.20). On a much larger scale, this same problem is known as the swing boat at the fair ground. The system consists of a platform that can rotate around a horizontal pin O, fixed in a frame, and having a body at each end, e.g. a boy and a girl sitting on the see-saw seats. If one of the bodies is pushed down, and then released, the system will begin to oscillate around its equilibrium position.

Fig. 2.20 See-saw problem

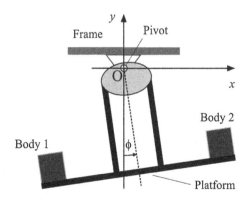

This problem is more complicated than those presented previously, as it consists of three interconnected bodies moving in the vertical plane. The model could be developed easily by treating the system as a physical pendulum. Instead, we consider it as a multi-body system and demonstrate how the model can be developed systematically by decomposition.

We start by defining the overall model structure (Fig. 2.21). The word models Body 1 and Body 2 represent the bodies on the platform. They each have a port that corresponds to the location where the body acts on the platform. The component Platform has four ports, two of which correspond to the places where the bodies act, a third for connecting to the Pivot component, and the last for the input of information on the rotation angle. The Pivot permits only rotation of the platform about a horizontal pin fixed in the Frame.

We assume that the bodies are firmly placed on the platform and move with it. Hence, we join the ports of the bodies to the corresponding platform ports by bonds. We assume the power flow sense from Body 1 and Body 2 through the Platform and Pivot to the Frame. Further, the information on the force at the pivot is of interest. Thus, we take the rectangular components of force F_x and F_y on the Pivot and feed them to the Display. Similarly, we extract information on the rotation angle *Phi* and feed it to the node. This information branches further to the Platform and to Display. Note, Display is output component in form of x-y plotter (see Sect. 2.6.2).

We proceed by developing models of the components. This requires defining the interactions taking place between them. Motion of the system is described in global co-ordinate frame *Oxy*, with origin *O* at the point of rotation of the platform in the vertical plane, axis *y* directed upward, and *x* to the right (Fig. 2.20).

The Body1 and Body2 models are shown in Fig. 2.22 Body 1 and Body 2. Separate effort junctions are used for the summation of the *x* and *y* components of forces acting on the bodies. The junction variables are the *x* and *y* components of the velocities of the bodies. The junctions are connected internally to their respective ports. The order of connection going from the left to the right is the *x* component first, then the *y* component,. This order of connection is also used for the other ports. Hence, Body1 and Body2 port variables are pairs of effort flow vectors $\mathbf{F}_1,\mathbf{v}_1$ and $\mathbf{F}_2,\mathbf{v}_2$, respectively.

Fig. 2.21 Overall structure of the see-saw

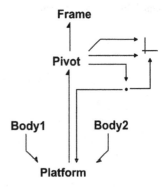

Fig. 2.22 Dynamics of
bodies attached to the see-saw
platform

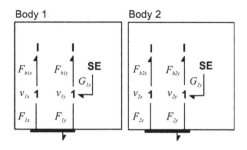

The inertial effects of the bodies in the x- and y-directions are represented by the
inertial elements **I** connected to the corresponding effort junctions, with power flow
directed into the inertial elements. The weights of the bodies, acting in the y direction only, are represented by source efforts connected to the y component
junctions, with power flow directed into the junctions. The equations of motion of
the bodies can be obtained directly from the bond graphs of Fig. 2.22. The masses
of the bodies are m_1 and m_2, and g is the gravitational acceleration. The relevant
variables are also shown in the figure.

Body 1:

$$\left.\begin{aligned}
\dot{p}_{b1x} &= F_{b1x} \\
\dot{p}_{b1y} &= F_{b1y} \\
p_{b1x} &= m_1 \cdot v_{1x} \\
p_{b1y} &= m_1 \cdot v_{1y} \\
G_{1y} &= -m_1 \cdot g \\
-F_{1x} - F_{b1x} &= 0 \\
-F_{1y} - F_{b1y} + G_{1y} &= 0
\end{aligned}\right\} \quad (2.72)$$

Body 2:

$$\left.\begin{aligned}
\dot{p}_{b2x} &= F_{b2x} \\
\dot{p}_{b2y} &= F_{b2y} \\
p_{b2x} &= m_2 \cdot v_{2x} \\
p_{b2y} &= m_2 \cdot v_{2y} \\
G_{2y} &= -m_2 \cdot g \\
-F_{2x} - F_{b2x} &= 0 \\
-F_{2y} - F_{b2y} + G_{2y} &= 0
\end{aligned}\right\} \quad (2.73)$$

The Frame simply fixes the pivot, about which the platform rotates, against translation and rotation (Fig. 2.23). The equations are:

Frame:

$$\left.\begin{array}{l} v_{Px} = 0 \\ v_{Py} = 0 \\ \omega_P = 0 \end{array}\right\} \tag{2.74}$$

The Pivot allows rotation only about the pin (Fig. 2.24). Two flow junctions are inserted to extract information on pin force components. Rotation is assumed frictionless, but friction can be added if required, e.g. by a resistive component R used instead of the source effort. The relative angular velocity of the platform is denoted by ω_r. The signal taken from the effort junction is integrated to get the platform rotation angle ϕ. The governing equations are again very simple:

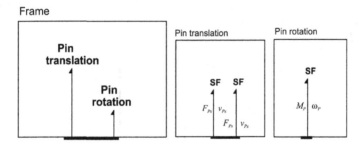

Fig. 2.23 Model of the see-saw frame

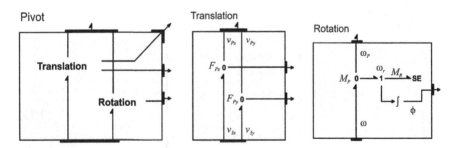

Fig. 2.24 Model of the see-saw pivot

Pivot:

$$\left.\begin{array}{r} v_{3x} - v_{Px} = 0 \\ v_{3y} - v_{Py} = 0 \\ \omega - \omega_P - \omega_r = 0 \\ M_P - M_R = 0 \\ M_R = 0 \\ \dot{\phi} = \omega_r \end{array}\right\} \tag{2.75}$$

The platform acts as a transformer of the velocities of the attached bodies. Simultaneously, the transformation of the reaction forces of the bodies also takes place. To develop the bond graph model of the platform we analyse the plane motion of the platform in the global co-ordinate frame Oxy (Fig. 2.25).

The position and orientation of the platform is defined by the body frame $C\bar{x}\bar{y}$ with the origin at its mass centre. The position vector of the origin C is described by a column vector of its global co-ordinates, i.e.

$$\mathbf{r}_C = \begin{pmatrix} x_C \\ y_C \end{pmatrix} \tag{2.76}$$

Orientation of the body is defined by the rotation matrix (see e.g. [7])

$$\mathbf{R} = \begin{pmatrix} \cos\varphi & -\sin\varphi \\ \sin\varphi & \cos\varphi \end{pmatrix} \tag{2.77}$$

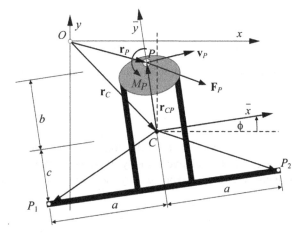

Fig. 2.25 See-saw platform plane motion

The vector of the position of a point P in the body with respect to the origin of the body frame can be expressed in the body frame by a vector of its coordinates

$$\bar{\mathbf{r}}_{CP} = \begin{pmatrix} \bar{x}_{CP} \\ \bar{y}_{CP} \end{pmatrix} \tag{2.78}$$

The position of the same point P with respect to the global frame is defined by the vector of its global co-ordinates

$$\mathbf{r}_P = \begin{pmatrix} x_P \\ y_P \end{pmatrix} \tag{2.79}$$

The relationship between these vectors is given by

$$\mathbf{r}_P = \mathbf{r}_C + \mathbf{r}_{CP} \tag{2.80}$$

Note that vector $\mathbf{r}_{CP}$ is the relative vector expressed in the global frame, i.e.

$$\mathbf{r}_{CP} = \begin{pmatrix} x_{CP} \\ y_{CP} \end{pmatrix} \tag{2.81}$$

The relationship between the vectors of (2.78) and (2.81) is given by the co-ordinate transformation

$$\mathbf{r}_{CP} = \mathbf{R}\bar{\mathbf{r}}_{CP} \tag{2.82}$$

Substituting the rotation matrix of (2.77) and evaluating the resulting expression yields

$$\begin{pmatrix} x_{CP} \\ y_{CP} \end{pmatrix} = \begin{pmatrix} \bar{x}_{CP} \cos \varphi - \bar{y}_{CP} \sin \varphi \\ \bar{x}_{CP} \sin \varphi + \bar{y}_{CP} \cos \varphi \end{pmatrix} \tag{2.83}$$

The velocity of a point P in the body can be found by taking the time derivative of (2.80), i.e.

$$\mathbf{v}_P = \mathbf{v}_C + \mathbf{v}_{CP} \tag{2.84}$$

which relates the velocity of the point P to the velocity of the origin C of the body frame and to the relative velocity of the point P with respect to the point C. These velocity vectors are expressed by their components in the global frame as

$$\mathbf{v}_P = \begin{pmatrix} v_{Px} \\ v_{Py} \end{pmatrix}, \quad \mathbf{v}_C = \begin{pmatrix} v_{Cx} \\ v_{Cy} \end{pmatrix}, \quad \mathbf{v}_{CP} = \begin{pmatrix} v_{CPx} \\ v_{CPy} \end{pmatrix} \tag{2.85}$$

and are defined by

$$\mathbf{v}_P = \frac{d\mathbf{r}_P}{dt}, \quad \mathbf{v}_C = \frac{d\mathbf{r}_C}{dt}, \quad \mathbf{v}_{CP} = \frac{d\mathbf{r}_{CP}}{dt} \tag{2.86}$$

Taking the time derivative of (2.82), and noting that $\bar{\mathbf{r}}_{CP}$ is a constant vector, we arrive at the expression for the relative velocity of point P:

$$\mathbf{v}_{CP} = \frac{d\mathbf{R}}{dt}\bar{\mathbf{r}}_{CP} \tag{2.87}$$

The time derivative of the rotation matrix $\mathbf{R}$ in (2.77) yields

$$\frac{dR}{dt} = \begin{pmatrix} -\sin\varphi & -\cos\varphi \\ \cos\varphi & -\sin\varphi \end{pmatrix} \cdot \frac{d\varphi}{dt} \tag{2.88}$$

The time derivative of the body rotation angle is the body angular velocity

$$\omega = \frac{d\varphi}{dt} \tag{2.89}$$

Thus, substitution of (2.88) and (2.89) into (2.87) yields

$$\mathbf{v}_{CP} = \mathbf{T}\omega \tag{2.90}$$

where $\mathbf{T}$ is the transformation matrix, given by

$$T = \begin{pmatrix} -\bar{x}_{CP}\sin\varphi - \bar{y}_{CP}\cos\varphi \\ \bar{x}_{CP}\cos\varphi - \bar{y}_{CP}\sin\varphi \end{pmatrix} \tag{2.91}$$

Compared with (2.83), this matrix also can be expressed as

$$\mathbf{T} = \begin{pmatrix} -y_{CP} \\ x_{CP} \end{pmatrix} \tag{2.92}$$

Equations (2.84) and (2.89)–(2.91) are the basic relations describing the kinematics of rigid body motion in a plane. Next, we consider the kinetic relationships relating the forces and moments applied to the platform.

A force $\mathbf{F}$ applied to the platform can be described by a vector of its rectangular components in the global frame, i.e.

$$\mathbf{F} = \begin{pmatrix} F_x \\ F_y \end{pmatrix} \tag{2.93}$$

The power delivered at point P is given by $\mathbf{v}_P^T\mathbf{F}$, where the superscript T denotes matrix transposition. From (2.84) and (2.90) we get

$$\mathbf{v}_P^T\mathbf{F} = \mathbf{v}_C^T\mathbf{F} + \mathbf{T}^T\mathbf{F}\omega \tag{2.94}$$

Evaluating the leading part of the second term on the right of (2.94) yields

$$\mathbf{T}^T\mathbf{F} = -y_{CP}F_x + x_{CP}F_y \tag{2.95}$$

We recognise this as the moment M_C of the force at P about point C, thus

$$M_C = \mathbf{T}^T\mathbf{F} \tag{2.96}$$

By substituting in (2.94), we finally arrive at the equations of power transfer across the body

$$\mathbf{v}_P^T\mathbf{F} = \mathbf{v}_C^T\mathbf{F} + M_C\omega \tag{2.97}$$

This equation can be read as a statement of *force equivalents*, well known from Engineering Mechanics (see e.g. [8]). That is, a force applied at a point P is equivalent to the same force applied at a different point C plus the moment of force about C. If at point P a torque also acts, its moment M_P should be added, too. Equations (2.96) and (2.97), jointly with (2.84) and (2.89)–(2.91), constitute the fundamental equations of rigid-body motion in a plane. To complete the dynamical equations we need to add the inertias of translation and rotation. These equations clearly show how to represent the dynamics of the platform (see Fig. 2.26 Platform).

At every point of application of a force (platform ports) we introduce a component 0 corresponding to summation of the velocities, as given by (2.84). These components contain two flow junctions. The corresponding junction variables are the x and y components of the force at the port (Fig. 2.26 0). The effort junction 1 is used to represent the angular velocity of the body, and the component CM describing the motion of the mass centre (Fig. 2.26 Platform and CM). We connect the junctions 0 to the angular velocity junction 1 by the components LinRot, and to the mass centre motion component CM.[3]

The LinRot components represent the *linear to rotation* transformations given by (2.90) and (2.96). The components consist of two transformers, which implement the transformations by matrix given by (2.91), and an effort junction that sum up the moments according to (2.96) (see Fig. 2.21 LinRot). The necessary

---

[3]Because of space limitation, only one of the **0** and **LinRot** components are shown. The others have a similar structure.

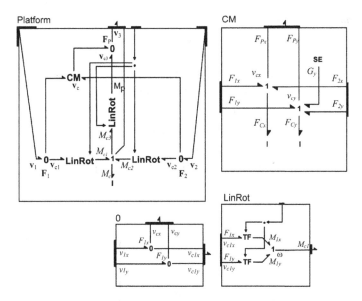

Fig. 2.26 Model of the platform

information on the angle of rotation of the see-saw platform is taken from the input port. In addition to these force effects, any moment at a port is transmitted directly to the rotation effort junction 1. An inertial element added to the junction represents the rotational inertia of the platform with respect to mass center. The body translation inertia with mass center is represented by component CM which consists of two effort junctions that add inertial elements corresponding to the x and y motion (Fig. 2.26 CM). The platform gravity is also added there.

The mathematical model of the platform can be written directly from the Fig. 2.26. Respective variables are given in the figure and parameters a, b, and c are dimensions shown in Fig. 2.25; m is the platform mass, and I_C is its mass moment of inertia about its mass centre.

The equations read:

Platform—left side:

$$\left.\begin{aligned}
&v_{1x} - v_{C1x} - v_{Cx} = 0 \\
&v_{1y} - v_{C1y} - v_{Cy} = 0 \\
&v_{C1x} = (a \cdot \sin\varphi + c \cdot \cos\varphi) \cdot \omega \\
&v_{C1y} = (-a \cdot \cos\varphi + c \cdot \sin\varphi) \cdot \omega \\
&M_{1x} = (a \cdot \sin\varphi + c \cdot \cos\varphi) \cdot F_{1x} \\
&M_{1y} = (-a \cdot \cos\varphi + c \cdot \sin\varphi) \cdot F_{1y} \\
&\quad - M_{C1} + M_{1x} + M_{1y} = 0
\end{aligned}\right\} \qquad (2.98)$$

Platform—right side:

$$\left.\begin{array}{l} v_{2x} - v_{C2x} - v_{Cx} = 0 \\ v_{2y} - v_{C2y} - v_{Cy} = 0 \\ v_{C2x} = (-a \cdot \sin \varphi + c \cdot \cos \varphi) \cdot \omega \\ v_{C2y} = (a \cdot \cos \varphi + c \cdot \sin \varphi) \cdot \omega \\ M_{2x} = (-a \cdot \sin + c \cdot \cos \varphi) \cdot F_{2x} \\ M_{2y} = (a \cdot \cos \varphi + c \cdot \sin \varphi) \cdot F_{2y} \\ - M_{C2} + M_{2x} + M_{2y} = 0 \end{array}\right\} \qquad (2.99)$$

Platform—upper side:

$$\left.\begin{array}{l} - v_{3x} + v_{C3x} + v_{Cx} = 0 \\ - v_{3y} + v_{C3y} + v_{Cy} = 0 \\ v_{C3x} = -(b \cdot \cos \varphi) \cdot \omega \\ v_{C3y} = -(b \cdot \sin \varphi) \cdot \omega \\ M_{3x} = -(b \cdot \cos \varphi) \cdot F_{Px} \\ M_{3y} = -(b \cdot \sin \varphi) \cdot F_{Py} \\ M_{C3} - M_{3x} - M_{3y} = 0 \end{array}\right\} \qquad (2.100)$$

Platform—mass centre motion:

$$\left.\begin{array}{l} \dot{p}_{Cx} = F_{Cx} \\ \dot{p}_{Cy} = F_{Cy} \\ p_{Cx} = m \cdot v_{Cx} \\ p_{Cy} = m \cdot v_{Cy} \\ G_y = -mg \\ F_{1x} + F_{2x} - F_{Px} - F_{Cx} = 0 \\ F_{1y} + F_{2y} - F_{Py} - F_{Cy} + G_y = 0 \end{array}\right\} \qquad (2.101)$$

Platform—rotation:

$$\left.\begin{array}{l} \dot{K}_C = M_C \\ K_C = I_C \cdot \omega \\ M_{C1} + M_{C2} - M_{C3} - M_P - M_C = 0 \end{array}\right\} \qquad (2.102)$$

The model consists of fifty-four very simple equations. No substitutions or other simplifications have been made, as we wished to develop the model strictly by describing every elementary component in terms of its variables and parameters.

Fig. 2.27 Single level model
of the see-saw

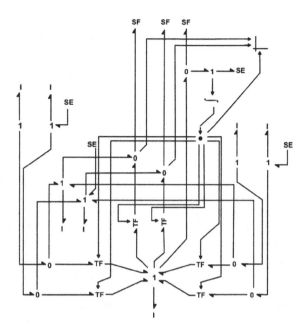

This procedure, based on the systematic decomposition, results in multi-level models. Such models can be developed and changed more easily, if necessary, than conventional "flat" models. Some of the components can also be reused in other models. For example, the `Platform` component can be used for problems dealing with the plane motion of rigid bodies. For comparison, a flat model corresponding to the model developed above is shown in Fig. 2.27. Such a model, however, is not easy to follow, particularly for people unfamiliar with bond graphs: There are many bonds, and it is not easy even to draw them correctly! It is thus more susceptible to errors and more difficult to change.

2.8 Causality of Bond Graphs

2.8.1 The Concept of Causality

The concept of *causality*, or *cause-effect relationships*, was introduced in the bond graph method to define the *computational* structure of the resulting mathematical equations at bond-graph level. Thus, the physical and computational structure of the model is defined in parallel during the modelling stage. It should be stressed that physical laws do not imply any causal preference: There is no physical reason to treat forces as the cause, and velocities of the body motions as effects; or voltages as

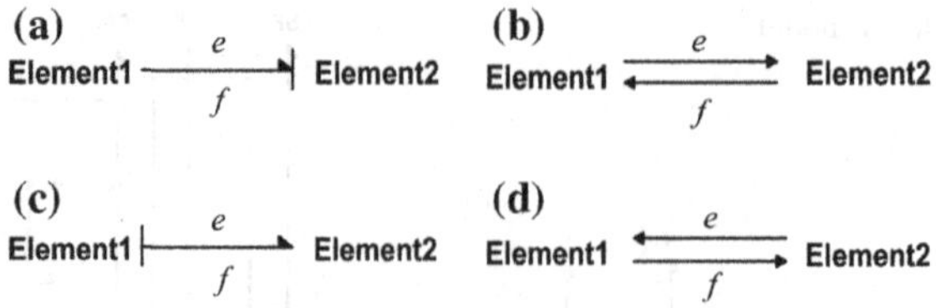

Fig. 2.28 Causality assignment: **a** and **c** possible stroke attachments, **b** and **d** meaning of the attachments

the cause, and currents in circuits as effects. The assignment of causality can be looked on as a convenient—but not an essential—part of the modelling task. Further it is arguable that it is convenient at all, in particular when using the object-oriented paradigm in simulation model building. We nevertheless briefly describe causality and its consequences in bond graphs because of their close connection with bond graph theory (see e.g. [1, 2]).

Causality means that, at every port of an elementary component, one of the power variables is the input (cause) and the other is the output (effect). Because bond lines in bond graphs connect the ports, the same variable is the input variable at one port and the output variable at the other connected port. Causal relationships between connected port variables are depicted in the bond graph literature by *causal strokes*. These are short lines drawn at one bond end (port) perpendicular to the bond (Fig. 2.28). This stroke denotes that the effort at the port is the input to the element and the flow variable is the output (Fig. 2.28a, b). At the other port just the opposite relation is valid; that is, the flow variable is the input and the effort variable is the output. Causal stroke assignment is independent of the power flow direction (Fig. 2.28a, c).

2.8.2   Causalities of Elementary Components

The causality assignment defines the input-output relationship of the elementary component constitutive relations. Possible types of causalities of elementary component ports are summarised in Fig. 2.29.

Source effort ports (Fig. 2.29a) can have only one possible type of causality, i.e. the effort is always the output, because flow at the input port is not defined. Similarly, at source flow ports the output is the flow, because the effort is not defined (2.29b). Thus, sources have *fixed causalities*.

The inertial component can have one of two possible causalities. If effort at the port is the input and the flow is output (Fig. 2.29c), the constitutive relations are then given by (2.2) and (2.5)

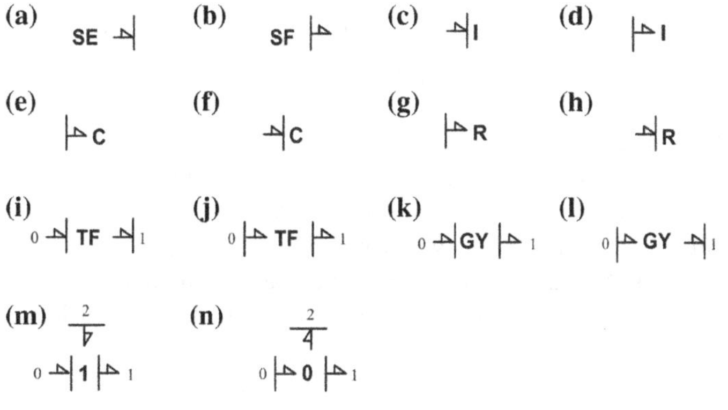

Fig. 2.29 Causalities of elementary components

$$p = p_0 + \int_0^t edt \atop f = \Phi^{-1}(p,par) \Bigg\}$$

(2.103)

Such causality is known as *integrating causality* because *integration* is used to calculate the output flow.

The other possibility is that the flow is the input and effort is the output (Fig. 2.29d). In this case evaluation proceeds by (2.4) and (2.1), i.e.

$$p = \Phi(f,par) \atop e = \dot{p} \Bigg\}$$

(2.104)

This type of causality is known as *differentiation causality* because *differentiation* is used to calculate the output.

Analogous causal forms exist for capacitive ports. If we take the flow variable as input and the effort variable as output (Fig. 2.29e), by (2.10) and (2.13) we have

$$q = q_0 + \int_0^t fdt \atop e = \Phi^{-1}(q,par) \Bigg\}$$

(2.105)

In this case we have integrating causality. On the other hand, if effort is the input and flow is the output (Fig. 2.24f), the calculation proceeds by (2.12) and (2.9), i.e.

$$\left. \begin{array}{l} q = \Phi(e, par) \\ f = \dot{q} \end{array} \right\} \qquad (2.106)$$

This yields differentiation causality.

Of these two possible causalities, integrating causality is *preferred* because integration is more easily implemented than differentiation. This is because integration works on the past values, whereas differentiation involves prediction.

For the resistor there are also two possible causalities. If the flow is the input and effort is the output (Fig. 2.29g), evaluation of the output is done using (2.18), i.e.

$$e = \Phi(f, par) \qquad (2.107)$$

On other hand, if the effort is input (Fig. 2.29h) and the flow is output, calculation is implemented by (2.19), i.e.

$$f = \Phi^{-1}(e, par) \qquad (2.108)$$

The first one is sometimes called *resistive causality*, and the other *conductive causality*. Preference of one over the other depends on which form is better defined, as some non-linear constitutive relationships are not invertible.

Transformers can also have two possible types of causality. If the effort at one port is the input, then at the other port the effort has to be the output; the same applies to the flows. For causality as expressed in Fig. 2.29i, the constitutive relations are given by (2.27), i.e.

$$\left. \begin{array}{l} e_1 = m \cdot e_0 \\ f_0 = m \cdot f_1 \end{array} \right\} \qquad (2.109)$$

On other hand, if causality is as in Fig. 2.29j, the constitutive relations are given by (2.28), i.e.

$$\left. \begin{array}{l} e_0 = k \cdot e_1 \\ f_1 = k \cdot f_0 \end{array} \right\} \qquad (2.110)$$

Two possible causalities for gyrators are shown in Fig. 2.29k, l, respectively. Inputs at the gyrator ports can represent either the efforts or the flows. For the case in which inputs are *efforts*, output flows are given by (2.30), i.e.

$$\left. \begin{array}{l} f_0 = k \cdot e_1 \\ f_1 = k \cdot e_0 \end{array} \right\} \qquad (2.111)$$

Similarly, if the inputs are *flows*, the output efforts are given by (2.29), i.e.

$$\left.\begin{array}{l} e_0 = m \cdot f_1 \\ e_1 = m \cdot f_0 \end{array}\right\} \tag{2.112}$$

Effort junctions represent the balance of efforts at the junction ports. Hence, one effort can only be the output at one port; all others must be inputs. For the effort junction in Fig. 2.29m, the effort at port 2 is taken as the output and all others are inputs. Thus, output effort e_2 is given by

$$e_2 = -e_0 + e_1$$

A similar statement holds for the flow junctions: Flow can only be the output at one port; all other flows must be inputs. For the flow junction in Fig. 2.24n, the output flow f_1 is given by

$$f_1 = f_0 - f_2$$

In the expression for output effort or flow, the sign of all input efforts or flows should be positive if the sense of the power flow is opposite to the sense of the output power flow. Otherwise, the sign is negative.

2.8.3 The Procedure for Assigning Causality

The causalities of junction, transformer, and gyrator ports are interrelated and thus imply constraints on the causalities of connected elements. The causalities of the complete bond graph can be assigned in a systematic way. The usual procedure is known generally as the *sequential causal assignment procedure* (SCAP) [1]. This procedure is summarised as follows:

1. Choose a source effort or source flow and assign causality to it. Extend the causality assignment, if possible, to the connected effort and flow junctions, the transformers, and the gyrators. Proceed in a like fashion until the causality of all sources has been assigned.
2. Choose an inertial or a capacitive element and assign to it the preferred (integrating) causality. Extend the causality assignment as in 1. Proceed until the causality of all such elements has been assigned. Otherwise, if the causality assignment of the all bonds is not achieved, go to the next step.
3. Assign causality to an unassigned resistor using any acceptable causality. Extend the assignment to the connected effort and flow junctions, transformers, and gyrators. Proceed until the causality of all resistors has been assigned. Otherwise, if the causality of all bonds is not already assigned, go to the next step.

4. Assign causality to any remaining bond. Extend the causality assignment to effort and flow junctions, transformers, and gyrators. Proceed until the causality of all bonds is assigned.

The bond graph to which causality has been assigned usually is termed a *causal* bond graph. Otherwise, it is termed an *acausal* bond graph.

The procedure can be illustrated on the simple body-spring-damper problem of Sect. 2.7.1. Other more complex examples can be found in, for example, [1]. Here we use the simplified bond graph of Fig. 2.16c, which is repeated in Fig. 2.30a.

We start with the source effort SE (step 1 of SCAP) and assign its causality as shown in Fig. 2.30b. We cannot extend the causality assignment immediately to the effort junction, as the connected port is an effort input port. There are no more sources, thus we proceed with step 2 of SCAP. We can choose to assign causality to either the inertial or the capacitive element. Let us choose to assign integration causality to the inertial element I (Fig. 2.30c). We now can extend the causality assignment to the effort junction, because the port connected to the inertial port is an effort output port, and all other junction ports must be effort input ports (Fig. 2.30d). This completes the causality assignment of the bond graph.

We have obtained integration causality of the capacitive element C, as well. If we start at step 2 by choosing the capacitive element instead of the inertial element, we would have to assign the causality of the inertial element before we could proceed to the effort junction. The first procedure is somewhat shorter.

The causal assignment of Fig. 2.30 defines the order of evaluation of the equations. This is shown by the block diagram of Fig. 2.31.

We start with the SE first. Next, we calculate the output flow of the inertial element. This is the input to the effort junction and the output of its all other ports. It

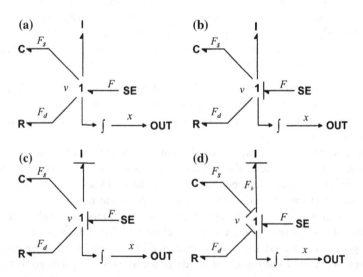

Fig. 2.30 Illustration of the causality assignment procedure

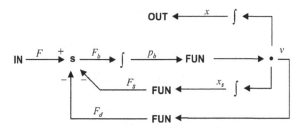

Fig. 2.31 Computational order of the bond graph Fig. 2.30d

is the input to the capacitor and the resistor used to calculate their outputs. These, together with the source effort output, are used to calculate the output of the effort junction, hence the inertial element input. Independently, it is used as input to the integrator to calculate body position.

2.9 The Formulation of the System Equations

The bond graph of a system completely defines its mathematical model. In Sect. 2.7 it was shown that the model could be generated directly from the bond graph by describing the elementary components, including junctions, in terms of their constitutive relations. This way of representing mathematical models is known as the *descriptor* form [9] and is widely used in electrical circuits. This is a non-minimum form because the equations are not expressed using the minimal number of variables. Some variables could be eliminated, e.g. by substituting into the equations of flow and effort junctions. This approach is in effect used in modified nodal analysis (MNA) of electrical circuits [10, 11]. This also is the case with certain approaches used in multi-body dynamics [7]. In Sect. 2.7 it was shown that the matrix of the equations is typically very sparse, and this can be used to advantage in their solution.

The descriptor form of equation formulation leads to the model in the form of systems of differential-algebraic equations (DAEs). The success or failure of the descriptor formalism depends to a large extent on the possibility of solving equations in DAE form efficiently and reliably. Solving such equations has a relatively long history and started with the famous DIFSUB routine of CW Gear [12] for stiff systems. The work reported in this book also has its roots in software that solves DAEs in a way that is based on the DIFSUB routine. From that time significant advances have been achieved in the theory of DAEs and their application [13, 14]. Today this is a viable approach to solving simulation models. We return to this again in Chap. 5.

Another common approach is to formulate the system in *state space* form. This technique uses a minimal set of independent variables to formulate the governing equations. It has its roots in the generalised coordinate methods of Analytical Mechanics [15], but it also is used widely, and is perhaps better known, from

Control Theory. The theory of state-space equations has been a topic of research for a long time and is well understood. This approach is followed not only in bond graph theory, but is also used in many continuous system simulation languages (Sect. 1.7).

The usual approach in continuous system simulation languages is to create a system of sorted equations that is solved sequentially. Such systems can be solved relatively easily. Unfortunately, in many engineering problems of practical interest it is not easy to put the equations in such a form.

The sequential causal assignment procedure of Sect. 2.8.3 was really designed as an aid to the generation of mathematical model equations in sorted form. From that the equations can be reduced to the state space form. The bond graphs with completed causality assignment can be put in such a form if inertial and capacitive elements have integrating causality, and if there are no algebraic loops [1, 16]. We illustrate this with the body-spring-damper system represented by the causal bond graph of Fig. 2.30d (or, equivalently, by the block diagram of Fig. 2.31). More elaborate examples can be found elsewhere [1].

We start with the source effort (sixth of (2.59)), following the order of causal assignment of Sect. 2.8,

$$F = \Phi(t) \tag{2.113}$$

The output of the inertial element **I** is given by (see (2.103) and fifth (2.59))

$$v = p_b/m \tag{2.114}$$

The variable v (by the effort junction) is used as the input to the capacitor C, resistor R, and the integrator. The order of evaluation of these elements is immaterial. From the first (2.105), written in derivative form, or first (2.59), we get

$$\dot{x}_s = v \tag{2.115}$$

Output of the capacitor is given by the equation (see (2.105) and second (2.59))

$$F_s = k \cdot x_s \tag{2.116}$$

Output of the resistor (see (2.107) or the third (2.59)) is

$$F_d = b \cdot v \tag{2.117}$$

Hence, all the inputs to the summator are found and we can calculate its output as

$$F_b = F - F_s - F_d \tag{2.118}$$

The output of the summator is the input to the inertial element. Thus, from the first equation of (2.103), written in differential form, or fourth (2.59), we get

$$\dot{p}_b = F_b \tag{2.119}$$

To these equations we add the output of the integrator written as (the last (2.59))

$$\dot{x} = v \tag{2.120}$$

This completes the generation of the system of sorted equations.

The equations above consist of differential equations (2.115), (2.119) and (2.120), and algebraic equations (2.113), (2.114), (2.116), (2.117), and (2.118). Hence, it is a differential/algebraic system of equations (DAE), but of a special structure. We classify all variables in these equations as being either differentiated or participating in algebraic operations only. The first are called *differentiated variables*, i.e. x_s, x and p_b. The others are *algebraic* variables; in the equations above these are F, v, F_s, F_d, F_b. All algebraic variables above can be expressed as functions of the differentiated variables and time. We see that the variables F, v, and F_s are already in this form (see (2.113), (2.114) and (2.116)). Eliminating v from (2.117) and (2.114) we get

$$F_d = \frac{b}{m} \cdot p_b \tag{2.121}$$

Finally, substituting from (2.113), (2.116), and (2.121) into (2.118) we obtain

$$F_b = \Phi(t) - k \cdot x_s - \frac{b}{m} \cdot p_b \tag{2.122}$$

We now substitute these expressions into the differential equations (2.115), (2.119), and (2.120). We, thus, obtain

$$\dot{x}_s = p_b/m \tag{2.123}$$

$$\dot{p}_b = \Phi(t) - k \cdot x_m - \frac{b}{m} \cdot p_b \tag{2.124}$$

$$\dot{x} = p_b/m \tag{2.125}$$

Note that (2.123) and (2.125) has the same form, but generally different initial conditions, because the first refers to the spring extension, and the second to the body position.

Equations (2.123)–(2.125) represent the model in the *state-space form*. Variables x_s, p_b and x constitute a minimal set of independent variables that completely define the state of the system. Solving these equations with suitable initial conditions, the all other variables can be found from (2.113), (2.114), (2.116)–(2.118).

In general, if all capacitive and inertial ports have integrating causality, then the corresponding differentiated variables, i.e. generalised moments and displacements of Sects. 2.5.2 and 2.5.3 can be looked upon as independent variables where accumulation of past histories of the efforts and flows take place. Such variables completely determine the future state of the system and usually are called *state variables*. All other variables can then be determined if the state of the system is known. If, in addition, there are no algebraic loops—that is, there are no implicit algebraic equations between variables—then all other variables can be eliminated from the governing equations. Thus, if all state variables are represented by a vector **p** and all external inputs (represented by the sources) by a vector **u**, then a change of system state can be described by a vector equation

$$\dot{\mathbf{p}} = \mathbf{\Phi}(\mathbf{p}, \mathbf{u}, t) \tag{2.126}$$

where $\mathbf{\Phi}$ is a suitable vector-function of the state, inputs, and eventually time. This is an ordinary differential equation that can be solved given the *initial state* of the system. Such an equation is termed the *state-space equation* of the system.

2.10   The Causality Conflicts and Their Resolution

The sequential causal assignment procedure (SCAP) of Sect. 2.8.3, in many cases of practical interest, leads to a causally augmented graph that cannot be described by equations in state space form [1, 17, 18, 19]. We illustrate this using the examples of Sect. 2.7.

We first analyse the electrical circuit of Fig. 2.18, but with the resistor replaced by a diode (Fig. 2.32a).

If we model the diode as a non-linear resistor, we get the equivalent causally augmented bond graph shown in Fig. 2.32b (see Fig. 2.19c). The problem here is that the diode is a non-linear element normally described in conductive form, i.e. the diode current is a function of the voltage across the diode. It is thus in conflict with the assigned causality that implies resistive causality. In order to resolve conflicts caused by non-linear elements, the relaxed causal assignment procedure was proposed in [20] and its modification in [21]. This procedure requires that, at step 2 of the SCAP (Sect. 2.8.3), propagation of causalities over junctions may not violate the fixed causality of non-linear elements. Thus, applying the causal assignment procedure again results in the augmented bond graph of Fig. 2.32c. The conflict caused by the fixed causality of the diode disappears, but a *causal conflict* appears at the effort junction because there is more than one output. Thus, the equation corresponding to the effort junction constitutes an algebraic constraint that the variables have to satisfy. The procedure permits casual conflicts at effort or flow junctions as an indication that the mathematical model is of the differential-algebraic equations (DAE) form, rather than of the state space form.

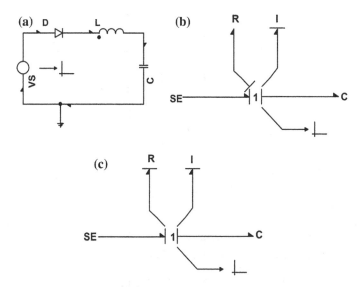

Fig. 2.32 Causality conflict in the electric circuit with a diode

In the see-saw problem (Sect. 2.7.3) there also is a causality problem. The see-saw is a single-degree-of-freedom mechanical system. The motions of the bodies depend on the motion (rotation) of the platform, which is represented in Fig. 2.26 by the `LinRot` transformers. To show this we apply the SCAP to the bond graph of Fig. 2.27. The resulting causally augmented bond graph is shown in Fig. 2.33. There is only one inertial element with integrating causality. All others have differential causalities.

This bond graph is rather complicated, so numbers are used to indicate the order of the causality assignments. Selection of the preferred (integrating) causality for one inertial element, e.g. the platform rotation, implies derivative causalities for all other inertial elements. Hence, there is only one state variable and all other generalized variables are non-state variables. The model again is a system of differential-algebraic equations.

There have been attempts to resolve causality conflicts by, for example, adding 'parasitic' compliances or inertias [22, 23]. This is not an acceptable approach, however, because, in the first instance, it is not clear how to do this without adversely changing the model behaviour. On the other hand, such modified models are not much easier to solve numerically than the corresponding DAE models because they are very stiff.

The causality assignment defines the model's computational scheme based only on the model's structure. In cases in which the model changes sufficiently, such a priori schemes can lead easily to a loss of efficiency and even failure of the equation solving routines. This is the case with models that have discontinuities.

Discontinuities are present in engineering systems in various forms, e.g. switches in electrical circuits, hard stops, clearances, and dry friction. For example,

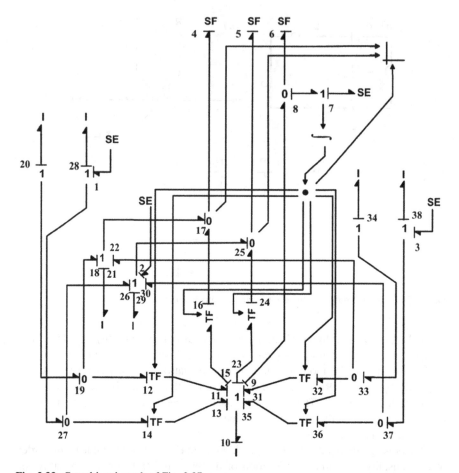

Fig. 2.33 Causal bond graph of Fig. 2.27

the diode represented in the circuit of Fig. 2.34a is modelled as a switch in Fig. 2.34b.

If the diode is forward-biased (conducting), then the switch behaves as a source effort implying a zero voltage drop across the diode (Fig. 2.34c). When, on the other hand, the diode is reverse-biased, the switch behaves as a flow source of small reverse saturated current (Fig. 2.34d). The model structure and causalities are apparently different for these two states. In the conducting regime the system has two state variables, while in the non-conducting regime it has only one.

There have been various attempts to solve causality problems with switches [24–28]. Overall, these procedures are not completely satisfactory in the general case. This is particularly true if the discontinuities are not confined to switch elements, but appear in the element constitutive relations, too.

Fig. 2.34 Change of causality pattern in the circuit with a switch

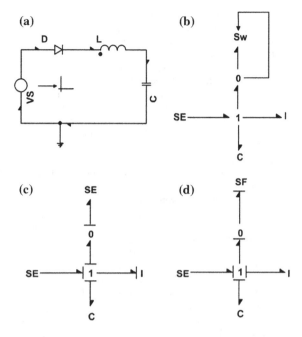

The concept of causality is generally not suitable for use in an automated object-oriented modelling environment. It is not only too restrictive with respect to the forms of models that can be used, but also puts restrictions on the design and usability of models libraries. A component that has one causality pattern in one system can have a quite different one when inserted in another system. We disregard causality issues when developing models of general engineering and mechatronic systems. Modelling is treated as a separate task from model simulation. The models will be generated in the form of DAE systems and solved as such.

References

1. Karnopp DC, Margolis DL, Rosenberg RC (2000) System dynamics: modeling and simulation of mechatronic systems, 3rd edn. Wiley, New York
2. Thoma J, Bousmsma BO (2000) Modelling and simulation in thermal and chemical engineering, a bond graph approach. Springer, Heidelberg
3. Borutzky W (2010) Bond graph methodology: development and analysis of multidisciplinary dynamic system models. Springer, Berlin
4. Breedveld PC (1982) Proposition for an unambiguous vector bond graph notation. J Dyn Syst Measure Control 104:267–270
5. Fahrenthold EP, Wargo JD (1991) Vector and tensor based bond graphs for physical systems modeling. J Franklin Inst 328:833–853
6. Analog-to digital converter. http://en.wikipedia.org/w/index.php. Accessed 22 Apr 2014

7. Haug EJ (1989) Computer-aided kinematics and dynamics of mechanical systems, Vol I: basic methods. Allyn and Bacon, Needham Heights
8. Beer FP, Johnson ER (1990) Vector mechanics for engineers, 2nd SI edn. McGraw-Hill Book Co., Singapore
9. Newcomb RW (1981) Semistate description of nonlinear and time variable circuits. IEEE Trans Circuit Syst CAS-26:62–71
10. Märtz R, Tischendorf K (1997) Recent results in solving index-2 differential-algebraic equation in circuit simulation. SIAM J Sci Comput 18:135–159
11. Vladimirescu A (1994) The spice book. Wiley, New York
12. Gear CW (1971) Numerical initial-value problems in ordinary differential equations. Prentice Hall, Englewood Cliffs
13. Brenan KE, Cambell SL, Petzold LR (1996) Numerical solution of initial-value problems in differential-algebraic equations, classics in applied mathematics. SIAM, Philadelphia
14. Hairer E, Wanner G (1996) Solving ordinary differential equations II, stiff and differential-algebraic problems, 2nd Revisited edn. Springer, Heidelberg
15. Goldstein H (1981) Classical mechanics, 2nd edn. Addison-Wesley Publishing Co., Reading
16. Rosenberg RC (1971) State-space formulation of bond graph models of multiport systems. Trans ASME J Dyn Syst Measure Control 93:35–40
17. Van Dijk J, Breedveld PC (1991) Simulation of system models containing zero-order causal paths—I. Classification of zero-order causal paths. J Franklin Inst 328:959–979
18. Van Dijk J, Breedveld PC (1991) Simulation of system models containing zero-order causal paths—II. Numerical implications of class 1 zero-order causal paths. J Franklin Inst 328:981–1004
19. Gawthrop P, Smith L (1996) Metamodelling: bond graphs and dynamic systems. Prentice Hall, Hemel
20. Joseph BJ, Martens HR (1974) The method of relaxed causality in the bond graph analysis of nonlinear systems. Trans ASME J Dyn Syst Measure Control 96:95–99
21. Van Dijk J, Breedveld PC (1995) Relaxed causality a bond graph oriented perspective on DAE-modelling. In: Cellier FE, Granda JJ (eds) International conference on bond graph modeling and simulation. Las Vegas, Nevada, pp 225–231
22. Margolis D, Karnopp D (1979) Analysis and simulation of planar mechanisms using bond graphs. ASME J Mech Des 101:187–191
23. Zeid A, Chung CH (1992) Bond graph modelling of multibody systems: a library of three-dimensional joints. J Franklin Inst 329:605–636
24. Borutzky W (1995) Representing discontinuities by sinks of fixed causality. In: Cellier FE, Granda JJ (eds) International conference on bond graph modeling and simulation. Las Vegas, Nevada, pp 65–72
25. Cellier FE, Otter M, Elmqvist H (1995) Bond graph modeling of variable structure systems. In: Cellier FE, Granda JJ (eds) 1995 International Conference on Bond Graph Modeling and Simulation. Las Vegas, Nevada, pp 49–55
26. Lorentz IF, Haffaf H (1995) Combination of discontinuities in bond graphs. In: Cellier FE, Granda JJ (eds) International conference on bond graph modeling and simulation. Las Vegas, Nevada, pp 56–64
27. Mosterman PJ, Biswas G (1998) A theory of discontinuities in physical systems models. J Franklin Inst 335B:401–439
28. Soderman U, Stromberg JE (1995) Switched bond graphs: towards systematic composition of computational models. In: Cellier FE, Granda JJ (eds) International conference on bond graph modeling and simulation. Las Vegas, Nevada, pp 73–79

Chapter 3
An Object-Oriented Approach to Modelling

3.1 Introduction

This Section describes the principles of object-oriented design based on the bond graph technique. The idea is to perform the modelling visually, without any coding; all the interaction between the modeller and the model will be through a visual development environment.

We do not develop a new modelling language, but instead design the classes that are used to create a structured model as a tree of linked objects. This will be stored as a set of linked files. It is not necessary to have the complete model in memory, as parts are loaded as required.

It is very important during the development phase to relieve the user of certain implementation details, such as how to create components, where and how to store data, etc. These operations are automated. The user communicates with the model in familiar terms, e.g. by the names of a project and of its components, how they are connected, and by defining component constitutive relations in terms of variables meaningful to the user. The developer thus concentrates on the problem. The modelling environment (framework) provides as much support as possible to fulfil this task.

3.2 The Component Model

The concept of *component model* plays a central role in our approach to modelling [1].[1] The model of a component is created in two steps (Sect. 2.2). First, a *component* object is created as a *word model*. Next, by "opening" the component, the associated *document* object is created. The document describes the component

---

[1]This is not to be confused with Microsoft's Component Object Model (COM) [2].

© Springer-Verlag Berlin Heidelberg 2015
V. Damić and J. Montgomery, *Mechatronics by Bond Graphs*,
DOI 10.1007/978-3-662-49004-4_3

Fig. 3.1 The *CComponent*
class

Attributes:
Type, position and extent
Component ID and filename
Textual data
List of the ports
etc.
Methods:
Constructors and destructors
Copy operations
In place editing
Port operations
Drawing operations
Storing and loading
etc.

model in terms of a bond graph. It contains lists of the word model components and interconnecting bonds that constitute a bond graph model of the component. The component and its associated document are intimately related and are used as the engine for systematic model development either by the top-down (decomposition) or bottom-up (composition) approach, or by a combination of both.

3.2.1   The Component Class

CComponent is the base class used for the creation of the component word model (Fig. 3.1). It holds textual information, including its name, ports that interconnect to other components, as well as information that define its position on the screen. Visually, the component is designed as a rectangle with a strip surrounding its name (Fig. 3.2). This strip is used to place the ports.

Creation of a word model component typically consists of

Fig. 3.2 The component
word model visual
representation

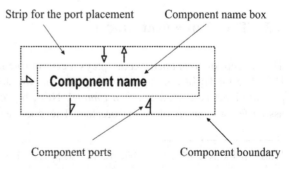

1. The selection of a position on the screen where the component is to appear.
2. The construction of the *CComponent* and initialising its position, type, and text font.
3. The creation of a unique identifier (*id*) and associated document file name, and storing them in the object.
4. Editing the name and storing it in the object; it is also possible to use a pre-defined graphical symbol.
5. Saving the bounding rectangle in the component object.
6. The creation of ports and adding them to the component object
7. The component word model is created within a suitable *document* object that manages the complete process of object creation (Sect. 3.2.2).
8. The component class supports simple text editing operations for the in-place editing of the object name (title). Typical of such operations are inserting or deleting of a character, new line creation, joining lines, jumping to the head or the end of the text, etc. The component name is only for the user's convenience; the framework refers to a component solely by its *id*. Thus, two different components can have the same name and appearance, but be different objects with different *ids*. The component *id* and document filename are unique in the workspace in which all model files are stored.

The Component class implements several methods for port creation and deletion, and for moving the ports around the component periphery. All ports that are created are stored in the component object as a list.

The component is responsible for its own visual appearance. This means that the component class has methods for drawing itself on the screen and for printing to a printer or to a file. The component is drawn as part of the drawing operations of the document object, which contains the component (Sect. 3.2.2).

The component class supports creation of a copy of a word model component. In addition to the usual copy constructor, a virtual method is provided for creating a copy of the component object. Because this is a new component object, a unique component *id* and a unique document file name are created and stored in the object. All other necessary changes are also made, e.g. in the ports (Sect. 3.4).

3.2.2 The Document Class

The document class associated with *CComponent* is termed *CBondSimDoc* (Fig. 3.3). This is a container class used to create and store the bond graph model of a component. The model is held as a list of document ports, components, and connecting bond objects.

Visually, a document is designed as an area framed by a rectangle that is used for placement of the components and drawing interconnecting bond lines (Fig. 3.4). Document ports are created in the surrounding strip and correspond to the ports of its word model. The ports serve as outside connectors of the components contained

Attributes:

> *Component name and document file name*
> *Links to previous document and the word model*
> *component*
> *List of document ports, components and bonds*
> *List of model parameters*
> *etc.*

Methods:

> *Create a new document*
> *Open saved document*
> *Closing and saving document*
> *Creation of document ports, components and bonds*
> *Creation of model parameters and their linking*
> *Remove document*
> *Copying document*
> *Support for mathematical model generation*
> *etc.*

Fig. 3.3 The *CBondSimDoc* class

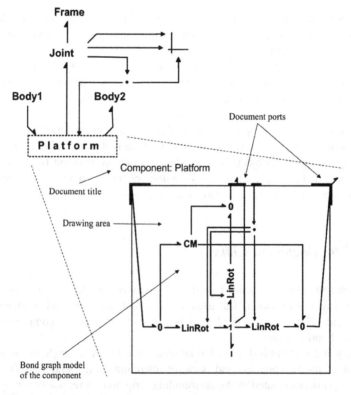

Fig. 3.4 The visual representation of a document object

in the document. Their positions correspond to the positions of the word model component ports. In this way it is visually clear which port correspond to which.

The document and its corresponding word model constitute the complete model of a component. Such an object typically is accessed through the corresponding word model. Hence, a word model serves as the interface to the document that contains its model. Important parts of this interface are the ports (Sect. 3.4). A new document object typically is created as follows:

1. Assuming that a document object (the current document) that contains a word model has already been created, a component for which the accompanying document is created is selected and a message is sent to the document to create a new document object.
2. The document method creates a new document object. This method uses a pointer to the word model object, from which the necessary data are taken, such as the name of the component, the filename where the new document object is to be saved, and the component ports.
3. A new document is created and its corresponding attributes are set. Document ports also are created. These correspond to the word model object ports (Sect. 3.4) and are added to the list of document ports. Otherwise, a new document is empty, i.e. without any component or bonds. The new document also creates links to the previous (lower level) document.

A new document object is not created if the ports of the corresponding word model component are not already defined. Such a document object would make no sense.

Once a document object is created, new word model objects of the contained components can be created inside the document working area. The creation process is managed by the document as already described (Sect. 3.2.1). The components created are added to the list of components that the document contains. In the same vein, the document manages the creation of the bonds that are used to interconnect components contained in the document, or to connect them to the document ports (Sect. 3.4). Model development can continue by creating a new document object for every word model in the document, thus developing higher levels of abstraction of the component model.

The document class manages saving the document objects to the document file. Before any a new document object is created, the current object is saved. All objects that the document contains are saved to the document file by calling the corresponding method. During this operation a copy of its word model object also is saved as a header, as well as the filename of the previous document and other pertinent data.

This saves the model as a tree of interconnected document files. All the files that make up the model are double-linked, i.e. forward through the components contained in the document, and backward by the filenames of the previous document. We return to this problem again in Sect. 3.9.

Document objects are *persistent* [3] because they exist even after they are closed and deleted from memory. To remove a component completely from a document it is not enough to remove it from the list of the components that the document object maintains and to destroy it. The accompanying document file has to be removed, too. Removing a component only creates dangling documents, i.e. documents of which nobody is aware. This means that it is necessary to open documents until the leaf documents are reached; then move backward by removing the document files until the component that is to be removed is reached. Only then can the component be disconnected from the container document and destroyed.

The document class is complex and supports many operations of the automated modelling framework. The drawing operations are executed in a visual environment in which the modelling system is implemented, e.g. the Windows system. During its execution the objects contained in the document are called to draw themselves, e.g. the document ports, the components and their ports and the bonds. Similar actions are performed when printing a document to an external printer or to a graphic file.

One important and much used operation that the document class supports is copying. This creates a copy not only of the document, but also of all documents in the tree. This operation starts from the document root and proceeds toward the document tree leaves. In this process the *ids* and filenames of all contained components are changed. This causes a new document tree to be created that does not share any component or document with the original tree. Document copying is implemented in the following way:

1. Open the document object from a file and set the new filename and path where the copy is to be saved.
2. For components contained in the document:

   - Get the original document filename
   - Create a new component *id* and document filename
   - Reset the component *id* and filename to the new values, saving the original filename to a temporary location. In the process, component *ids* are changed in the component ports, as well as in the bonds connected to these ports.

3. Save the document under the new filename
4. For every component contained in the document:

   - If there is an existing document file repeat step 1

5. Close the document

This operation is called when a copy of a component model is needed. In this case a copy of the word model component is made first as explained in Sect. 3.2.1, and then a copy of its accompanying document is performed as described above. When this operation is called directly (Sect. 3.9), a copy of the word model component stored in the document header is made first, then a copy of the document is made, as explained above.

The document class also supports many other operations, such as creation of the mathematical model parameters defined at the level of a document object. These parameters are visible to all components contained in the document. Linkage of these parameters is designed so that a parameter defined in a document overrides (hides) a parameter of the same name that is defined in a lower-level document. The document class also has methods for mathematical model generation. This is addressed in Chap. 5.

3.3 The Component Class Hierarchy

The *Component* class is the base class from which more specialised component classes are derived. Figure 3.5 illustrates the hierarchy of the component classes.

One type of specialisation is visual appearance. The component can be depicted, for example, by graphical symbols, not only by its name. In electrical engineering such practise is widely used and has been standardised (e.g. by ANSI, DIN). It is also used in mechanical engineering, but not to such an extent or with such versatility. Hence, to simplify modelling with bond graphs, several derived classes are defined that partly support such schemes.

In the first branch on the left of the hierarchy tree of Fig. 3.5 shows classes that represent some of the more common electrical components. These, of course, don't imply any specific model, for they are really word model classes represented differently. The model should be defined in the accompanying document object (see e.g. Fig. 2.18). Using these components and connecting them by bonds gives the bond graph an appearance very close to the usual electrical schemes. An important difference from the latter is the half-arrows used to indicate power flow directions in the circuit.

Mechanical engineering uses various schematics to depict, for example, systems in vibrations, hydraulics, and pneumatics. Here there are defined classes to represent simple mechanical components only, such as bodies in translation and rotation, springs, dampers, etc. (the second branch of the tree of Fig. 3.5).

Classes of fundamental importance involve derived classes that represent the elementary components discussed in Sect. 2.5. These are the classes in the third branch of the hierarchy tree of Fig. 3.4. In accordance with bond graph practise, these components use predefined textual symbols that can readily be changed.

It should be noted that some simple electrical and mechanical components can be defined as a specialisation of certain elementary components, such as ground potential, and electrical or mechanical junctions.

Elementary components differ fundamentally from the base class component in that they do not have an accompanying document; they are entities in themselves. The question, then, is where to put the variables and constitutive relations of these components. The natural answer is: in their *ports*. Derived elementary classes

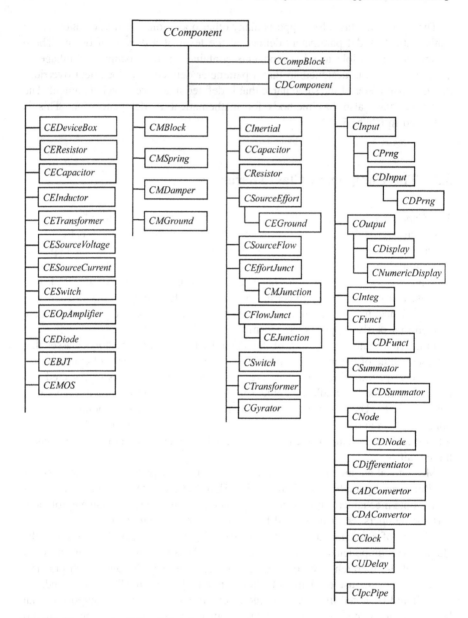

Fig. 3.5 The component class hierarchy

support editing of the constitutive relations in the form given in Sect. 2.5. The elementary component classes also support creation of locally defined model parameters; that is, those valid only in the component.

Elementary component classes define various class-specific methods that over-ride the base class methods. They also have an important role during mathematical model creation, and in simulations, too.

The last group of components deals with the block diagram operations of Sect. 2. 6. Components in this group are shown as the last branch of the hierarchy in Fig. 3. 5. They are similar to the elementary components discussed previously, but support only control ports. These components also use their ports to store the variables and input-output relations. An exception is the *COutput* class that serves for collecting the simulation output and displaying it on the screen.

There are two specializations of this last class: *CDisplay* and *CNumericDisplay*. The first serves to represent the output component in the form of the familiar *x*-*y* plotter symbol as has already been discussed in Chap. 2 (see e.g. Figs. 2.18 and 2. 21), and the other displays the output in form of a numeric counter.

The components in this last group describe both the continuous- and discrete-time components. While continuous-time components can be freely mixed with components having power ports, the discrete components are an exception to this rule. Therefore, to support development of discrete models a discrete subclass of word model component the *CDComponent* is defined as shown at the top of 3.5.

Some of discrete components are just specialization of the corresponding con-tinuous time components, e.g. input, function, node, summator. The same holds for the pseudo random number generator classes *CPrng* and *CDPrng*. On the other hand there are also classes used for modelling intrinsically discrete (digital) pro-cesses such as A/D and D/A conversions, digital clock, unit delay components, and others which do not have continuous-time counterparts.

There is one other interesting component class: *CIpcPipe*. This is a class that supports creation of *anonymous or named pipe* components used for *inter-pro-cesses communication* (*IPC*) between the *BondSim* program environment and an external program.

Finally, there is a derived class that differs from the classes discussed above. This is the *CCompSet* class, a container class for a group of component objects. It is described in Sect. 3.8.

The Document class, as previously described (Sect. 3.2.1), manages creation of component objects. Because there are no virtual constructors, the technique of the object factory is employed [4] to construct objects of the correct type. Similarly, every derived class overrides the method for creating a copy of a component, insuring that a copy of the object of the correct type is created.

3.4   Port and Bond Classes

The component model introduced in the last two sections doesn't specify com-pletely the interconnections between components. It is necessary to work out the port interconnections, too. One approach is based on the notion of *multibonds* [5, 6], a generalisation of the concept of bonds to the multidimensional case. We accept

Document object: Component

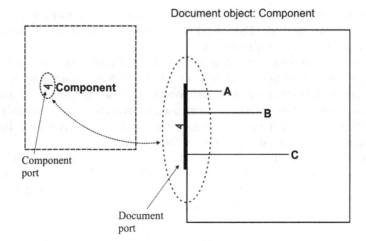

Fig. 3.6 A component and the accompanying document port

that another approach, based on the concept of *compound* ports, better fits the object-oriented philosophy. Thus, bonds are taken as simple objects that define only which port is connected to which; everything else is the responsibility of the ports. We define the necessary port classes, but first describe what we ask of them.

Looking at the component object and its accompanying document object (Fig. 3.6), we identify two types of ports: component ports and document ports.

The first type connects *external* components. Such a port "knows" that it belongs to a certain component and that it is, or it is not, connected by a bond. The port doesn't need to know what is on the other side of the bond. This is the responsibility of the bond. The document port, on the other hand, belongs to the document object and serves for the internal connection of the document's components. This kind of port knows what, and how many, bonds are connected to it.

These two types of ports are used to describe connections looking from two sides of the same component, from outside and inside. Looking from the outside—that is, at the component port side—it would be helpful to know how many bonds are connected inside. This we call the *dimension* of the port. On the other hand, looking from the inside, it would be useful to know if the port is connected on the outside by a bond or not.

To simplify the interchange of information between these two types of ports, we define the component port class *CPort* as the base class and the document port class *CDocPort* as a derived class (Fig. 3.7).

Creation of these objects is the responsibility of their containers, *CComponent* and *CBondSimDoc*, respectively. A possible scenario is as described below:

1. When a user issues a command to the component object to insert a port, the component method responsible for port creation is called. It, in turn, calls the base class constructor, which supplies the port type requested, and its position

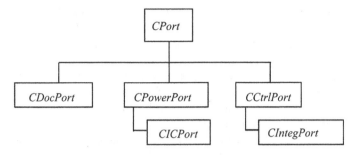

Fig. 3.7 The port classes hierarchy

on the component boundary. The port constructor creates a port object and assigns to it the component and the port *ids*, as well as the port type. The object initially is unconnected, both externally and internally (dimension 0). Depending on it's type and position on the component boundary, the necessary visual data are created that are used for drawing. The object subsequently is added to the list of the ports that the component maintains.

2. The document object is responsible for the creation of document ports. These are created during document object creation, based on the component object ports (Sect. 3.2.2). The base class part of the document port object is just a copy of the accompanying component port object translated to a position on the outside of the document drawing area (Fig. 3.6). The position is calculated such that it corresponds to the position of the accompanying component port in the strip surrounding the component name. Thus, they share the same information.

3. When a bond is connected to a component port, its *id* is sent to the port and stored there. The accompanying document port is updated at the same time. Similarly, when a bond is internally connected to the document port, the bond *id* is inserted in the list of bonds that the document port maintains. The position in the list corresponds to the position of the connected bond line in the connection rectangle. The dimension of the port is changed, as well as that of the corresponding component port. Thus, every change at one side of the interface affects the other side.

The ports belonging to elementary components are different because they have no document port counterparts. These ports serve mainly for storing data about the component model constitutive relations as "seen" at those ports. The necessary information is different for the power and control ports. Thus, two port classes, *CPowerPort* and *CCtrlPort*, are defined that are derived from the *CPort* base class (Fig. 3.7).

The power port class defines the port effort and flow variables, as well as the element constitutive relations. Similarly, the control port class defines the control port variables (signal). The constitutive relations are defined only at the output ports.

There are two other classes, *CICPort* and *CIntegPort*, derived from the *CPowerPort* and *CCtrlPort* classes, respectively (Fig. 3.7). These define differentiated (state) variables that the Inertial, Capacitor, and Integrator components need (Sects. 2.5 and 2.6).

Port object creation is managed by the corresponding elementary component objects in a similar way to the creation of ports of the base components, as described above. The variables are defined with default names and default constitutive relations. These depend on the type of the elementary component. Subsequent to creation, the names of the variables and the specific form of the constitutive relations can be changed. Details of the syntax used for the description of the constitutive relations are given in the next section.

Port classes, like component classes, define methods for operations on the port objects, such as construction and deletion of the object, copying, saving, loading, and drawing.

The final object required to close the bond graph is the bond itself. onds are simple objects that indicate which port is connected to which. A *CBond* class is defined with attributes that hold the bond identifying label (*id*), the starting and the ending component, and the port *ids*, as well as data necessary for the visual representation of the bonds. The bond class defines methods necessary for the creation and destruction of bond objects, copying, saving, loading, and drawing. Bonds, as objects, are contained in documents. The procedure for the creation of bonds is as follows (Fig. 3.8):

1. Select a starting component port by clicking it. Drag the mouse cursor with the left mouse pressed to the next intermediate point and release it. Move the mouse cursor with the key released it until it is over a port of the other component we wish connect to and click the mouse. The starting or ending ports could be the document ports as well.
2. The document class is called to create a *CBond* object. The *ids* of the starting and ending components and ports are set to the object. The coordinates of the starting and ending points of the bond line are stored into the bond object as well. Add the bond object to the list of bonds in the document object.

Fig. 3.8 The creation of a bond line

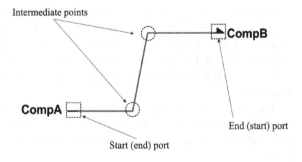

3. We can modify the bond drawn by simply picking a point on the bond line and dragging it (similarly as we can drag a rubber string) and release the mouse. When the mouse is released the additional points are added to the collection of the points stored in the bond object, which define the bond line, and the bond is redrawn.

3.5   Description of the Element Constitutive Relations

The constitutive relations of elementary components are defined at the component ports in the form of symbolic expressions. These expressions are of the form

$$variable = algebraic\ expression$$

where the algebraic expression is formed of variables, numerical constants, parameters, and the operators.

The acceptable variables in the expression are the port variables and time t, which is a reserved symbol, i.e. it cannot be used for port variables or parameters. Other port variables of the corresponding component can also be used. There are also some restrictions on the implied form of the constitutive relations of specific components. These are described in Sects. 2.5 and 2.6.

The standard forms of integer and floating-point constants are acceptable, e.g. 12, -1020, 2.7612, $-1.36e-5$. Integer constants are internally converted to floating-point constants. Constants can be defined symbolically in the form of parameter expressions formed from numerical constants and other symbolically defined constants using the operators that are described later. Symbolically defined constants can be freely used as parameters in algebraic expressions. These can be defined at the level of the component or at some lower level, (see Sect. 3.2.2).

The common arithmetic, relational and logical operators (Table 3.1) are permitted in algebraic and parametric expressions. The exponentiation operator ("^") can also be used, as well as function calls. To describe discontinuous relations, e.g. in Switch

Table 3.1 Operators supported in expressions

Operator	Meaning
()	Function call
!	Logical not
+ −	Unary plus and minus
^	Exponentiation
* / %	Product, division and mod
+ −	Addition and subtraction
< > ≤ ≥	Relational operators
& |	Logical AND and OR
? :	Arithmetic if (conditional)

elements, the *if-else* type of control statement can be expressed by C/C++ like operators "?" and ":". These can be nested. Table 3.1 lists the supported operators. The operators follow the common priority rules, as implemented in C/C++.

All common elementary mathematical functions are supported, such as *sin*, *cos*, *tan*, *log* (natural base), etc. Sometimes the functions are not known in analytical form but as tabular data. For example, tables might result from finite element analysis or experiments. Such data can be interpolated by polynomials and used in the expressions. The *BondSim* program supports an interface that accepts one- and two-dimensional tabular functions. Functions are interpolated by B-splines [7] and are referred to in an expression by a user-assigned name, e.g. *flux(i,x)*.

3.6   Modelling Vector and Higher-Dimensional Quantities

The power ports of elementary components define a pair of effort-flow variables. Similarly, a control port defines a single control variable. Using components—not necessarily elementary ones—it is possible to represent more complex variable structures. We describe these structures for the power ports, but the description holds for the control ports, too.

Thus, if several elementary components that are contained in a component are connected internally to its document port, externally the accompanying component port can be looked as representing a pair of effort-flow vectors (Fig. 3.9a). This relationship can be seen more clearly if we substitute the component word model with its document, i.e. transform it from a multi-level to a single-level representation (Fig. 3.9b). The port of Component A (Fig. 3.9a) holds a list of bonds, each

Fig. 3.9 Representation of a pair of effort/flow vectors:
a Component representation.
b Single level representation

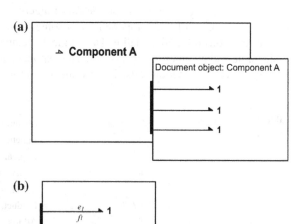

Fig. 3.10 Representation of complex effort/flow structures: **a** Component representation. **b** Single level representation

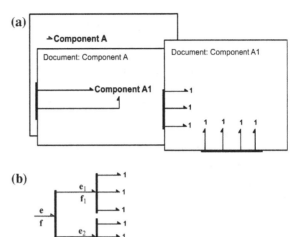

of which points to the elementary ports where the component effort-flow pairs are defined. Hence, such a port represents a pair of effort and flow *vectors*

$$\mathbf{e} = (e_1 \quad e_2 \quad e_3)^T \tag{3.1}$$

and

$$\mathbf{f} = (f_1 \quad f_2 \quad f_3)^T \tag{3.2}$$

where superscript T denotes vector transposition. This approach can be used for 3D representation of a force applied at a body point and the corresponding velocity (for 2D representation, see Fig. 2.22).

The dimension assigned to a port indicates the number of ports connected internally. In the case shown in Fig. 3.9, the port dimension is 3; but it could, of course, be any number. The connected ports are the ports of contained components. These also can be of higher dimension than 1. Quite complex structures of port effort-flow variables can, thus, be constructed.

As an example, Fig. 3.10a shows a component port to which two other ports are connected internally. These are the ports of another component that contains seven elementary components. Two groups of three and four components, respectively, out of these seven elementary components are internally connected to the corresponding document ports.

The port of Component A thus can be viewed as representing a pair of *block (partioned) matrix* containing two row blocks (Fig. 3.10b), i.e.

$$e = \begin{pmatrix} e_1 \\ e_2 \end{pmatrix}, \quad f = \begin{pmatrix} f_1 \\ f_2 \end{pmatrix} \tag{3.3}$$

Note that the blocks are of different dimensions. Thus, we have

$$e_1 = \begin{pmatrix} e_{11} & e_{12} & e_{13} \end{pmatrix}^T, \quad e_2 = \begin{pmatrix} e_{21} & e_{22} & e_{23} & e_{24} \end{pmatrix}^T \tag{3.4}$$

and similarly for flows f.

In general, the dimension of the connected ports can be different. Hence, a component port can be looked at more properly as a tree of connected component ports, with elementary ports as its leaves. A component port, thus, can be interpreted as representing a pair of tree-like, higher dimensional effort-flow objects.

3.7   Port Connection Rules

An important question that must be answered is what component port connections are allowed. We have already discussed this in Sect. 2.3.

We impose additional restrictions to the permissible connection of elementary ports. This is motivated by physical reasons already discussed in Sect. 2.5.7. The ports of elementary components should be connected to junction ports only. For example, ports of inertial and capacitive components should not be directly interconnected, except to an effort or a flow junction.

The rule permits direct interconnection of two junctions of the same type, i.e. an effort junction to another effort junction, or of a flow junction to another flow junction. Such junctions can be interpreted as a single junction. The ports used for the interconnection of such components are treated as internal and are not counted in junction balance equations (Sect. 2.5.7). The connection of junctions of the same type is usually not permitted in the classical bond graph modelling approach. Nevertheless, we find this quite useful in modelling based on the component model approach. This can be easily seen from the example of Sect. 2.7.2.

We permit, however, the ports of word model components to be directly connected. It is necessary, however, to take into account the fact that such ports are, in general, compounded. That is, they correspond to a node of the tree of connected component ports. We require that the structure of connected ports is the same, which means that the port connection trees on each side of the bond line are symmetrical (Fig. 3.11a). In this case, it is easy to find out which port is connected to which. Ports are interconnected if they have the same position in their trees. Thus, in Fig. 3.11a, the port of component A1 is connected to the port of component B1; likewise, port "a" of component I is connected to port "b" of component e. We further require that document ports be not interconnected directly; instead, we can simply bypass the component!

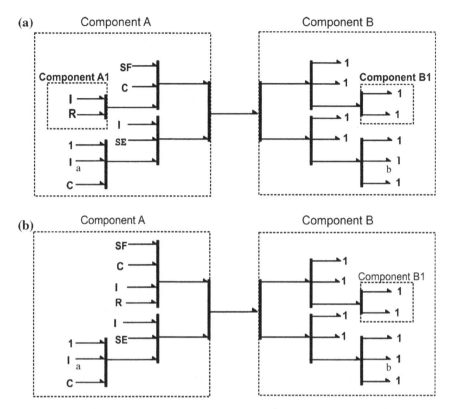

Fig. 3.11 The proper connection of the component ports. **a** Symmetrical structure. **b** Unsymmetrical structure

This assumption is a reasonable one. If we remember that connecting component ports by a bond means equating the effort and flow variables on both sides of the connection we cannot do it if the structures of these variables are different. Thus for two matrices to be equal they must have exactly the same structure, i.e. the same number of rows and columns.

It is also possible to permit more flexibility in the structure of the ports connected by a bond line. Thus, in Fig. 3.11b, component A1 (Fig. 3.11a) was substituted with its document object, and the document ports were integrated without changing the bond connection order. Thus the process does not change the interconnection of the elementary component ports of components A and B. Now, however, it is much more difficult to find out which elementary ports are connected to which, as there is no similarity in the structure of the connected component ports. It is thus necessary to find out the equivalent linear lists of elementary components ports using ordering of the bond connections in the corresponding document ports (Sect. 3.4). By comparing indexes of the ports in such lists, it is possible to discover which port is connected to which.

Fig. 3.12 The procedure for finding the port connected to a given port

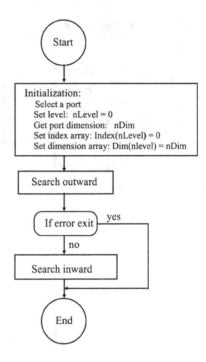

We do not allow the connection of ports with different structures, as it is more natural (not to mention also proper) to correlate ports of the same structure and, hence, the variables of the same structure. We also think such connections are more transparent and therefore easier to understand.

We next describe a procedure to find the port connected to a chosen component port (Fig. 3.12). This procedure treats only those ports that are branch ports—that is, ports where more than one bond is internally connected—and leaf ports. Ports where there is no branching bonds are taken as simple connector ports and are skipped. To describe this procedure, several variables are defined:

- nLevel current port level
- nDim current port dimension
- nIndex index of the current bond in the list of document port connections
- Index array for storing bond line indexes during port tree traversal
- Dim array for storing port dimensions during the tree traversal

Starting at level 0 the procedure searches outward, then inward.

At every step of the outward search we first check if the port has a connection. If it does not, the search is stopped and a suitable error message is displayed. Otherwise, it continues along the bond line from the chosen port to the document port to which it is connected. It then moves out of the component and on to the next document port. The search continues in this way until a port is found that is either unconnected or is not connected to a document port, but to a port of another component. At every step of the search we find the dimension of the document port and the index of the bond connection. The level is incremented and these values are stored in the corresponding index and dimension arrays. These actions are skipped if the port is a simple connection port (of dimension 1).

The inward search proceeds step-by-step, from the outside into a component, to find the connected port. At every step the dimension of the current port is found first. It then is compared with the dimension of the array corresponding to the current level. If it is OK, the index of the bond is taken from the index array corresponding to the current level and the level is decremented. This action is skipped if the port is a simple connector (dimension 1). If the dimensions don't match, the search and the complete procedure are aborted. To enter the component the accompanying document must be opened first. Next, the corresponding document port is obtained and then, using the index taken from the array, the corresponding bond is found. Finally, the port at the other side of the bond is found and the process is repeated. If the level reached at the end of the search is the same as that at which the search started, the port that was found indeed is the one for which we were searching, and the process ends.

The procedure can be used for checking the component's port connections. This can be implemented in a way that visually shows component port connections over various levels of the document. It can also be used during generation of the system's mathematical mode.

3.8 The Component Set Classes

In addition to the component and document classes of Sect. 3.2, two other classes are useful for operations on a group of components. During model development, for example, it can be useful to copy, move, or delete not only single components, but also a set of components and their interconnecting bonds (Fig. 3.13). The set can even be unconnected. To support such operations we introduce a helper class *CSelSet* (Fig. 3.14).

CSelSet defines a list of components that makes up the set and a list that contains their internal bonds (i.e., the bonds that interconnect the components in the set.) There also is a list of external bonds to components not contained in the set. We use this class to create an object that contains a list of a document's components and their bonds. This is done in the following way (Fig. 3.13):

- Create a rectangle encompassing a group of components
- Find all the components enclosed in the rectangle and add them to the list.
- Check the bonds connecting the ports of selected components

If the bond connects the components from the set, add it to the list of internal bonds; otherwise, add it to the list of external bonds.

Fig. 3.13 Selecting a set of
the components and bonds

Component: Platform

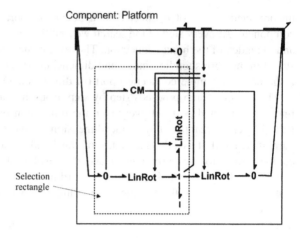

Selection
rectangle

Fig. 3.14 The helper class
CSelSet

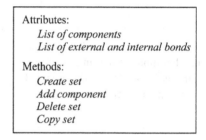

Attributes:
> *List of components*
> *List of external and internal bonds*

Methods:
> *Create set*
> *Add component*
> *Delete set*
> *Copy set*

We also can add a component to the set and check its bonds. If the component already is in the set, it is then removed from the set. In this case, any bond of the component already contained in the list of internal bonds will be removed and stored in the list of external bonds.

The set of components is *free* if its list of external bonds is empty. Such a set can be moved or deleted. Otherwise, it is fixed and can only be copied (if it is not opened). Moving the set means that every object in the set—i.e. the components and their bonds—is displaced by the same amount.

Delete and copy operations involve creation of an object of the other type that contains all information on the components and bonds in the set. *CSelSet* objects are lightweight components that contain only pointers to the components and bonds in a document. They are not persistent objects, but can be used to create persistent objects. When an object is destroyed, the lists only are destroyed, not the component and bond objects

We define a class that is used for creation of a persistent object using information from *CSelSet*. This is the *CCompSet* class we met when discussing the component class hierarchy in Sect. 3.3 (Figs. 3.5 and 3.15).

Fig. 3.15 The component
group class *CCompSet*

Attributes:
    List of components
    List of bonds
 	
Methods:
    Constructors and destructor
    Copy operation
    Create from
    Storing and loading

CCompSet is a kind of *CComponent*, but it differs from the latter in that it does not have a separate document to hold its model. It is a container for a set of components and their interconnecting bonds. It doesn't have a separate visual representation, as other components, because it is not used for model building.

CCompSet has one thing in common with *CComponent*: It can be stored and loaded like other component objects. The main use of this class is to create objects that contain a set of components and their bonds that are deleted or copied from a document. Such objects can be held in buffers, from which they can be reused (Sect. 3.9). In performing these operations, there is a close co-operation between *CSelSet* and *CCompSet*.

The components contained in a *CCompSet* object can be inserted into a document using *Create from* method (Fig. 3.15), which creates a copy of all components, bonds, and associated documents, attaches them to the document where they should be inserted, and creates a corresponding *CSelSet* object.

3.9   Systematic Top/Down Model Development

The previous Sections described the concept of the component model, the concept that lies at the core of the systematic top/down modelling philosophy developed for complex engineering and mechatronics systems. This Section describes an object-oriented environment that implements this modelling philosophy.

Physical model development normally begins at the *system level* (Fig. 1.2), a step that identifies the system and defines its interaction with its environment. The model at this level represents a world-view of the problem under study and is represented by a document object without the ports. That is, this is the system model root document.

We use the term *project* to mean the model development. We introduce a separate project class *CProject* (Fig. 3.16) to serve as the starting point of model development. Every project has a project object with a unique identification label (*id*) and filename that indicates where the project's root document is to be saved. This class is, in essence, a reduced version of the component class (Fig. 3.1).

Fig. 3.16 Project class
CProject

> Attributes:
>> *Project ID*
>> *Document filename*
>
> Methods:
>> *Constructors and destructor*
>> *Data access functions*
>> *Storing and loading*
>> *etc.*

Fig. 3.17 *CBondSimApp*
class—modelling support part

> Attributes:
>> *Index of projects*
>> *Collection of IDs created*
>> *Root document object*
>> *Current document object*
>> *List of document opened*
>> *Buffer of undo objects*
>> *Waste bin buffer*
>> *Component libraries*
>> *Project storage*
>> *Export/import storage*
>> *etc.*
>
> Methods:
>> *Constructors and destructor*
>> *Initializing and closing application*
>> *Maintenance of IDs collection*
>> *Create new project and Open project*
>> *Copy, rename and delete project*
>> *Undo buffer operations*
>> *Waste bin buffer operations*
>> *Library operations*
>> *Project storage operations*
>> *Export/import operations*
>> *Other operations*

Another class, *CBondSimApp*, manages the complete operation of the bond graph based project model development and simulation. It works in close co-operation with a suitable visual environment (Chap. 4). This is a quite complex class. The part that deals with the most important operations of model development is shown in Fig. 3.17.

The application class contains an index of the projects of interest. These generally are projects under development or currently being studied. Other projects are held in a separate storage (Sect. 3.10). Project indexes can be stored as a map (dictionary) to permit rapid access using the project name as the key.

At the start an application object is constructed and initialised. As a part of the initialisation the index of current projects is loaded. A new project may be started as follows:

1. Send a message to the application object to create a new object.
2. Enter the title of the new project. (The name should be unique.)
3. After the name is supplied and checked in the index of projects, a unique project identifier (*id*) is created, along with a file name under which the project's document will be saved. A new project object is constructed using the *id* and the file name as parameters; it then is added to the project index using the new project's name as the key.
4. A new document object is created.

The document object that is so created takes as its name the project name. This is the document from which model development starts; it thus represents the root of the project's document tree. Further model development proceeds as described in Sect. 3.2.2.

A project is opened (or reloaded) by selecting its name in the index of projects. The project object to which the index refers is found and the corresponding document is obtained. The project document then is loaded, as in step 4 above.

The application class supports other important operations on projects that parallel those for components. Thus, to rename a project it is enough to change the key entry. To make a copy of a project, the procedure is:

1. Get the project object that is to be copied from the index of projects. Get the file name of the original document. Define the name of the project copy.
2. Create a new *id* and filename for the copy. Construct a new project object using the new *id* and the new filename as parameters. Add the new object to the index of projects using the name of project copy as the key.
3. Make a copy of the project's document, as explained in Sect. 3.2.2. The new document is saved under the new file name created in step 2.

To simplify model development the application object maintains a *Component* buffer that serves a function similar to the Windows clipboard. It also supports operations similar to the cut, copy, and paste operations of the Windows environment. Here, these operations are component based. This means that a component contained in a document can be selected and copied, as explained in Sect. 3.2.2, then added to the buffer. Similarly, a component can be cut—disconnected from the document—and added to the buffer. Components from the buffer likewise can be copied back and inserted in the same, or some other, document. These actions can be performed on a selected set of the components, as explained in Sect. 3.8.

Another useful service of the application object is the *Waste Bin* buffer. This buffer functions in a similar manner as the Windows Recycle bin. The project can be deleted from the project index using methods from the application class. Instead of removing all of the underlying document files, however, only the project entry is removed from the index; the corresponding project object is sent to the waste bin buffer. Similarly, other objects—e.g., a component or group of components and bonds—that are deleted from a document are disconnected, can be added to the buffer.

Projects or components in the Waste Bin buffer can be either restored or removed from the application. When restoring a project, the information object is moved from the buffer and inserted back into the project index. Similarly, when restoring a component, it again is moved from the Waste Bin buffer back to the Component buffer. From there, the component can be inserted into any document. On the other hand, removing a component or project from the buffer means removing the object and all underlying documents. This can be achieved as discussed in Sect. 3.2.2.

When closing the application the computer frees computer memory by destroying all objects and saving these data to files. The index of projects, Component and Waste bin buffers are saved too. When restarting the application, the buffers are restored. Hence, the models in the Component—and Waste bin buffers are available across modelling sessions. It is advisable, of course, to empty these buffers from time to time to conserve memory.

Finally, we return to the mathematical model parameters already mentioned in several places. These can be defined at the document level or at the component level. In the former case, they are visible in all contained components of this level and higher. Parameters defined in a component port, however, are visible only in that component. In this way, common parameters—such as gravity, and some other physical and mathematical constant—can be defined in a lower-level document, or maybe at the project root document level. Parameters that are specific to a process should be defined in the corresponding elementary component. Such flexibility comes with some dangers. If, for example, we delete a parameter defined in a certain document, this could create a problem if the parameter also is used by some higher-level components. The corresponding mathematical relationship then would be incompletely defined, thereby resulting in an undefined parameter. It is quite difficult to monitor all such changes. One remedy is to postpone final checking until the phase in which the complete mathematical model is built.

3.10 Component Libraries and Model Reuse

The idea of code reuse is very old and very appealing, but well known not to be an easy problem to solve. Code reuse significantly improves the efficiency and quality of development and, hence, of problem solving. There are many approaches and techniques developed and are in wide use today that are based on code reuse, such as COM and DCOM technology. OOP languages are developed with code reuse as a goal. The following four fundamental reuse problems can be identified [8, 9]:

- Finding components
- Understanding components
- Modifying components
- Composing components

As used in this sense, a component is any output of the solution process, e.g. code components, on-line documentation, specifications, etc.

Here we are interested in reuse of component models and projects. The concept of component oriented modelling, as discussed at the beginning (Sects. 1.2 and 1.6), is introduced not only to enable systematic model development, but also to support reuse of components for building system models. Component reuse can significantly improve the quality and efficiency of model building. Similarly this applies to projects as well. Here we describe approaches and methods used to that goal.

Component and project files are held in the library, a separate section of the workspace. The library is divided into two segments

- Project repository
- Component libraries

The project repository serves as a convenient storage place of complete projects. Any project can be moved into the project repository and removed from the main workspace section (models section). Such a project can be reused at any time. To simplify the search for a particular project, the repository can be hierarchically organised according to application areas, or by some other criterion. This is not considered further here, for it is a separate problem. Currently, only the basic mechanisms of project managment is provided.

Component libraries are divided into three sets:

- Word model components
- Electrical components
- Mechanical components

The first serves for components represented by word models and are for general use. The electrical library stores electrical components represented by electrical circuit symbols. This library can store models of, for example, electrical resistors, coils, semiconductors, and electrical motors. The third is a library of mechanical components represented by suitable graphical symbols, e.g. springs, bodies in translation or rotation, connectors, etc. These libraries can be further specialised.

It is relatively easy to organise components into libraries. Project and component files of interest are put from the model section into a library section of the workspace simply by copying. The application object maintains an index of project and component files put into libraries. In this way, storing important components and project models separately from the model workspace is more secure; there is less chance that they will be accidentally modified or removed.

Projects or components of interest can be found by searching library indexes. To reuse a particular project it needs simply to be copied back into the model workspace. After that it can be opened and used as any other project.

Components from the library are inserted into the model workspace. It can be done as follows:

1. Find component in the library index
2. Open the document and extract a copy of the component object held in its header
3. Select the position in the target document working area where the component is to be inserted. Change the component visualisation data to reflect this selection. Create a new component *id* and document filename, and then reset the corresponding attributes in the component object.
4. Add the component to the document component list and save the document.
5. Copy the library document files the to the model workspace

The other three reuse problems stated in the beginning of the section are solved by the component modelling approach. Any component or project in the library can be opened to analyse its structure, constitutive laws and parameters, and how it may be used. As with other coding methods, understanding of a project or component model depends also on how the model is constructed. If developed logically and by strictly applying systematic decomposition, the models will be transparent and more easily understood. After their insertion into a document they can be modified easily. They can be further connected to other components and the connection checked as discussed in Sect. 3.7.

The component model concept presents another possibility for component reuse: Component use is not confined to the application in which they are created; different applications can exchange components.

To exchange models they are exported as (compressed) files in the form of *ept* files. These can be sent by e-mails, but can be transmitted in other ways as any file. Likewise, the projects and components can be imported into another application.

The exchange of models is a very useful means of supporting collaboration between people engaged in solving similar problems. A complex modelling project can be divided into separate development tasks. After components are developed, they can be integrated into an application. We do not wish to imply that this is a simple task. For components to work properly, they must have ports designed to comply with the requirements of Sect. 3.7. This is not much different from the situation that arises when dealing with real components. Every such component can function properly only in an environment specifically designed for it. The approach developed here can help in understanding this problem and aiding design of real engineering components and systems.

References

1. Damic V, Montgomery J (2003) Mechatronics by bond graphs: an object-oriented approach to modelling and simulation. Springer, Heidelberg
2. Williams S, Kindel C (1994) The component object model: a technical overview. http://msdn. microsoft.com/library/techart/msdn_comppr.htm
3. Booch G (1991) Object-oriented design with applications. Benjamin Cummings, New York
4. Stroustrup B (1998) C++ Programming Language, 3rd edn. Addison-Wesley, Reading

5. Breedveld PC (1982) Proposition for an unambiguous vector bond graph notation. J Dyn Syst Measure Control 104:267–270
6. Fahrenthold EP, Wargo JD (1991) Vector and tensor based bond graphs for physical systems modeling. J Franklin Inst 328:833–853
7. de Boor C (1998) A practical guide to splines. Springer, New York
8. Biggerstaff T, Richter C (1989) Reusability framework, assessment, and directions. In: Biggerstaff TJ, Perlis AJ (eds) Software reusability: concept and models, vol 1. Addison-Wesley, Reading
9. Walpole RA, Burnett MM (1997) Supporting reuse of evolving visual code. In: Proceedings of 1997 IEEE symposium on visual languages, Capri, Italy, pp. 68–75

References

Chapter 4
Object Oriented Modelling in a Visual Environment

4.1 Introduction

This section describes *BondSim,* a program which offers a visual environment for the modelling and simulation of engineering and mechatronics systems, based on bond graphs. The general concept of the program is shown in Fig. 4.1.

It offers several services to a user. The two basic services are Modelling and Simulation. The first supports model development tasks and represents implementation of the ideas and methods of Chap. 3. The program also supports model data base maintenance and library support. It also supports collaboration using model import and export as well as inter-process communications (IPC).

The simulation subsystem uses models developed by the modelling subsystem to study system behaviour. This chapter focuses mostly on the modelling part of the program. The simulation part is the topic of Chap. 5.

The operating system for which the application has been developed is Microsoft Windows 7 and later. The program has been developed using the Microsoft integrated development environment Visual Studio 2012, the VisualC++ and Microsoft Foundation Class Library (MFC) [1]. The program *BondSims 2014* version is an extension of the program described and included in Ref. [2]. Use was also made of the Zlib library for exported data compression.[1] We greatly appreciate the permission of Zlib's authors for the free use of the library. *BondSim 2014* is freely available through the author's web site. Readers are encouraged to use it in conjunction with the text.

---

[1]ZLib is a free data-compression library developed by Jean-Loup Gailly and Mark Adler, http://www.info-zip.org/pub/infozip/zlib/.

© Springer-Verlag Berlin Heidelberg 2015
V. Damić and J. Montgomery, *Mechatronics by Bond Graphs,*
DOI 10.1007/978-3-662-49004-4_4

Fig. 4.1 The visual modelling and simulation environment

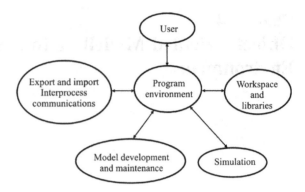

4.2   The Visual Environment

The main application class *CBondSimApp* was derived from the MFC class CWinAppEx and is used to construct a Windows application object, which in turn is used to implement a modelling and simulation environment (Sect. 3.9). The application object is declared as global and is constructed at the start of *BondSim*. During its initialisation the main application window is created and appears on the screen as shown in Fig. 4.2.

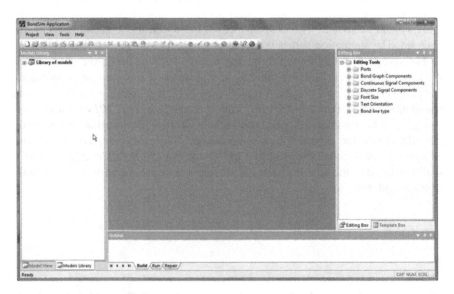

Fig. 4.2 The BondSim main window

The main window follows the *MS VisualStudio form*. It is divided into several parts. The upper part contains the Program title, menu bar and tools bar, which are typical for a Window program. The area below these bars is divided into four main parts. In the middle is a large window which plays a central role in the model development. On the left is a window which contains two tabs: *Models view* and *Models library*. The first serves to show the structure of the model currently under the work in the main window. The models library tab enables easy access to the models library (see Sect. 4.9). Similarly, to the right of the main window is another tabbed window, which contains *Editing* and *Template* tool *boxes*. These boxes contain tools that are used during model development (Sects. 4.6 and 4.7). Finally, at the bottom is an area which is used for messages sent during model building, simulation run or the program repair phases.

The main application window is based on the *CMainFrame* class. This class is derived from the MFC Library's *CMDIFrameWndEx* class, which provides the functionality for a multi-document interface (Sect. 4.5). The class also defines some specific methods needed by the modelling and simulation environments. These belong to two groups. The first controls messages sent to the main window and gives information on the current status of the operations. Such messages are displayed in the status bar at the bottom of the main window, e.g. *Ready*, *Create a new project*, etc. Attributes and methods are also provided for the creation and operation of a progress control bar located to the right of the status bar. This provides the developer with feedback on the completion percentage of some lengthy operations, such as are encountered during simulation runs. The other methods are used for the distribution of messages sent to the window during various phases of modelling and simulation, such as during the creation of bond graphs, operations on libraries, etc.

The methods accessible at the application level are organised in the *Project*, *View*, Tools and *Help menus*. There is also a row of toolbar buttons for the most important commands. Some of the most often used commands can also be accessed by keyboard shortcuts.

The *Project* menu contains commands for modelling operations implemented along guidelines given in Sect. 3.9. The first two are *New* and *Open*.

The *New* command is used to create a new project. When this command is chosen, a dialogue window appears with an edit box into which the user inputs the name of the new project (Fig. 4.3).

The dialogue window also contains a list box with the names of projects that already exist. This information is held in the project index maintained by the application. The name of the new project is accepted, provided it is unique. The new project document is then opened in a suitable window. This is the starting point of the model development process (Sect. 4.6).

The *Open* command is used in a similar way. This command opens a modelling project already in the projects list. It uses a similar dialogue window as the *New* command (Fig. 4.4). The name of the project can be selected from the

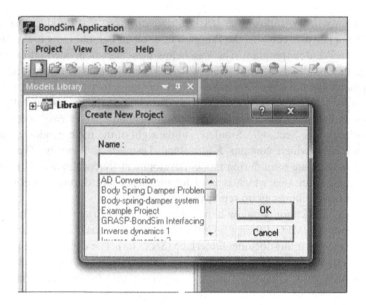

Fig. 4.3 The *New* modelling project dialogue window

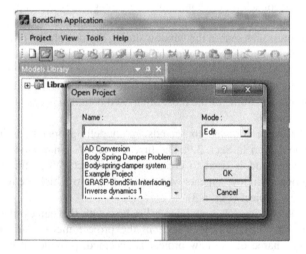

Fig. 4.4 The Open a new project dialogue

accompanying list box. It is possible to open a project in the *Edit* or *Read only* mode by selecting one or the other from the *Mode* box. The *Edit* mode is the normal mode for opening projects, in which models are created or modified. The *Read only* mode, on the other hand, can be used only for reviewing. In this mode, project documents cannot be changed, thus protecting the model's original form. The document project is read from the project root document file and displayed in the central window.

The next group of commands contains project manipulation routines, such as *Delete*, *Copy*, and *Rename*, each of which uses a similar dialogue. The *Delete* command removes the project from the project data base and puts it into the *Waste Bin*. It is also possible to delete the project completely by bypassing the *Waste Bin*. The *Copy* command creates a copy of an existing project under a new name. This command invokes the copy operations of all documents that make up the project, as described in Sect. 3.9. The *Rename* command changes a project's name.

The other useful command is *Copy to Library*, which moves a copy of a project to the projects library. Once copied to the library, the project can be removed from the project workspace.

The commands *Export* or *Import* serve for exporting and importing a project respectively to or from another *BondSim* application. The *Export* command generates a file in compressed form having the file extension "*.ept*", which contains a project data that the user wishes to export from the current program environment. The file is stored in a place selected by the user using standard Windows *Save as* file dialog. Similarly, the *Import* command uses a similar *Open* file dialog to access a BondsSim exported project file, which we are importing. At the end of the operation the program asks for the name under which it will be stored in the program workspace.

The *Repair Projects* command repairs the model database in case of corruption. During this operation all projects that are not correctly linked are collected and could be restored into the database or rejected.

The *Waste Bin* command provides access to the buffer that stores information on previously removed or repaired projects or components. These can be completely deleted from the application workspace, or restored back into the workspace. It plays a similar role as *Recycle Bin* in Windows. At the end there is the standard *Exit* command. The *View* menu contains commands that the user can apply to change the toolbars, docking widows, status bar or the application look.

The *Tools* menu contains commands that can be useful during the work. Thus, it is possible to access the *Component Buffer*, (which has similar role as windows *Clipboard*[2]), to inspect its contents or to remove components that are not needed anymore. There is also a *Functions* sub menu, which contains the commands for creating a new user function, opening an existing one, their deleting or exporting. In addition there is a *Symbolic Operations* submenu with currently only one command

---

[2]Note, however, that the components stored in the *Component Buffer* are retained between the *BondSim* sessions.

Differentiation, which enables a symbolic evaluation of derivative of a function defined analytically. Finally, there is *Display Plot* command that enables import and displaying external functions using text files as the input, similarly as in *Microsoft Excel*. The *Component buffer, Functions* and *Symbolic Operations* are also accessible from the *Tools* menu when projects are opened (Sect. 4.10).

The Help menu contains *Help Topics, BondSim on the Web* as well as *About BondSim*. There is also a command for registering a BondSim application.

4.3   The Component Hierarchy

The component classes discussed in Sects. 3.2 and 3.3 are derived ultimately from *CObject*, the MFC's base class. This provides basic support for dynamic creation of the objects and object persistence (serialization), as well as other services [1]. Objects derived from the *CObject* class are responsible for their storing to, and loading from, a persistent medium (typically a hard disc).

The hierarchy tree for the model classes of Sect. 3.3 is given in Fig. 4.5. In addition to the functionality already discussed in Sect. 3.2.1, the base class, *CComponent*, also defines an internal state attribute and methods, which change the visual appearance of the component objects in the development environment.

The possible states are:

- *Normal*
- *Select*
- *Open*
- *Text*

We will discuss some of these. When a component is created it is internally set into *Text mode* and appears on the screen as a *caret* (a blinking line or block). Thus, the component is in the editing mode, and when the user starts typing in its name (title) appears on the screen in a specific colour, e.g. red. To finish editing the user may click to a place outside of this text. This closes the editing mode and changes the component state into the *Select* mode in which a bounding rectangle appears around the component name in the specified colour, e.g. red, and the text changes its colour to the normal black colour. Clicking again deselects the component and the bounding rectangle disappears. The component is now in the *Normal* mode. Finally, if we *open* the component, as described in Sect. 4.7, a rectangle appears around the component name indicating that its document is opened in a separate window.

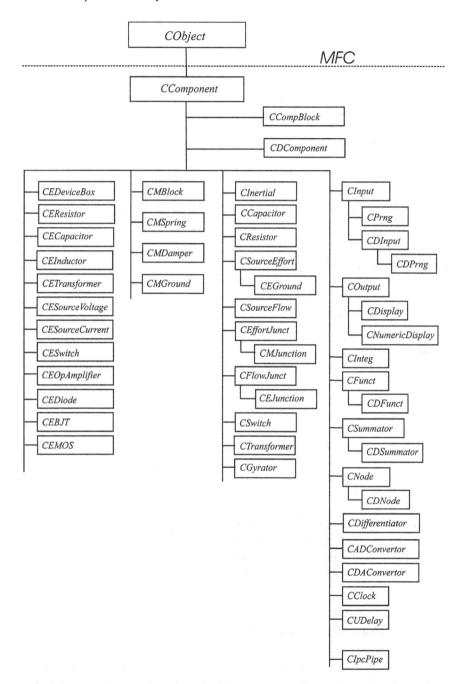

Fig. 4.5 Component class hierarchy as extension of *MFC CObject* class

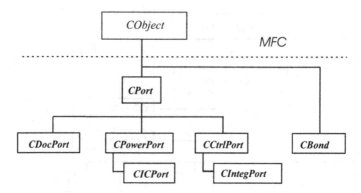

Fig. 4.6 The port and the bond classes hierarchy as extension of MFC library

4.4 The Port and Bond Classes Hierarchy

Similar to the component classes, the port and bond classes of Sect. 3.4 are also derived from the *CObject* class. These also use *CObject*'s dynamic creation and serialization support. The complete class hierarchy is given in Fig. 4.6.

The *CPort* and *CBond* classes also define its internal state similarly to the *CComponent*. Thus, when created they are in the *Selected* mode, which is shown by the ports and bonds drawn in red. When we deselected them, by e.g. clicking outside of these objects, they assume the *Normal* mode in which they appear in the normal colour, typically black. There are also other modes often used in logical combinations.

Power In and *Power Out* modes are used by power ports, and are inherited by the corresponding document ports, to visually represent the ports by half arrows directed into, or out of, the corresponding component, respectively. Similarly, *Control In* and *Control Out* modes are used by the control ports, and also inherited by the corresponding document, to visually represent them as full arrows directed into, or out of, the corresponding component, respectively. In the same vein *Discrete Port* mode is assigned to the ports belonging to the discrete components, and similarly, the trigger and activation ports are normal control ports to which it was assigned *Trigger Port* and *Activate Port* modes when they were created.

4.5 The Document Architecture

MFC supports the creation of document windows based on three associated classes: *CDocument*, *CView*, and *CFrameWnd*, or the classes derived from them [1]. The process is coordinated by a class derived from the *CDocTemplate*.

The frame class is used to display documents, and has a title bar and a border within which the view class displays the contents of the document. The view class handles events generated by keyboard input and mouse actions. Typically, the document template object is constructed in the global application object at initialization (Sect. 4.2). When a command to create a new document is issued, the document template object dynamically creates document and frame objects. The latter creates a view object that displays an empty document. When opening a document that exists in a file, the procedure is similar, but this time the document is read from the file and displayed inside a window frame. Reading, as well as storing, a document is done by an archive object that supports serialization. Details of this and of related processes can be found in [1].

We accepted this architecture but use the *CScrollView* derived class, a view class that supports scrolling large documents, and *CMDIChildWndEx* multi-document child window derived class. They are coordinated by the *CMultiDocTemplate* class. In this way we can open multiple document windows on the same project. However, these classes do not satisfy completely our needs and we thus introduce classes derived from these classes as shown in Fig. 4.7.

We define two derived document classes: *CBondSimDoc*, and *CBondSimSSDoc*. These are necessary because we wish to radically change the document architecture of the program.

Windows documents, as implemented in the MFC library, are single-level documents, i.e. they are all created at the modelling project level. Because we wish to support hierarchical multilevel model development, only the project root document, which describes the basic system level model, should be opened at the project level. All others need to be opened within the previous document using word model objects as the interfaces. In addition, all the documents should be linked backwards as well. This is the main task of the CBondSimDoc class.

Another important change is the use of a *structured storage file system*. This is a storage scheme introduced by Microsoft for dealing with the OLE (Object Linking and Embedding) technique. Thus, instead describing the model of a project by a multitude of the files, a single *compound* (structured storage) file is used. The compound files contain *storages*, which correspond to directories in the common file system. The storages may contain other storages and the *streams*, which are

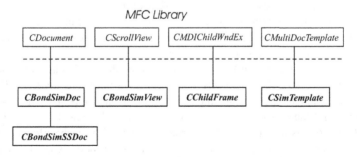

Fig. 4.7 The *CDocument* and the associated classes

analogous to the common files. Thus, a compound file is a file system inside a file. In our approach, we use the word model components in the current document to create corresponding storages, and store the corresponding documents as the streams in these storages. Each document, which describes a component model, is stored in the corresponding component storage, and thus the word model component and the corresponding model document are closely coupled indeed.

Hence, to 'open' a component model we open the accompanying document in a frame using the corresponding stream (instead of a file). And similarly, by closing a document we store it as a stream in the corresponding component storage. Only the lowest level document, i.e. the documents at the system level, do not have the corresponding storage, because they are already at the project level. Thus, by storing this document we store the project file to the disk. This means that until the project is stored, all changes are only temporary stored in the previous document stream and could be rejected. This storage scheme is implemented by the *CBondSimSSDoc* class, where 'SS' means the 'structured storage class'.

To illustrate this concept remember that the program *BondSim* as discussed in Sect. 4.2 contains a tabbed window on the left side of the *Model structure*, which shows the hierarchical structure of the current project. Figure 4.8 shows the *See-saw* project (Sect. 2.7.3) opened in the central window and on the left is given its model structure in a *Windows Explorer* form. By expanding a branch, e.g. *Body1*, it opens the corresponding document containing its model and shows it in a separate tab in the central window (see the document tab named *Component:Body1* in Fig. 4.8). By expanding or collapsing the branches we can walk through the model. (As shown later this can be done also in other ways).

Some operations, however, do not need the complete document-view-frame architecture. Any document object contains the complete description of the component model. Thus, for operations on models, the document object can be treated simply as a C++ object and directly work with them. This is also the case when objects at both sides of a component interface—i.e., a word model object and its document object—must be updated. Such operations typically run in the background, hence, it is inefficient to use the windows resources and the corresponding processor time.

The size of the document displayed on the screen is limited to fit into a page (A4) when printed. The page can be displayed vertically or horizontally. This really does not impose any restrictions on the size of model that can be developed because, using the component model technique, a large document can always be represented by the smaller ones. We hope that such a restriction on the document size encourages component-wise model development.

The view class *CBondSimView* is derived from the MFC class that supports scrolling views. This class supports two modes: *Normal* and *Scale to Fit*. The models are edited in the *Normal* mode. The scroll bars appear if the document extends beyond the size of the frame window. A new document is opened in this mode. This mode is also used when a project is opened from the file and the *Edit* mode is chosen (Sect. 4.2). Otherwise, the *Scale to Fit* mode is used. This mode displays no scroll bars and the document is fitted to the frame window by scaling it

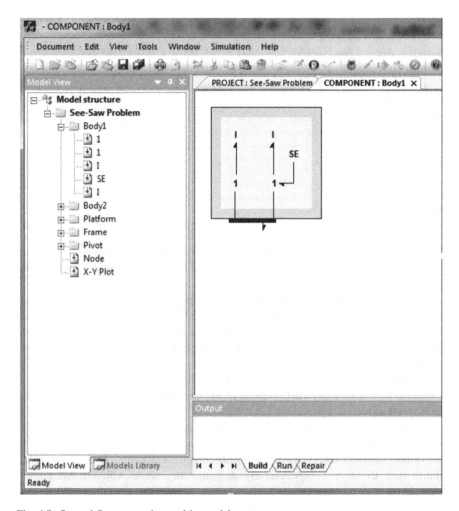

Fig. 4.8   Opened See-saw project and its model structure

up or down. No operations that change the model appearance are permitted in the *Scale to Fit* mode.

This class manages drawing the document in response to the paint commands from the Windows system. The corresponding drawing method forces all objects contained in the document to draw themselves. It also defines methods for handling commands forwarded to the view in response to the mouse or keyboard. The response depends on the mode chosen. Thus, the class implements an internal state attribute used to determine which action to execute.

The most important states are:

1. *Normal*
2. *Create component*

3. *Move component*
4. *Insert component*
5. *Insert library component*
6. *Edit text*
7. *Create port*
8. *Drag port*
9. *Move port*
10. *Create bond*
11. *Connect ports*
12. *Change bond*
13. *Size document*
14. *Size port*
15. *Set selection*
16. *Move selection*

We explain here some of the actions that deal mostly with selecting and moving model objects. Those concerned with the creation and editing of model objects are explained in Sect. 4.6.

In the *Normal* mode the view is not set to any specific editing mode; it is ready to accept commands. A component can be selected simply by clicking it.[3] The component reacts by changing its visual appearance. The selected component can be opened using the command from the menu (Sect. 4.6.2). The component can also be opened simply by double-clicking it, or by the keyboard shortcut. Other commands can be executed on the selected component, as well. The component is deselected by clicking anywhere outside of the component rectangle and returns to the *Normal* mode.

We can drag a component around the document drawing area provided it is not connected to other objects by the bonds. This is accomplished by pressing the left mouse button when the cursor is within the component name. This changes the editing state to *Move Component*. By dragging the mouse with the left button pressed, we can move a component rectangle around the document. When the mouse button is released, the component reappears in the new position and the view mode changes to the *Normal* mode, and its component state to *Selected*.

A component port can be selected by clicking the port. In a similar way as with components, we can move a port around the component periphery, provided it is not joined by a bond. Simply put the cursor over a port, then press and hold the left mouse button. This changes the editing state to *Move Port*, and the cursor changes to the shape of the port. As we drag the cursor around the component periphery, the cursor shape changes to reflect the correct port shape. For example, a power-in port should point every time into the component. When the mouse button is released, the editing mode returns to *Normal* and the port reappears in the new position in the selected mode. Again, the port is deselected by clicking outside its boundary.

---

[3]We will use terms like clicking or pressing mouse to mean using the left mouse button. Otherwise the button used will be stated.

Similarly, we select a bond by clicking it, or deselect the bond by clicking outside of the bond. We can also move the bond by dragging it. This changes the bond shape in a manner similar to stretching a thin rubber band, the ends of which are fixed. If we press the left mouse button when the cursor is close to, or on, the bond, the editing mode changes to *Change Bond*. This creates a new intermediate point that is inserted into the array of bond points (Sect. 3.4). Thus, by dragging the cursor with the mouse button pressed, the coordinates of the point under the cursor changes, as does the shape of the bond line. By releasing the mouse button, the view again returns to the *Normal* state and the bond is redrawn in the new changed shape and is selected.

The operations described above are, in effect, single-object selection operations, i.e. we select or move or apply a menu command to a component, a port, or a bond. It also is possible to select a set of objects. Thus, if we put the mouse cursor outside of any object, then press and drag it, a rectangle appears. When we release the mouse, all components inside this rectangle are selected, and all of the bonds between these components as well (Fig. 3.13). This action creates a temporary *CSelSet* object that contains a list of pointers to the components contained within the rectangle, as well as of the bonds joining them (Sect. 3.8). The view object changes its state to *Set Selection*, and all components and bonds in the set change their state to selected ones.

We can also add a component to the selection by holding down the *Ctrl* key and clicking on the component outside the selection. Similarly, we deselect a selected component by clicking on it. If the components in the selected set are not connected to other parts of the bond graph, they can be moved jointly within the document area, as was the case with a single selected component. To do this, we put the cursor somewhere within the rectangle enclosing all the selected components, and then press the left mouse button. This changes the view state to *Move Selection*. By dragging the mouse while the button is pressed, all selected objects move as a block. Other operations can be applied to a set of selected components (Sect. 4.10). We can remove the selection simply by clicking outside of the rectangle encompassing all selected components. This returns the states of the objects to *Normal*, the temporary *CSelSet* object is destroyed, and the view object state also returns to *Normal*.

CChildFrame (Fig. 4.7) is derived from the MFC's *CMDIChildWndEx* class, which supports multi-document frame windows. Finally, we come to the document template class, *CMultiDocTemplate*, from which we derived a document template class, *CSimTemplate*. This class inherits all functionality of its parent class. It is used for minor adjustments of menu items during normal document- view-frame creation. For example, it adds the *Last* command in the Windows control menu, which opens the previous document. It also defines a method for creating a document-only object, i.e. a document object without the associated frame and view objects. This is used when executing certain background tasks, as discussed at the beginning of this section.

4.6 Editing Models

When a new project is created (Sect. 4.2), an empty document is created in a
window with the name of the new project in the title bar. The menu changes to
reflect commands accessible at the document level. We discuss these commands in
Sect. 4.10. Here we explain how bond graphs are developed systematically.
Figure 4.9 shows a new project titled *Example Project*.

An empty project document appears inside a frame with the title *Example
Project*. Note that the model structure on the left contains only the project title. On
the right is seen a tab *Editing Box* containing the editing tools. We will familiarize
ourselves with this box, which plays the central role in model editing.

4.6.1 The Editing Box

The *Editing Box* (Fig. 4.9) tools consist of three main branches: *Bond Graph
Components*, *Continuous* and *Discrete Signal Components*, which are used to
create the corresponding components. The *Port* branch serves to add the ports to the
components. The *Font* is used to select the size of the component text, for component name editing, and the *Text orientation* to set either to *horizontal* (default) or
vertical. The *Bond Line Type* enables selection of the visual representation of the
bonds either as a *Single bond* (default) or *Multi bond*. Note that this is only the bond
visual appearance. As discussed in Sect. 2.3 the multi-bond technique is not

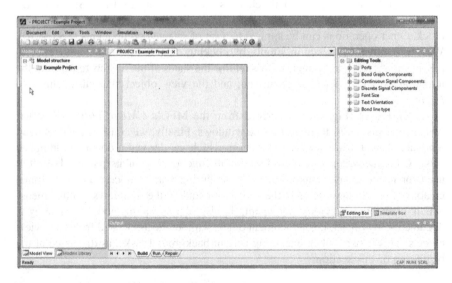

Fig. 4.9 Creating a new project document

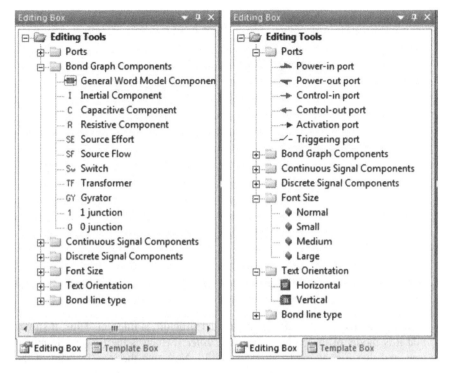

Fig. 4.10 Editing Box with expanded Bond Graph editing tools

applied, and thus it may be used to indicate only that the connected ports are not simple. Figure 4.10 shows the main Bond Graph model development tools.

To create a component the drag and drop technique is used. Thus, to create a word model component in an empty document window Fig. 4.9 we can proceed as given bellow (Fig. 4.11):

- Drag the *General Word Model Component* from the *Bond Graph Components* tools (Fig. 4.10 left) and drop it into the rectangular area in the *Example Project* window, but not too close to its borders.
- A caret appears and we may type in the title, e.g. "CompA" and click outside of it to end the text editing of the component title.
- The CompA appears having two power ports bounded by a red rectangle (denoting that it is selected). Note that when a component is created it already has the default ports pre-created. We can retain them, remove some or add new ones.
- To add a control-in port, e.g., drag it from the *Port* tools and drop it at the component boundary. Note that when we drag the port over the component a red bounding rectangle appears. We need to drop the port in the space between the component title and the boundary rectangle. When we drop it a control-in port appears at the component boundary in the selected state.

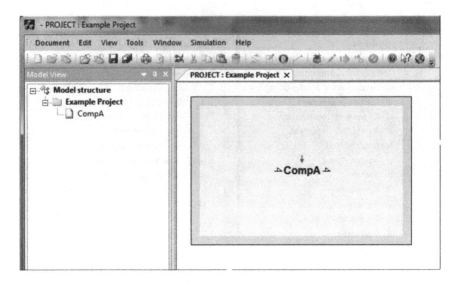

Fig. 4.11 Creating a component and adding a control-in port

- Note in Fig. 4.11 on the left that the model structure is filled in. Thus, it shows that it is model of the project *Example Project*, which is the root of the model tree, and that it contains *CompA* as a child. Note that the icon in front of the *CompA* branch is in the form of an empty page because the component is empty; there is no document contained, just the word model component alone.

We may add other components as well. Thus by dragging and dropping we add a *Source Effort* (*SE*) component (Fig. 4.12). This component appears also as a leaf in the model tree as can be seen in Fig. 4.12 on the left.

Note that the component is not created if it is placed too close to the rectangle boundaries. Thus the best way is to create it somewhere away from the rectangle boundaries and then drag it and put in a suitable place.

Every created component is a different object in the computer memory, or in the persistent medium, even if two have the same names. The component name is for user convenience and is not used by the program to identify a component. To uniquely identify every component object the program assigns to it a unique *identifier* (ID).[4] Thus, we can freely select the component name (title). There cannot be a name clash. We can even change standard Bond Graph element names, e.g. we may rename the I component name as Inertia. The components created are unique by its design.

---

[4]The component *ID* is based on Microsoft *GUID*, a globally unique identifier. Each generated GUID is not guaranteed to be unique, but the number of possible GUIDs are so large (2^{128} or 3.4×10^{38}) that the probability of generating the same number twice is very low. It is used in many software products not only by Microsoft, but also by Oracle, Novel, Intel and others [3].

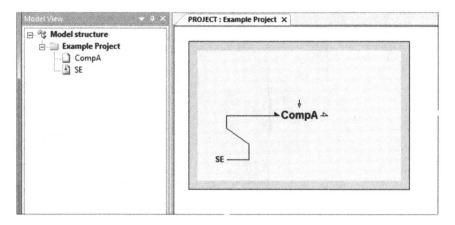

Fig. 4.12 Adding **SE** component and connecting its port to the component port by the bond

Now we may connect the power-out port of SE and the power-in port of the
CompA by a bond. The procedure is:

- Pick the power port of SE.
- Move (with the key pressed) mouse cursor to the right and click, then move up
  and click again, then move to the left and click, and continue until it is over the
  port of the correct power sense (power-in in this case) of CompA. This is
  signaled by the port becoming selected.
- End drawing the bond by clicking on the port below the mouse cursor and the
  bond line appears.
- If we wish to correct a bond line we can put the mouse over the bond, pick it and
  drag it like an elastic string, and drop it in a new position.

4.6.2   Developing Bond Graph Models

The first step in the project development is editing the bond graph model at the system
level. If the model at this level contains word model components, such as CompA in
Fig. 4.12, we need to define its model. This is done by *opening* the component.

We may *open* a component in several different ways, as is the usual approach in
windows programs. The easiest way to do it is by double-clicking the component
title in the document window, or in the model tree. We may expand the component
branch in the model tree, or click using the right mouse button the component title
in the model tree and select the *Open* command from a dropped down menu. Or, we
can select the component in its document window and click the corresponding
toolbar button. To see which one to choose we may move the mouse cursor over the
toolbar buttons until the tooltip windows that appear shows *Open Component*. We

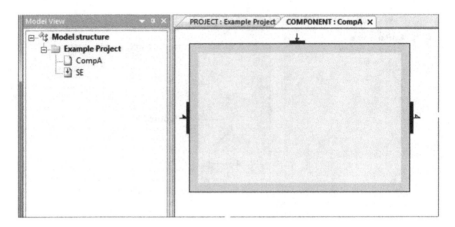

Fig. 4.13 CompA: Opened empty model document

can also choose the *Open Next* command in the *Document* menu, or simply press the PGUP (Page up) short-cut key.

If the model of the component has not been defined already, as the case is in Fig. 4.12, this operation tries to create a new document object. The program, first, asks the modeler if she or he wishes to edit a new model of the component. If the answer is 'yes' the program creates a new document object and shows it in the corresponding frame window that has the same title as the component. Figure 4.13 shows this window as a tabbed window. The document is empty. There are, however, three document ports corresponding to the ports of the component (Fig. 4.12).

We can proceed with component model development similarly as earlier by constructing new components by dragging and dropping the corresponding component tools from the toolbox (Fig. 4.10).

Thus, we can drag a *1 junction* and drop it near the left document port (Fig. 4.14). This operation creates a 1-junction with three ports. We will pick the bottom port and drag it around the component and drop it at the top. Next, we will pick the I (Inertial) component and drop it above the 1-junction. After clicking outside of the component, the component appears on the screen with the predefined I symbol and the power-out port on the left. We will again pick this port and move it to the bottom (against the corresponding 1-junction port). Now we can connect these two ports. If the line is not fine looking, we may reposition the I component and its port. Firstly, we must disconnect the ports, by selecting bond and pressing the *Delete* key to remove it. Then, we may pick the component by the title and drag to a suitable place. We may also select the component and use the cursor keys to fine-position it. Then we reconnect the ports again. We can next create a CompB, add to it a control-in port and interconnect its ports as shown in Fig. 4.14.

Now we may close the CompA document and store it in the previous document, within the *CompA* storage. We can close the document simply by clicking its 'x'

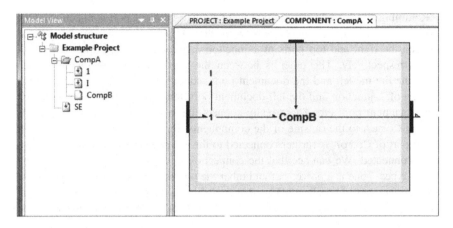

Fig. 4.14 Model of CompA

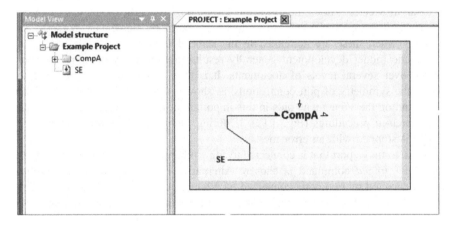

Fig. 4.15 The project model structure after the CompA document was closed and stored

button. (We can also do it in other ways, which parallels that of *Open* operation.) When the program asks we may save the document. Now the model structure of the project changes as Fig. 4.15 shows.

Comparing with Fig. 4.12 we see that the icon in front of the *CompA* branch changes to a folder form and '+' connector. This indicates now that *CompA* storage contains its model. By clicking '+' we can expand the branch, which effectively opens the component model document using the contained model data.

Finally, we finish the project model development by closing the project window. The program again asks if we wish to store the document. This time this amounts to storing the model documents to the disk as a compound file. If we decide to say 'no' all changes made after the last storing of the project document are lost.

Returning to Fig. 4.14 we see that there are two types of port connections. One correspond to internal connections between the components in the model, such as between the right and top ports of 1-junction and corresponding ports of CompB and I, respectively. The other is between the ports of the components, which constitute the model, and the document ports, such as the connection between the left port of 1-junction and the left document port, the right port of CompB and the right document port, etc. These really connect the inside components through the ports of CompA to the outside of the components. In Fig. 4.12 we see that the left power port of CompA is further connected to the SE power port, and the other two are unconnected. We can see that the connection is proper because both have the correct types—one is a power-out and other the power-in port—, and both have the same dimensions. The elementary ports has dimension one, by the construction. The port of CompA internally has only one connected bond, and this is inside connected to a 1-junction, i.e. to a port of dimension 1. Thus, the port of CompA has also the dimension one.

Great care should be taken when connecting ports by bonds. The rules discussed in detail in Sect. 3.7 should be followed closely; otherwise, the model can easily be incorrect. It is relatively easy to connect the ports of elementary components properly when these are contained in the same document. As a rule, however, systematic model development generally results in elementary components connected over several levels of documents. It is then of paramount importance to insure the symmetry of port connections, as shown in Sect. 3.7. The *Show Joined* command on the *View* menu aids in this important task by invoking the search-out and search-in procedures (Fig. 3.12). If during the search an error is found, the search is stopped with an error message.

Thus, to find a port that is connected to some other port, the port is selected, then the *Show Joined* command is chosen. Alternatively you may press the toolbar button depicted by two ports connected by a bond, or use the *Ctrl+J* shortcut. The search displays all ports and connecting bond lines, starting with the selected port and ending at the other port. During the search the states of the ports and connecting bonds are set and coloured. We illustrate the procedure with the example of the *See-saw* model of Sect. 2.7.3 (See Fig. 2.21).

We open the Body1 component and select the power-out port of the left junction as shown in Fig. 4.16 (indicated by the circle), i.e., the x-force junction port. We wish to find the other port connected to it. By clicking the *Joined Ports* toolbar button we start the search procedure, the result of which is shown in Fig. 4.17. The search starts at the selected port, go out of the Body1 component continue along the bond to the Platform, then go inside the Platform (Fig. 2.26, see also the model tree on the left in the Fig. 4.17). It follows then the bond until the component 0 is reached, goes inside, and finally ends at the x-force component of the flow junction (denoted by the oval). Thus, the connection is valid, i.e. the x-force component applied at the Body1 really is the x-component of the reaction force of the Platform. The port connections are symmetrical.

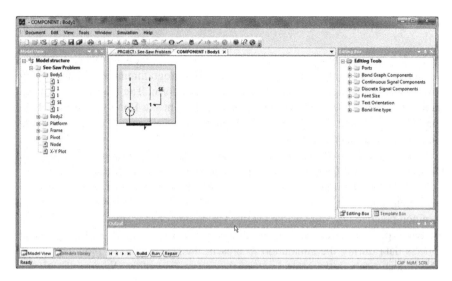

Fig. 4.16 Selection of Body1 port (inside the *circle*)

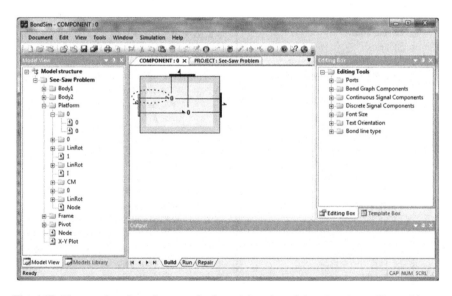

Fig. 4.17 The search ends at the port of *x*-force 0-junction of the Platform (denoted by an *oval*)

4.6.3   Developing Block Diagram Models

Block diagrams are traditionally used for developing models in different fields as was already discussed in Sect. 1.4. The *BondSim* programming environment supports also developing and using models in such a form. The main reason for the inclusion of this type of model here is to supplement the Bond Graph models by continuous-time and discrete-time block diagrams when complex physical model are developed in mechatronics.

Figure 4.18 shows the expanded *Continuous Signal Component* (on the left) and the *Discrete Signal Component* (on the right) branches of the *Editing Box* tab.

The continuous-time component tools support the creation of components that implement block-diagram operations discussed in Sects. 2.6.2 and 3.3. Following common practice the *Node* is implemented as a component depicted as a dot. There are two types of output component. One is in the form of an *X-Y Display*, which collects data during the simulation to present them as *x-y* or *x-t* plots in separate tabbed documents in the central part of the screen. The other, *Numeric Display*, is

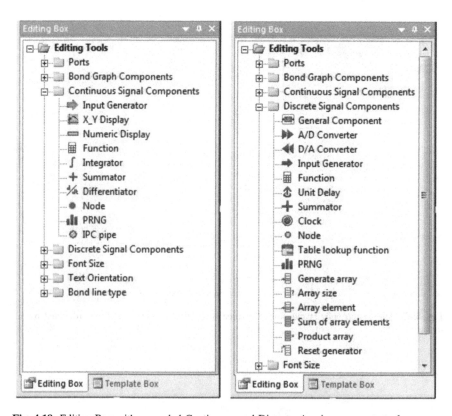

Fig. 4.18 Editing Box with expanded Continuous and Discrete signal components tools

implemented as a running counter. There is also a tool for generating noise in the system by a pseudo random number generator. The tool in the form of a ring, or a pipe cross-section, *IPC pipe*, serves for creating of an object based on the *named pipes* [4, 5] for inter-process communication with an external program.

The discrete-time signal component tool box is shown in Fig. 4.18 on the right and will be discussed shortly.

The development of block diagram models is very similar to that of Bond Graphs given in the previous sections. To illustrate this we will show a typical use of such components in Bond Graph models. We often add a control-out port to a flow (0) or effort (1) junction to pick up the common signals. In the first case it is an *effort* signal, and in the second a *flow* signal. These signals can be used for monitoring change in the corresponding variables over time during the simulation.

We will start the *BondSim* program and open the *Example Project*. We open the CompA, e.g. by expanding its branch in the model tree. The CompA document opens and appears in the central part of the screen (Fig. 4.19). We will add a control out port at the bottom of the 1-junction. To do this we may open the *Ports* branch in the *Editing Box* tab on the right of the main window, pick the *Control-out port* tool, and drag it over the *CompA* document window. Finally, we will drop it at the bottom of the 1-junction. The control-out port appears now as shown in Fig. 4.19.

Now we have created an access to the junction common flow (the body velocity if the I represents body inertia). But, we cannot take the signal out of the component because there is no the corresponding control-out document port. To create one we must add the corresponding port to the component object. Thus, we can click the *Example Project* document tab to activate the previous document (Fig. 4.15), and add a control-out port at the left-bottom of the CompA periphery (but inside the bounding rectangle) in a similar way as we did with 1-junction.

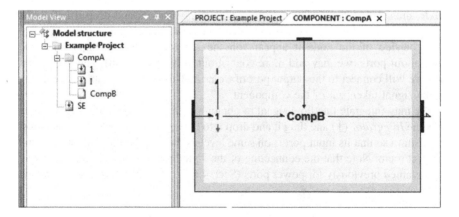

Fig. 4.19 Adding a control-out port to 1-junction

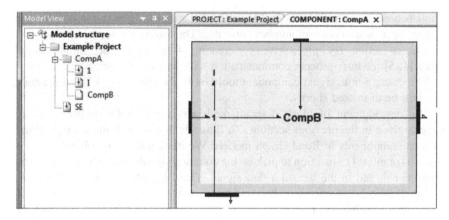

Fig. 4.20 Taking out the flow signal

When we return back to the CompA document (by clicking on the corresponding tab) we see that we now have the document port we need.[5]

If the port is not positioned at the periphery of the document as we would like, e.g. just opposite of the 1-junction control port, we can adjust it before connecting. However, we must do it at the component side, by dragging the port by mouse, or moving it by the cursor arrows keys. We can also widen the document port by moving the cursor over the port. When the port becomes selected and the cursor changes shape to a two-side arrow, we may press the left mouse button and drag the port edge.

After connection the *signal line* appears as in Fig. 4.20.

Now we can close the *CompA* document and save it (Fig. 4.21). We will next add a node into the system level document below the CompA. To do this we will expand the *Continuous Signal Component* branch in the *Editing Box*, pick the *Node*, drag it into the *Example Project* document window and drop it below the CompA and near (but not too close) to the document bounding rectangle. After we click outside of the node it appears on the screen. It has one input and two control-out ports (we may add more control-out ports if we wish). The node input port we will connect to the output port of CompA. This way we establish branching of the signal taken out of the component.

We may integrate this flow signal to obtain e.g. the displacement. Thus we may pick the *Integrator* ($\int$) and drag it and drop it to the right of the node. We can adjust its position so that its input port is on same level as the node's right output port, and connect them. Note that the connecting of the signal ports is done in the same way as explained previously for power ports (Sect. 4.6.1). The only difference is that the

---

[5]Note that this operation silently, in the background, saves the component document in order to ensure the correspondence between the component and document ports just created.

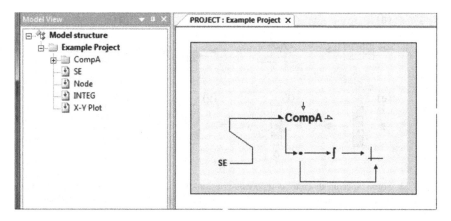

Fig. 4.21 The final view of the project model

connecting line is the signal line and not the bond line; a signal line basically transfers a single signal, and a bond line a pair of the signals.

The final component that we will create is the X-Y Display at the right of the *Integrator*. It has two input ports predefined (we may remove one or add more if we wish). We may adjust their positions so that the left port is at the level of the integrator output. Finally we connect the left port of the display to the output port of the integrator, and the bottom port to that of the node. The final form of the model is shown in Fig. 4.21 (at the right). The *Model structure* tab shows now the structure of the *Test project* model. Now we can close and save the model.

4.6.4   Modelling Discrete-Time Processes

We return now to the *Discrete Signal Component* tools (Fig. 4.18 right). The first entry in the tool serves to create a *General Component*, which parallels the *General Word Model Component* in the *Bond Graph Components* branch (Fig. 4.10 left). However, this one supports only discrete control input and output ports and can contain only discrete-time components. Thus, continuous-time and discrete-time models are separate, but can communicate. The same is valid in the physical world as well. An embedded microprocessor cannot directly acts on a motor. They work in different physical domains. It is, thus, necessary to convert the signals from one domain to the other and scale them appropriately.

The conversion can be achieved by *A/D* and *D/A converters* described in Sect. 2. 6.3. These components can be created by the drag and drop technique as any other component using the corresponding tools. These two components, however, have to be created in the main project root document.

In the same way we can create other components. These components can be put in the system level document or inside a discrete component. There are many tools,

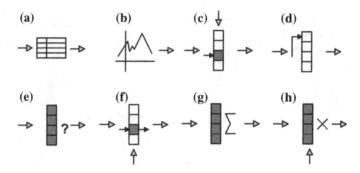

Fig. 4.22 Some special discrete components: **a** lookup table function, **b** PRNG, **c** buffer generation, **d** reset operation, **e** buffer size, **f** buffer element, **g** buffer summation, **h** element wise buffers multiplication

which serve to purpose. The next six components were already discussed in Sect. 2. 6.3. We explain the others.

The *Table lookup* function tool serves to create a corresponding discrete component (Fig. 4.22a). Currently, the *lookup table function* supports evaluation of *sine* and *cosine* functions using a table of pre-calculated sine and cosine values. The value generated by such a function uses linear interpolation of the input value between the nearby table values. It is often used when a great number of such calculations are taking place in a processor. Using such a model function it is possible to closely simulate the behavior of such equipment both in terms of the efficiency and accuracy.

The *Pseudo random number generator* (*PRNG*), is similar to its continuous-time counterpart (Fig. 4.18). It is a component, which is often used to simulate the noise present in digital systems. Its graphical representation is shown in Fig. 4.22b.

The last six tools serve to create the corresponding components that model the *buffered* operations in digital systems. Their graphical representations are shown in Fig. 4.22c–h.

The first of these components the *Buffer* (Fig. 4.22c) generates a storage buffer in the form of an array, whose size may grow. There is also an internal index pointing to the first free position in the buffer, which is initially set to zero. At every sampling instant the value received at the left input port is stored in the next free position in the buffer and the index is incremented. The port at the top serves to reset the *Buffer* by setting the buffer index to zero. The buffer is reset when this port receives a nonzero signal. The third, the output port, gives the value of the *buffer handle*. It is a number indicating the position of the buffer in the list of buffers that the program internally manages and is used to access the buffer to apply the operations on its elements by the other components. It, thus, plays a role of the buffer *id*.

The buffer reset operation can be generated directly, or by the *Reset* component shown in Fig. 4.22d. This component is intended to be used inside a trigger component, a general discrete component with a trigger port (Fig. 2.11e). If the trigger port is switched 'on' the *Reset* component generates at its output value *1*, otherwise *0*. If it is not inside a trigger component it unconditionally generates the value *1*. This component needs the input port in order to properly define the calculation sequence and thus define when the reset is applied.

The *Buffer Size* component shown in Fig. 4.22e serves to find the current number of the elements contained in the buffer whose handle was received at the input port. It can be used, e.g., to stop collecting data in a buffer when its size reaches some value and to start their processing.

The next component, the *Buffer Element* (Fig. 4.22f), generates at its output the value of the element of a buffer, whose handle is set at the left input port. The element corresponds to the zero-based buffer index supplied to the bottom input port.

The *Buffer Sum* (Fig. 4.22g) component calculates and outputs the sum of all elements contained in a buffer whose handle was set at the input port.

The last component is the *Buffers Product* (Fig. 4.22h), which calculates the element-wise product of two buffers. This function has two input ports, which should be connected to the corresponding *Buffers* output ports. The function creates a new buffer and when it is called it calculates successively the products of the elements of the buffers defined at its inputs and successively fills in the internal buffer. At the output it generates the handle of this product buffer.

The buffered components can be used e.g. to evaluate in a compact way an integral measure of a signal, as is often used in frequency analysis of signals. The other discrete components could be defined as well, but the components shown are taken as the basic and permit simulation of fairly complex processes.

4.7   Generating Electrical and Mechanical Schemas

When editing bond graph models of electrical devices, using word model components in the usual (textual) form described in the previous section is perhaps not too attractive to electrical engineers who are more used to the standard electrical symbols. A similar situation exists in other fields, e.g. mechanical vibrations. To enable generation of the Bond Graph components in more familiar forms in addition to the *Editing Box* there is also a *Template Box* (Figs. 4.2, 4.10 and 4.18). The template box currently contains only two template branches (Fig. 4.23): *Electrical Components*, and *Mechanical Components*. The third branch *Component Orientation* serves for the selection between horizontal (default) and vertical orientation of the generated components.

4.7.1   Developing Electrical Circuits

To illustrate a typical procedure used to model an electrical circuit by the templates we consider a simple RLC circuit. We apply the procedure described below (Fig. 4.24):

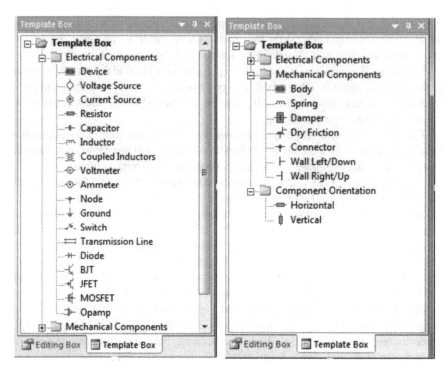

Fig. 4.23 Template Box with expanded components branches

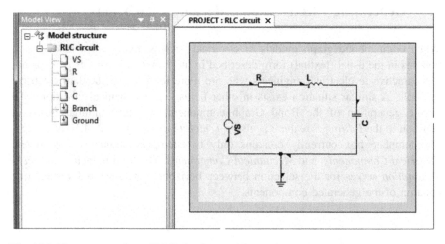

Fig. 4.24 The representation of RLC circuit

- Start the *BondSim* application. Open a *New* project and input the project name, e.g. *RLC Circuit*. Click *OK*. Now an empty document appears in the central part of the screen.
- Select the Template Box in the right tabbed window, and expand the *Electrical Components* and *Component Orientation* branches.
- Drag the *Voltage Source* template to the document window and drop it near the left edge and at the middle height. Each time when dropping a component, click outside of it to end text editing.
- Drag and drop the *Resistor* and place it near the upper document rectangle edge and to the right of the *Voltage Source VS*. Connect it to the voltage source by drawing a bond between the corresponding ports. Reshape the bond line by dragging and dropping to obtain a rectangular shape.
- Drag and drop the *Inductor* and drop it to the right of the resistor *R*. Adjust its position so that its left port is horizontally on the same height as the right resistor port. Connect the ports.
- Before inserting the capacitor, select *Vertical* in the *Component Orientation* template. Drag the *Capacitor* and drop it to the right. It appears as a vertically drawn capacitor *C*. Connect its upper port to the left port of the inductor. Reshape the bond so that it forms a right angle. Click *Horizontal* in the *Component Orientation* template to return to the default orientation.
- Before adding the Ground we add first a node by dragging and dropping the *Node* component. However, its power port senses are not correct. We expect that power flows from the capacitor through the right port and from the left port into the voltage source; or out of the bottom port to the ground. To change the power flow sense of a port we may select it and click the toolbar button which has two half arrows (or select the *Change Power Sense* command in *View* menu). After setting the proper power ports sense of the node connect it to the voltage source and capacitor.
- Now we need more space at the bottom of the node to add the ground component. We put the mouse cursor on the bottom edge of the inside bounding rectangle. When it changes shape to the up-down arrow, press it and drag the edge downward until we obtain enough space and then release the mouse.
- Now we can drag the *Ground* and place it below the node and connect it to the node. We thus obtain the Bond Graph model in as Fig. 4.24.

Now the model has the shape of the ordinary electrical circuit. Only the half arrow indicates that it is a bond graph model. The model structure of the circuit is shown in Fig. 4.24 at the left. We see that it has four word model components represented by symbols: *VS*, *R*, *L* and *C*. However, these are empty components, without underlaying model documents. Thus it is not the complete model. However, we can develop the models in the usual way as was explained in the previous section.

Note that *Branch* (node) and *Ground* are leaves of the model tree, or in Bond Graph terminology the elements. The electrical nodes physically represent branching of the currents flowing through them, thus it is really a 0–junction (see component hierarchy in Fig. 4.5). Similarly the *Ground* is really a SE (*Source Effort*), which defines the ground potential (typically zero).

More on modelling of electrical circuits will be given in the second part of the book.

4.7.2   Developing Mechanical Circuits

In a similar way we can develop models of mechanical systems. To illustrate the procedure we consider a body hanging on a spring the other end of which is fixed to a horizontal wall. If such body is dragged down and released it will oscillate vertically, but its amplitude will become smaller and smaller until it comes again to rest. Thus, there is damping in the system and such systems are typically modelled as the *Body-spring-damper system*. We will develop a model of such a system using the mechanical template tools (see Fig. 4.23 right).

The procedure is similar to the previous case (Fig. 4.25):

- Start the *BondSim*. Open *New* project and input the project name, e.g. *Body-spring-damper system*. Click *OK*. Now an empty document appears in the central part of the screen.
- Select Template Box in the right tabbed window, and expand *Mechanical Components* and *Component Orientation* branches.
- Select *Vertical*, and then drag and drop *Wall right/up* tool at the top. We obtain top wall oriented upward, i.e. with the port at the bottom. (If we select the other wall, it will be oriented downward. We can easily remove it by selecting it, if it is not already selected, and then press *Delete* key. In the dialogue windows that opens click the button *Remove completely*.)

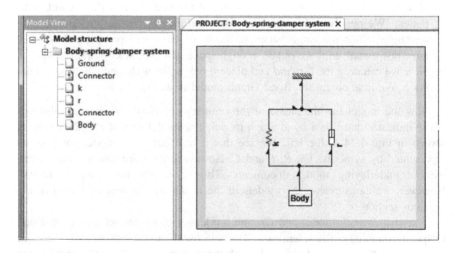

Fig. 4.25 Model of Body-spring-damper system

- Drag *Connector* and drop it below the top wall. A mechanical node appears having three ports: the left, right and top. We assume the power flow direction from the bottom (body) to the wall. Thus, we need to change the sense of the right port by selecting it and clicking double half-arrow toolbar button. Now we may connect the node to the top wall by drawing the bond.
- Next, we will drag *Spring* and drop it bellow the connector node and to the left. Then, drag *Damper* and drop it to the right of the node. The spring and damper are drawn vertically. We position them approximately at the same heights. We need to change direction of power flow through these components. Thus, we select the spring and damper by clicking and then click the toolbar button to change the power senses of their ports. Then connect the top ports to the connector node then reshape the bonds so that they have right angle shapes. Select the *Horizontal* orientation to return to the default orientation.
- Add the bottom connector below the bottom ends of the spring and damper. We must change the power sense of the left port, and remove the top port (by selecting it and pressing the *Delete* key). We also have to insert the port at the bottom. Hence, we need to return to the *Editing Box* by clicking its tab, and expanding the *Port* branch. We drag and drop the *Power-in port* to the bottom of the connector node. Return now back to the *Template Box*.
- Similarly as in the last problem we enlarge the document working area by dragging the lower edge of the inside bounding rectangle downward.
- Finally, we drag *Body* from the template box and drop it below the bottom connector node. A Body component appears, bounded by a box, and having the power port at the left side. We will drag the port to the top, and change its power flow sense. We will also drag the component to place its port exactly below the bottom *Connector* port and draw the bond connecting these ports.

The model of the Body-spring-damper system is shown in Fig. 4.25. The central window depicts the Bond Graph model, and on the left is shown its Model structure. We can see that the model consists of word model components: *Ground* (corresponding to the top wall), k (corresponding to the spring with stiffness constants k), r (corresponding to the damper with linear damping constant r) and Body. There are no underlying component models, but they can be easily added. The *Connectors* are leafs of the model tree (Fig. 4.5). They are the points of common velocity, hence the 1-junctions.

4.8 Editing Elementary Components Constitutive Relations

4.8.1 Component Port Dialogues

The port variables and the constitutive relations of elementary components are stored in their port objects (Sect. 3.5) and, ultimately, in the component. The same holds for block diagram components, with the difference that the constitutive relations are stored in their output ports; input ports only store the input variables. In addition, there are parameters that are defined at the document level, or locally in the component. The constitutive relation can have different forms depending on the type of component, as discussed in Sects. 2.5 and 2.6. The corresponding expressions should conform to the rules of Sect. 3.5.

Default variable names and constitutive relations are defined at the component port creation. These correspond to simple linear relations. The necessary parameters also are defined and stored in the component. Hence, it is necessary to implement methods that enable changing the variable names, parameters, and the constitutive relations. We address the problem of the editing of variables and constitutive relations first. The parameters are discussed later.

Suitable dialogues (Fig. 4.26) are used to support modifying the variables and the constitutive relations of the element ports in a user-friendly way. These are invoked by double-clicking the ports of elementary or block diagram components. In response, the component method is called to open the component port. The method constructs a dialogue object and transfers data to it, which is then displayed on the screen. Variable names and constitutive relations are edited using fields provided by the dialogue. When the editing is finished and the data are accepted (by clicking the *OK* button) the dialogue is closed and the data are stored back into the components ports.

To enforce specific types of variables and constitutive relations, a separate dialogue is defined for every type of component port. Figure 4.27 shows a dialogue corresponding to an inertial component power port. Such a port dialog can be opened e.g. by double-clicking I the component port in Fig. 4.20. Similar dialogues are used for the other types of elementary port. The top part of the dialogs has fields that are used for defining the constitutive relation, of the form discussed in Sect. 3.5. Next are the fields that define the port variables. On the left are the current port variables, and on the right a list of other port variables can appear if the component contains other ports as well. There is also a button that can be used to change the values of, or define new, parameters (constants).

Most fields can be changed simply by selecting it followed by editing using the keyboard. The fields are instances of MFC's *CEdit* class, which implements

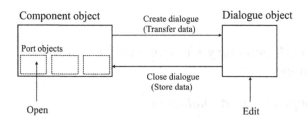

Fig. 4.26 Operation when editing the constitutive relations

Fig. 4.27 The inertial port dialog

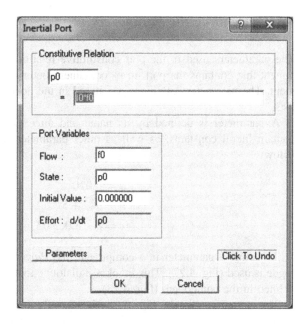

common text editing operations. Fields that should not to be changed are set as "read-only", e.g., the *Effort* field (Fig. 4.27). Anytime during editing we can undo the last change by clicking anywhere in the dialogue box.

When the editing is finished, the changes are accepted by clicking *OK*, or rejected by clicking *Cancel*. If the *OK* button is chosen, the data are validated before being sent back to the component port.

Validation is done by parsing and syntactical analysis of the expressions. A valid variable starts with a letter, its length is not limited, and is case sensitive. The symbols used in the constitutive relation can be port variables, the time symbol *t*, which is a global variable, or numerical and symbolic constants. The syntax of the constitutive expression should conform to the rules discussed in Sect. 3.5. If an error is found, a message box appears with information about the error. The error must be corrected before the data in the dialogue can be accepted, or else the user must dismiss the dialogue by clicking the *Cancel* button.

Another check is also made before the data are transferred back to the component port. When port variable names are changed, these new names are checked against the names used in the constitutive relations of the other ports contained in the same component. If a match is found, a dialogue is opened that prompts the user to change to a unique variable name. When all such corrections are made, the new data are accepted and stored back into the component ports.

4.8.2 Defining the Parameters

The parameters used in the port constitutive relation can be defined in the component that contains the port, in its container document, or in a lower-level document. The parameter definitions are stored in the document or components as the lists.

A parameter is defined by its name and an expression. The expression can contain literal constants, as well as other parameters already defined, as shown below:

$$a = 0.010$$
$$b = a^2 + 5 \qquad\qquad (4.1)$$
$$c = b + 5.0e{-}03$$

To define a parameter in a component, the *Parameters* button of the port dialogue is used (Fig. 4.27). This invokes a dialogue that displays a list of parameters defined in the component (Fig. 4.28).

The parameter list in Fig. 4.28 shows only one parameter *I0*. This is the parameter that was defined by default at the construction of a I (inertial) component port and appears in the constitutive relation in Fig. 4.27. To add a new parameter, e.g. the first parameter in Eq. 4.1, we type-in the parameter name in the editing box, and then click the *Insert* button. A new dialog appears now as shown in Fig. 4.29. It contains two editing boxes arranged in the form of an equation. The left side is already defined and contains the parameter name that we have typed in previously, and the right edit box is ready to accept a corresponding parameter expression. It shows the default value of 0.0. We change this to 0.010, the value of parameter *a*. When the *OK* button is pressed, the expression on the right side is parsed to determine if it is syntactically correct and well defined. The dialog box closes and the new inserted parameter appears in the parameter list. The parameter list is now

Fig. 4.28 Parameter editing dialog

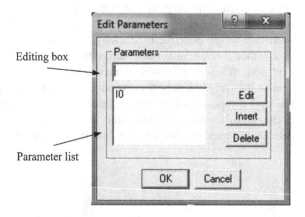

Editing box

Parameter list

Fig. 4.29 Editing the
parameter expression

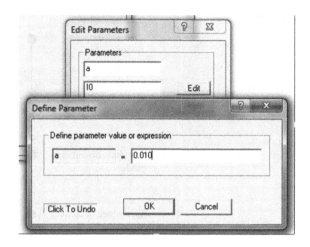

ready to accept the next parameter. Note that parameter names are case sensitive, thus e.g. "Mass" and "mass" are different parameter names.

The new parameter is added at the end of the list. However, we may change the position of a parameter in the list by dragging it before or after some other parameter in the list. A parameter already defined in the list can be changed by selecting it in the list first, then pressing the *Edit* button. The same dialog appears again, but this time showing the current value of the parameter. To save the changes made we need to close the parameter dialog by pressing the *OK* button.

The parameters can be defined at the document level as well. To that end we may click the toolbar button denoted by *P* (Fig. 4.11), or use the *Model Parameters* command from the *Edit* menu. In response the same dialogue appears; but this time it displays parameters defined in the active document. The procedure of editing parameters is of course identical to that described above, but the created or modified parameters are stored in the current document.

The parsing and syntactical analysis of parameter expressions are similar to that implemented for the constitutive relations. The parameter expression is accepted if the expressions are correct and all parameters are well defined, i.e. they are in this parameter list, or are defined in some other lower level document.

These parameter lists are ordered in the same way as the components and documents. Thus, the parameter definitions can be looked at as the "list of lists", or as a parameter tree. Each node of the lists corresponds to a list contained in a component, or in a lower level document, and finally in the application (Fig. 4.30). Some of the nodes may be empty.

The lists are searched in the direction from the component to the root document and the application. Thus if, during the parsing of a constitutive expression in an elementary component, a symbol is met that is not a port variable, the time or literal constant, then the parameter lists are searched to see if the symbol is already defined as a parameter. The search starts at the component parameter list, then the parameter list defined in the container document, then the last document is searched, and the

Fig. 4.30 Parameter list tree

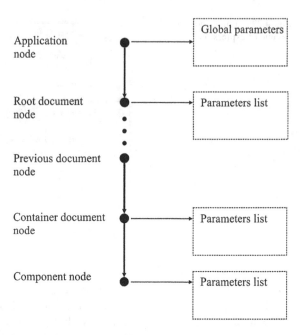

search is continued until the root document and global parameter list defined in the application are reached (Fig. 4.30). The first occurrence of the symbol is taken as its definition. All possible lower level definitions are ignored.

In this way, a parameter defined at some level hides the same parameter defined at a lower level. This may be termed the *parameter hiding*, and is similar to hiding of the local variables in the C/C++ language. The hierarchical structure of the parameter definition has several important impacts on modelling.

The often used constants such as the important mathematical and physical constants, the program constants and the alike are defined as the named parameters. The list of such constants defined in the *BondSim* program are termed the global parameters and are shown in Table 4.1.

Due to the search procedure defined above these constants are visible in any document and the contained elementary components. In the same vein parameters specific to some project can be defined in the project document (the root).

The typical values of parameters of a component could be e.g. defined at the level of its component model. However, the more specific values of the same parameters can be defined closely to its detailed model, i.e. at higher levels. In this way we can start to study by using a general model based on typical parameters, and when the model is refined, a new definition of the same parameter is introduced that hide the previous typical values. We use this procedure when dealing with practical problems in the second part of the book, e.g. when dealing with semi-conductor models.

Table 4.1 Global parameters

Parameter	Value	Note
g	9.80665	Gravity constant (m/s^2)
PI	3.141592653589793	π constant
BOLTZMANN	1.380662e−23	Boltzmann constant (J/K)
ECHARGE	1.6021892e−19	Electron charge (C)
EGSI	1.12	Energy gap in Si (V)
TNOM	300.0	Nominal working temperature (K)
BG_EPSILON	4·DBL_EPSILON	BondSim epsilon
BG_MAX	DBL_MAX	BondSim maximum value
MEXP	50.0	BondSim maximum exponent
BG_e	2.718281828459045	Basis of natural logarithm

DBL_EPSILON = 2.2204460492503131e−016, DBL_MAX = 1.7976931348623158e+308

Finally, a few words about deleting parameter. This is accomplished by selecting a parameter in the list, and clicking the *Delete* button (Fig. 4.28). This removes a node where the parameter is stored together with the parameter definition. This is a potentially dangerous operation because the parameter may be used elsewhere. It is the developer's responsibility to decide whether or not to delete a parameter.

4.9   Library Operations

The program space where the work is conducted and the data are stored is divided in BondSim into two main subspaces. The *workspace* is a part where the main development work is done and *models* are stored. Another is the *library*, which serves as a repository for projects and components. All the previous operations take place in the workspace. But, the *Library* is important, as well.

To simplify access to the *Library* its contents are shown using an Explorer style tree in the tabbed window *Models Library*. This window together with the *Model structure* tab is contained in a separate window frame at the left of the main window. An expanded part of *Models Library* window is shown in Fig. 4.31.

It is divided into several parts:

- Project
- Word model components
- Electrical components
- Mechanical components

We describe next these library branches and how they are used when developing models for a project.

Fig. 4.31 Contents of
Models Library

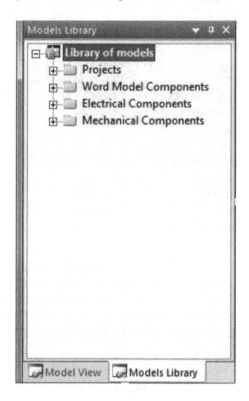

4.9.1   Library Projects

By expanding the Project branch we obtained a list of project files stored in the library (Fig. 4.32).

Most of these are the projects that we will develop in the second part of this book, when practical applications are considered. To see what we can do with these projects click by the right mouse key any of these projects, e.g. *Body Spring Damper Problem.* A drop down menu appears as shown in Fig. 4.33.

The menu contains several commands. The *Open* command is used to open the library project. The project opens in the main window, but the project title is now *View project: "project name"* (see Fig. 4.34). This indicates that opened project is in the viewing mode. Thus, we can walk through the project, but changing nothing. Simultaneously the *Models Library* tab switches to the *Model View* tab and the structure of the library project model is shown. We can walk through the project by double-clicking documents in the central window or expanding the branches in the Model structure (or in other ways as we already have discussed).

The *Remove* command enables the user to remove some of the project from the library database. We can insert a project into the library by copying a project from

Fig. 4.32 List of library projects

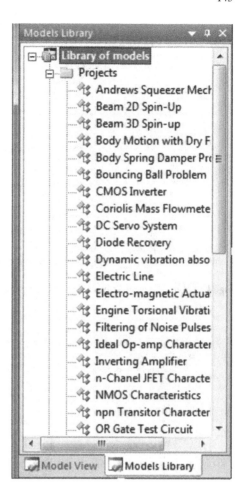

the workspace. For that purpose under the *Project* menu there is a command *Copy to Library* as already discussed in Sect. 4.2. Similarly, we may copy a project from the library into the workspace using the command *Copy to the Workspace* (Fig. 4.33).

Finally, there is a *Repair* command, which is used to repair the library subspace. The command is similar to that of the *Project* menu (Sect. 4.2).

4.9.2 Library Components

The components are organized in a similar way. Figure 4.35 shows an expanded *Word model components* branch. It contains different components, all of which are stored in the corresponding containers as files. If we right click any of these

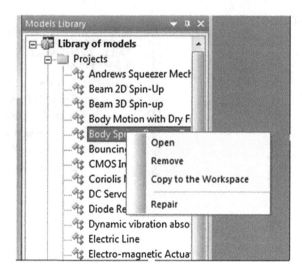

Fig. 4.33 Library menu activated by mouse right click on a project

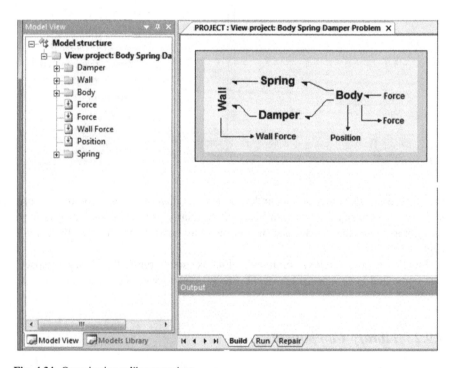

Fig. 4.34 Overviewing a library project

Fig. 4.35 Word models
component library

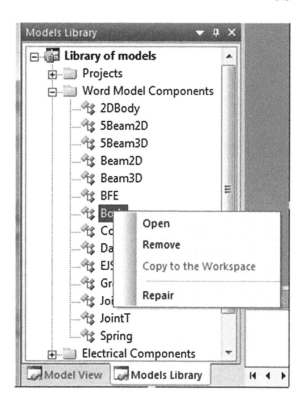

components a drop-down menu appears, which has the same form as in the previous subsection, with a difference that *Copy to the Workspace* command is now unavailable. Therefore, we may open a component as we did with projects, or remove it from the library. To use a component during model development we simply use the "drag and drop" technique as we did with the Editing and Template boxes. Thus, we pick a library component with the mouse and drag it into a document window. The application copies the corresponding word model component and all underlying documents.

The *Electrical* and *Mechanical Components* are organized in a similar way. Figure 4.36 shows the expanded *Electrical Components* branch.

The component library is further divided into specific types, such as *Resistor*, *Inductor*, *Diode*, *Opamp*, and others. Each of these sub branches contains specific components. Thus, e.g. the *Resistor* type contains only one component denoted as *R*. This is a specific component that we can open to see its description in terms of a Bond Graph model. We can also use it by dragging and dropping it into a corresponding document. The library can contain the other components as well, e.g. models of commercial resistors, etc. Similarly this holds for *Mechanical Components*.

Similarly as with projects we can also insert a component by copying it from a project document into the library. The *Document* menu, which is active when a

Fig. 4.36 Electrical
Components library

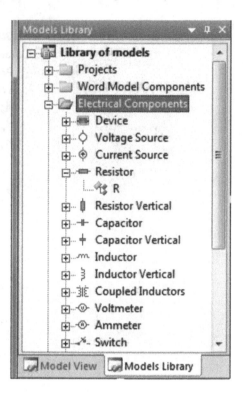

project is opened, contains the *Copy to Library* command (similarly as we have when dealing with the projects in Sect. 4.9.1). This time the command applies to a component that we selected in a document window. By applying this command the selected component is copied together with all its documents into a container document, which is inserted and stored in the library. If the component is a word model it is stored among the *Word Model Components*. But, if this is an electrical or mechanical component of a specific type, it is inserted and stored in the corresponding component library under the corresponding type.

4.10   Important Operations at the Document Level

This section reviews some of the most important document-level operations. These commands are collected under menus that appear when the first document—the project root document—is opened. These are: *Document*, *Edit*, *View*, *Tools*, *Window*, *Simulation*, and *Help*. Some of these commands can also be invoked by pressing the appropriate tool bar button, or using the keyboard shortcuts.

Many of these commands we have already met. We will describe here some others that concern model development. These are found in the first four menus:

Document, *Edit*, *View*, and *Tools*. They implement methods already discussed in Chap. 3.

The *Windows* menu contains standard Windows commands, such as *Cascade* and *Tile*, used for arranging windows on the screen. The menu also contains one specific command: *Refresh*. This is used to update (redraw) the active window, e.g. during editing of the bond graph model. There is also a list of all opened documents at the bottom. These are used to activate the corresponding frame windows. Note that there is no *New Window* command in the Window menu, as is customary in Windows applications. Every document is shown in just one document frame.

Commands in the Simulation menu deal with simulation tasks. We postpone discussion of these commands until Chap. 5. The *Help* menu contains commands that link to the program documentation, as already described at the end of Sect. 4.2.

4.10.1   Open, Close and Save Commands

First on the *Document menu* is the *Open Next* command. This command is applied if we wish to open the currently selected component. We have already met this command, as well as the others to open the next document (Sect. 4.6.2).

The opened document contains a link (a pointer) to the previous opened document. Hence, it is possible to return to the previous document, and to activate it by using the *Last* command on the *Document* menu.

An opened document can be closed in the usual way for Windows by clicking the '×' in the title bar. There is also a *Close* command on the *Document* menu. Because every opened document contains a link (pointer) to the previous document, it is necessary to close all upper-level documents first. Thus, a *Close* command invokes a dialogue asking if the user wish to close all upper-level documents. If the answer is no, closing is aborted. Otherwise, all upper-level documents are closed before the current document. When the document is closed, the state of the corresponding component object changes to *Normal*.

There are two other *close* commands in the *Document* menu. The *Close Component* command is used if, in the current window, there is an opened component. To close it, it is necessary to select the component first, and invoke the *Close Component* command. There is also a *Close Project* command in the *Document* menu. This command is used to close all currently opened documents and, thus, the modelling project.

When closing, an alert box appears to advise the user if the document about to be closed has to be saved. There are several possibilities (Fig. 4.37). We can choose simply to *Save* the current document to the corresponding component storage. Or, select *Save to all* meaning that all documents down to the system (project) level document are saved. We can also choose to *Don't save* document, or *Cancel* the closing operation.

Similarly to *close* command, there are several *save* commands on the Document menu: *Save*, *Save to Project*, *Save Project*. The first one, *Save*, we have already

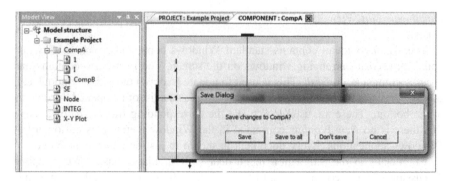

Fig. 4.37 Close alert dialog

met. This command stores the current document in its component storage, or disk if it is a project level document.[6] In the process, all objects contained in the documents, such as components, ports, bonds, and others, are called to store themselves in a document archive before the document is stored.

The *Save to Project* command successively stores documents from the current document to the project level document. In terms of the model tree structure it means storing data from a branch down the tree until and including the root. This command, thus, assures that the project contains the all changes. However, if we issue the command *Save Project* the complete project is stored including its all open documents.

4.10.2   Page Layout and Print Commands

The document orientation is important for printing, but it also influences the maximum width and height of the document. The default document is oriented horizontally, but this can be changed to the vertical by the *Page Layout* command in the *Document* menu. This command invokes a dialogue from which horizontal or vertical page orientation can be chosen.

The program also implements printing and print previewing of the documents based on the MFC Library support. The first page of the printout contains the bond graph diagram displayed in the current document. The following pages contain all parameters defined at the document level, as well as the constitutive relations and parameters of the elementary and block diagram components.

There also is a command *Print to File* for printing graphic data (Bond Graphs) in the *Enhanced Window Metafile* format (emf). The programs opens a *Save As* dialog and asks the user to define the name of the file and the place where it should be

---

[6]Usually the structured storage data are held on disk in temporary working locations, which are released after final store operation is completed.

saved. The extension is predefined as '.emf'. After the dialog is closed by clicking *OK*, the command invokes methods very similar to those used for drawing to the screen, but uses the MFC Library metafile context to draw to the specified file [1]. The emf files can be imported by most word processors and graphic programs, such as MS Word, CorelDraw, and Adobe Illustrator.

4.10.3   The Delete, Copy, Cut, and Insert Operations

To simplify model editing, several commands are implemented for deleting, copying, cutting, and inserting in the *Edit* menu.

Thus, a component port can be deleted and removed from the model by selecting it first, then using the *Delete* command from the *Edit* menu. The *Delete* key may also be used for this purpose. The port can be deleted if it is not externally and/or internally connected (Sect. 3.4). Deleting a port does not only remove the port in the component, but also in the corresponding document. This means that the accompanying document is opened in the background, the document port removed, and the document saved and again closed before the component port is removed. In a similar way, it is possible to delete a bond. This removes it from the document, and the bond data are removed from the ports that it connects.

Deleting a component is more involved. The component should be disconnected from the others and must be closed. To delete a component it first must be selected, and then the *Delete* command is chosen from *Edit* menu, or *Delete* key pressed. A dialogue then appears asking if we wish to move the component to the *Waste Bin* or delete it directly. There are three possibilities: *Move to Waste Bin*, *Delete*, or *Cancel*. If the first option is selected then the component is removed from the document and moved into the *Waste Bin* buffer (Sect. 4.2). This action does not really remove the component object, but remove the component (its pointer really) from the component list maintained in the current document and is inserted into the *Waste Bin* component list. The next time the screen is painted the components disappear from the screen. The *Waste Bin* buffer can be accessed from the document level, as well. If *Delete* is chosen the component and its documents are completely removed from the memory without sending it to *Waste Bin* first. By selecting *Cancel* the delete operation is aborted of course.

Three other commands—*Copy*, *Cut*, and *Insert*—work between the document and the internal *Component Buffer*, which the application maintains. This buffer plays a similar role as the *Clipboard* in Windows. However, different from the last the *Component Buffer* retains its contents between *BondSim* sessions. To cut a component not interconnected to others, we first select it, and then use the *Cut* command from the *Edit* menu, or the *Shift+Delete* key shortcut. This removes the component from the document—in the same way as the *Delete* command—and puts it into the *Component Buffer*.

We can also copy a component from the document to the *Component Buffer*. The component may be connected, because we are not moving it, but its copy. To copy a component to the buffer we first select it, and then apply the *Copy* command, or *Ctrl+C* shortcut. Copying means, as has been already stated, creating a new component object that is a replica of the one being copied. The copy has a new *id* and filename. Thus, the document of the original component, as well as those of any contained components, must be copied. The pointer of the newly created component is put into the *Component Buffer*.

The components in the *Component Buffer* can be inserted back into a document by applying the *Insert* command from *Edit* menu, or *Ctrl+V* shortcut (equivalent of *Paste* command). This opens a dialogue with a list of the components that are in the buffer. Selecting a component from the list and clicking the *OK* button closes the dialog and sends a message to the active view, which changes its state to the *Insert Component* mode. The status bar displays a message informing the user to pick a place in the document to insert the component. In the same way as when creating a component (Sect. 2.6.2), as soon as the cursor is moved into the document area it changes to a cross, which is surrounded by a rectangle corresponding to the bounding rectangle of the component that is being inserted. We need to place the cursor so that this bounding rectangle is completely inside the document working area. Clicking the left mouse button inserts a copy of the component into the document.

All the above commands can also be applied to a set of selected components (Sect. 4.5). We create a separate object to copy or move a selected set of the objects from the document to a buffer, and back (Sect. 3.8).

The components in the set can be *deleted* or *cut* from the document if they are not connected to other components not belonging to the selected set. In this case, we simply proceed as explained in Sect. 3.8. Hence, we create a component set object, remove the pointers of the selected objects from the document and add them to this object. The object can then be added to a buffer (the *Waste Bin* or the *Component Buffer*). When adding the component set to the buffer, we must append a suitable reference name. It is not necessary that this name be unique. Any name conflict is resolved easily by automatically adding a version indicator, e.g. *Body*, *Body.1*, *Body.2*. (Whether or not the names are the same, they represent different objects.)

We copy a selection in a similar way. In this case, the components in the set need not be disconnected from the other components. We again create a component set object and add to it copies of all selected components and bonds (Sect. 3.8). Copies of its documents and of all contained documents are made, as well. The new component set object is then added to the buffer.

Inserting a set of components from the buffer is made in the same way as when inserting a component. Copies of all components and bonds in the set are created and added to the document. Their positions are changed to correspond to the position chosen on the screen. Copies of all documents are also made. Next, a temporary selection set object is created, pointers to objects inserted in the document are added to the selection object, and all of the objects in the set are selected. The selected components can now be moved across the screen, if required.

As we already mentioned in Sect. 4.2 we can access the *Component Buffer* thorough *Tools* menu. It has the Component Buffer submenu with two commands: *Open* and *Delete*.

The *Open* command can be used to review components (or component sets) placed in the buffer. Selecting a component from a list opens a separate document window. A copy of the component object is inserted into this document and serves as the interface to the underlying documents. Editing is not permitted in this mode.

The components (or component sets) in the buffer can also be removed if, for example, they are no longer needed. The *Delete* subcommand is used to remove a component from the buffer and place it into the *Waste Bin*, or delete it directly, in the same way as the *Delete* command describe previously.

4.11   Inter-process Communications

Bond Graphs basically describe systems in a rather abstract way. For example a ball bouncing on a horizontal surface can be modeled by Bond Graph models in combination with signal processing components and display plots. Using the simulation the motion of a ball can be presented in the form of plots of the position of the ball's center in 3D space over the time, the changes of its velocity components, reaction forces at the impact with the surface, etc. It would be useful if we simultaneously visualize the ball motion in 3D space in a similar way as when we watch the ball motion over the floor. 3D geometrical modelling and visualization is in general very useful when dealing with mechanical systems, such as robots, clouds of particles, etc.

It is rather complicated to have all this in one place. The approach that we take here is to concentrate in *BondSim* on the dynamical side of the problem and to enable communication with other applications that look on the problem from a different point of view, e.g. using 3D geometric models and visualization. To that goal the BondSim application supports *inter-process communication* (*IPC*), Fig. 4.38. With modern computer technology the two systems can appear as one complex system. In one window we can have many different plots giving details of the system dynamics, and in the other on the same or different screen we can visualize the system motion in a 3D scene.

The communication between the Visualization and Simulation can be realized in different ways [6]. It must ensure that the movement of the geometric object such as the bouncing balls, robot arms etc. over a 3D scene are smooth for a viewer, without

Fig. 4.38 Concept of inter-process communication

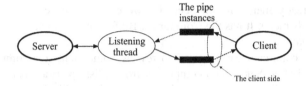

Fig. 4.39 IPC communication using the named pipe

"freezing" of the objects or their sudden jumps. To that end the scene should be fleshed for a short time (typically 1/24, 1/25 or 1/30 s) and then immediately replaced by the next one. Illusion of the objects moving is usually explained by persistence of the vision, which blends the picture frames together.

The data should be exchanged in a proper order, and also in a strict time sequence. The data packets exchanged are not large; typically they are of the order of 50–100 bytes. Thus e.g. if six joint angles of a robot were exchanged using the floating point precision this is only $6 \times 4 = 24$ bytes. The packets are, however, transmitted with a rather high frequency, e.g. every 30–50 ms. There is thus nearly a continuous flow of the data. After the data receiving the Visualization application recalculates the new positions of the objects and redraws the scene. The possible mechanisms for the IPC that are considered were the *TCP/IP* and *Pipes*.

The TCP/IP is the well-known set of communications protocols used for Internet and similar networks. Its basic protocols are the Transmission Control Protocol (TCP) and Internet Protocol (IP). IP works by exchanging the pieces of information called packets between the computers on the Internet. The TCP provides a communication service at an intermediate level between the application program and the Internet Protocol (IP). TCP ensures that all the bytes received are identical to the bytes sent and that are in the correct order. It is optimized for the accurate rather than timely delivery. Thus, during the transmission there are often relatively long delays due to waiting for out of order messages, or resending the lost messages. The sizes of the packets are often increased by collecting the data before the transmission.

The TCP/IP ensures the correct transmission of the bytes sent, but its timing is not guaranteed. Thus, it is not suitable for the IPC we need. After initial experimentation with the TCP/IP solutions, finally the *named pipe* mechanism was chosen. It enables that two applications running on the same or a different computer connected by a local net establish a two-way communication. It is also more responsive and can ensure the proper timing of the packets sent and received. A scheme of the *IPC* based on the named pipe is shown in Fig. 4.39.

There are two sides of the pipe, the server's and client's. The intercommunication between the processes starts on the server side, which creates a thread[7] that is used to listen for the massages received from the client. The *Server* then sends a

---

[7]A thread is an execution stream within a process. A single process may be broken up into multiple threads. The threads allow the concurrency. Thus, one tread may execute while the others are busy.

request to the *Client* to connect. When the *Client* connects the inter-process communications starts. In our case the *Server* is the visualization process, and the *Client* is the simulation process.

After the start and regularly during the simulation, e.g. after every 50 ms, the quantities that the visualization process needs are generated. They are packed in an order in which the *Server* expects them and written into a message, which is sent to the *Server*. The *Server* reads the message, processes it, reevaluates and redraws the visual scene. Eventually it answers by sending its own message back to the *Client*. The data that are sent back are what the *Client* expects from the *Server* and are used inside the simulation process. The complete process lasts until the *Client* closes the communication. Both processes interconnected by the pipe behave as one process showing the problem from both sides—the results of the simulation e.g. in the form of the plots and the visualization of the mechanism motion over space—in the separate windows.

It should be pointed out that there is a problem of different time scales in the simulation and the visualization processes, because they are the separate processes. On the simulation side the real (wall clock) time in which the system simulation is executed is generally different from the simulation time. Thus if we simulate the motion of a system during e.g. 5 s of simulation time, the computer on which the simulation is running can execute the complete simulation in much less times, e.g. 100 ms. On the other hand we expect that the visualization time corresponds to the motion of the real object. However, if the process is too quick, e.g. if the complete robot motion lasts 2 s, to visualize the motion in the scene it can be useful to slow it down, or replay it. To attack these problems the different approaches could be used. Currently the simulation is delayed for a fixed time each time it has sent the data to the server. The delay time can be set at the start of the simulation. How the both processes works together will be discussed in Chap. 9 on example of a robot motion.

The *BondSim* implements the client's side of the named pipe (Fig. 5.39) following [4, 5]. To that purpose it is necessary first to construct an *IPC pipe* interface component at the project's root document. It defines the client's side of the pipe. The *Edit Box* (Fig. 4.18 at left) under the *Continuous Signal Components* branch contains the *IPC pipe* tool. By dragging and dropping this tool into the project root document the corresponding client *IPC* interface component is created. Only a single pipe component is permitted.

Figure 4.40 shows a simple example of use of the *IPC* client interface. The system consists of three input generators (Fig. 4.40) named as Rate1 to Rate3, which generates the corresponding the angular velocities. These velocities are integrated to obtain the corresponding angles, which we wish to send to the *Server*. To that end we create a client *IPC* interface component by dragging and dropping the *IPC pipe* tool from the *Edit box* as has been explained earlier. After its creation it appears as a ring (a pipe cross-section) having by the default four input ports. The input ports serve to collect the signals whose values are sent to the *Server*. We may add more control-in ports if there are more signals we needed to transmit. Likewise,

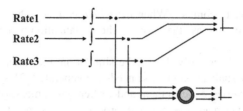

Fig. 4.40 Example of client side pipe

we may add the control-out ports to access the information received from the Server.

In this example we will remove one of the control-in ports and drag the other three to the left side of the *IPC* interface component and connect them to the corresponding nodes as shown in Fig. 4.40. Next, we double-click the output ports of each integrator and rename their variables to $q1$, $q2$, or $q3$. In the same way we change the names of the corresponding *IPC* component input ports. In this way we specified that the variables $q1$, $q2$, and $q3$ generated by the simulation will be collected and sent to the *Server*. At the right side of the *IPC* component, the three output ports are added (using the corresponding the *Port* tools in the *EditBox*) and its output variables are renamed to x, y and z. Hence, these are the signals received from the *Server*. These ports in this example are simply connected to an X-Y

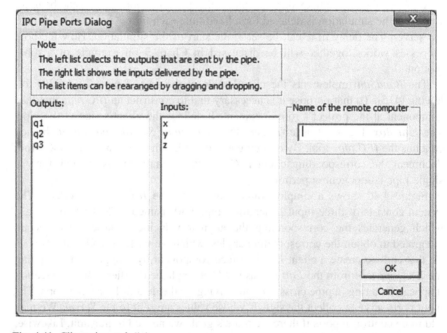

Fig. 4.41 Client pipe end dialog

`Display` component, and its ports variables are renamed accordingly. If we double click the *IPC pipe* component a dialogue window opens which contains two lists and an edit box (Fig. 4.41).

The output list contains the signals collected at the input ports, and which are sent by *IPC pipe* to the *Server*. The ordering of these signals should be as the *Server* expects them. If they are not properly ordered we can reorder the list simply by dragging and dropping the corresponding items. Similarly, the input list contains three signals received from the server, e.g. *x*-, *y*- and *z*-coordinates of the corresponding point in the mechanism. Their order depends on the server. If its order in the list is not the correct we can reorder the list it by dragging and dropping its entries.

In addition to the signal lists there is also an edit box in which the name of the remote computer on which the *Server* is running may be typed in. If it is left empty it is understood that both the *Client* (BondSim) and the *Server* are running on the same local computer.

References

1. MSDN Library (2012) Development tools and languages, Visual Studio 2012. http://msdn.microsoft.com/en-us/library/dd831853(v=vs.110).aspx. Accessed 13 May 2014
2. Damić V, Montgomery J (2003) Mechatronics by Bond Graphs—An object-oriented approach to modelling and simulations. Springer, Berlin
3. Microsoft Wiki (2014) Globally unique identifier, Microsoft.wikia.com/wiki/Globally_Unique_Identifier. Accessed 27 May 2014
4. Multithreaded Pipe Server, msdn.microsoft.com/en-us/library/windows/desktop/aa365588(v=vs.85).aspx. Accessed 12 Oct 2013
5. Named Pipe Client (2013) msdn.microsoft.com/en-us/library/windows/desktop/aa365592(v=vs.85).aspx. Accessed 12 Oct 2013
6. Vuskovic M (2013) Operating systems, inter-process communications. http://medusa.sdsu.edu/cs570/Lectures/Chapter9.pdf. Accessed 12 Oct 2013

Chapter 5
Generation of the Model Equations and Their Solution

5.1 Introduction

In the previous chapters the systematic component-based approach was developed that enables development of mechatronic system models in a formal way. An important part of this is the description of the element constitutive relation symbolically using a relatively simple language. Thus, not only non-linear relationships, but also piecewise expressions can be used. This is important in modelling discontinuous mechanical processes and in electronics and makes simulation of complex systems not only feasible, but also a challenging task. This chapter describes the generation of system mathematical models and their solution.

We start by describing methods used to build the mathematical model implied by the project model component structure. The equations are machine generated in the form of differential-algebraic equations (DAEs). We have already discussed some of the features of these equations. We continue here with the problem of their numerical solution. We chose the well-known backward differentiation formula (BDF) as the solution method. This is suitable for index-1 problems, but can also be extended to index-2 problems. For higher index problems the solution is much more demanding; and we have left this for further study. The problems of starting values and discontinuities are also discussed.

The methods we use depend to a great extent on computational algebra support. This adds flexibility to the modelling. It is also used in the numerical solver e.g. for evaluation of functions, generation of matrixes of partial derivatives, and their subsequent solution. It also enables generation of feedback to the modeller supplying information on the generated models. These capabilities can be extended.

This chapter closes the first part of the book, which deals with the fundamentals of a component-based bond graph approach to modelling mechatronic systems. This is implemented in the *BondSim* program that is freely available from the author's web site (see Appendix).

© Springer-Verlag Berlin Heidelberg 2015
V. Damić and J. Montgomery, *Mechatronics by Bond Graphs*,
DOI 10.1007/978-3-662-49004-4_5

5.2   General Forms of the Model Equations

In Sects. 2.5 and 2.6 we gave a fairly general description of the constitutive relation of components used as the building blocks for developing models of systems. Using the techniques of Chaps. 3 and 4, such models are generated and stored as a tree of component models. These models consist of relations defined in elementary components that are the leaves of the model tree and which are constrained by relations describing the interconnections of components that are the branches of the tree. The models are made more general and, unfortunately, more difficult to solve by allowing the mixing of bond graph based models with operations on signals. Here we discuss the general procedure for building system mathematical model of a project under the study.

5.2.1   Generating the System Variables

The first step in building the mathematical model is defining a unique set of system variables. When an elementary component is created the corresponding power ports are created too. The number of such ports depends on the type of the component. In some components, e.g. 1 or 0 junctions, a port can be removed or a new one added. Typically the elementary component ports are power ports, but could be control-in or control-out ones as shown in Sect. 2.5.8. This holds also for the block diagram components, but in this case, the ports are control-in or control-out only. Every created component is unique and is identified by a unique ID (Sect. 4.6.1). Similarly, every component port is unique as well and is identified by its label, which represents the ordinal number of the created port.

As already disused in Sects. 2.5 and 2.6 the processes seen at the elementary components power ports are described by *effort* and *flow* power variables and the corresponding constitutive relations in term of these variables and the parameters. During the creation of the ports default names of these variables are defined, and default constitutive relations as well (see Sect. 4.8.1). These names are local in the component in question and serve only to define the corresponding constitutive relation. These variables and the corresponding constitutive relation can be modified by the user. Some components in addition to the power variables use also the internal *state* variables. Thus, the inertial components use the generalized momentums, and the capacitive ones uses the generalized displacements. This holds also for the control ports of the block diagram components. In this case, however, there is only one port variable—input, or output—the constitutive relation is defined in the control-out port only.

In order to support generation of the system variables, to every port variable name a corresponding global variable is defined. These variables are not seen by the user and are internal to the *BondSim* program. They are employed to define a unique set of the system variables, which are used for defining the mathematical model of the current project. These global port variables are rather clumsy and consist of a leading symbol, the port ordinary number and the component ID. Thus, e.g. an effort global port variable can have a form

$$\text{EffortVar} = e0\_\#YzifgzgnKgzfe4w05gvi3iqc2f$$

Connecting two component ports by a bond means that the respective global efforts and flows of the connected ports are represent by the *same* system variables. Thus, these two efforts, or two flows, can be represented by their respective *system variable*. We need to develop a technique for defining a unique set of system variables that represents variables of all elementary and block diagram ports. We do this by means of the symbol tables.

A symbol table consists of pairs of the port global and system variables. The system variables can be described by integer indices running from 1 and up. The index zero (0) is reserved for time t. The symbol table is simply a map (dictionary) of the port global variables as the keys into the map and system variable indices as their values. The keys are a unique and are used to search the map for their values. It is similar to the ordinary language dictionaries. We use a word as the key into the dictionary to find its meaning (the value). The map structures are used because of their shorter search times, which is important during mathematical model creation.

Because we can have two distinct types of physical process—continuous-time and discrete-times—we use the separate symbol tables for each of these. We start with the continuous-time processes which encompasses Bond Graphs and continuous-time the block diagrams.

The procedure for the creation of the symbol table consists of traversing the components, starting from the lowest level and going upwards. During this procedure a port variable is put into the table along with its corresponding system variable index. We start with the effort and flow junctions, and then continue with other elementary and block diagram components.

The effort junctions are characterised by the fact that all of their port flows are the same variable, i.e. the common junction variable (Sect. 2.5.7). Similarly, this holds for the flow junctions, with the common junction variable being the effort. The procedure for assigning the system variables is as follows (Fig. 5.1):

1. Increment the index of the system variables by one. Insert in the symbol table the common junction flow (effort) variable symbol as the key, and the current system variable index as the value.
2. For every junction power port find first the value of the system variable index corresponding to the junction flow (effort) variable. Then search then for the port of the elementary component to which it is connected. If the component is of the same type, i.e. if the effort junction port is connected to a port of another effort junction or the flow junction port is connected to a port of another flow junction, treat the ports then as the internal ones and the junction as the same junction, and proceed further. If, on the other hand, it is a port of another

Fig. 5.1 The assignment of
the common junction variable

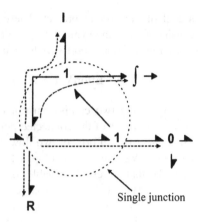

Single junction

component type, get the port flow (effort) variable and insert it into the symbol table using the current system variable index.

3. If the port is a control-out port, the procedure is slightly different. Find the control-in port to which it is connected. Get the corresponding input variable and insert it into the symbol table using the current value of the system variable index. The search is stopped if the connected to port is the input port of an *Integrator*, a *Differentiator*, or an *Output* component. Otherwise, if it is a *Function* or a *Summator* component, continue at the component output port. Break the search if the output variable is already defined in the symbol table. Otherwise, increment the index of the system variables. Insert this value in the symbol table using the output port variable as the key. Continue the search. Finally if it is a *Node* input port, find the port connected to each of its output ports and continue the search because all node ports share the same variable.

4. Continue with the search and assignment of variables until all power or control ports of the junction and of other junctions of the same type that are connected to it are traversed.

5. Continue with the next junction until all junctions are visited.

The procedure outlined above ensures spreading out the common junction variables to all ports connected directly to it, or across other junction components of the same type (Fig. 5.1). It also creates variables of input-output components connected to it. It is important to do this during effort or flow junction variable assignment, for it is one of the common connection points between bond graphs and the control signal paths. We stress that this procedure ensures that all directly connected junctions of the same type behave as a single junction.

After finishing with the effort and flow junctions, we proceed to the other components. The procedure is similar to the one described above, but is slightly simpler.

For every unprocessed elementary component port variable—i.e. an effort, flow, or control—we increment the system variable index by one and insert the variable symbol-system variable index pair into the symbol table. We then find the port to which it is connected and add its corresponding variable symbol and the current value of the system variable index into the symbol table. If the port is an output control port, we proceed as explained above (item 3). The elementary component control input ports, e.g. resistive component inputs or transformer input ports, are dealt with when assigning input-output (block-diagram) component port variables. Internal variables of inertial and capacitive components are also added to the symbol table.

Processing of the block-diagram components is similar. We start at the output ports of *Input*, *Integrator* or *Differentiator* components, increment the system variable index and insert it into the symbol table using the output variable symbol as the key. Next we find the input control port of the component to which it is connected, and continue in the same way as explained in item 3 above. The search is stopped when the connected port is the input to an *Integrator*, *Differentiator*, *Output* component, or a Bond Graph elementary component.

The discrete-time processes are treated similarly but using the discrete variables symbol table. The search starts at the output ports of *A/D converter*, *Input*, or *Delay* components, and continue at the output port of the connected to component, and stops at the *D/A converters* input ports. We end the discrete path by *D/A converters* because in Mechatronics we typically use discrete (digital) signals for either control of analog devices (e.g. by actuators), or for monitoring (displays).

During the generation of the system variables and populating the symbol tables, the correctness of the components interconnection is checked too. First, if some ports are not connected the procedure is stopped and the reason is conveyed to the user by a suitable message displayed in the *Output* widow under the *Build* tab (Figs. 4.2, or 4.8). Similarly, if the connection on each side of a component does not fit—e.g. because of incorrect dimensions of the bonds (Sect. 3.7)—the procedure is stopped and a message is posted. Finally, the elementary components are checked to ensure that the connections are valid (Table 5.1). The elementary components that are not the junction components—i.e. effort or flow junctions, Transformers, or Gyrators—can only be connected to the junction components. The junction components, on the other hand, can be connected to any component. We rule out direct interconnections of Transformers, or Gyrators, as these combinations are not necessary, e.g. TF-TF is just TF, as is GY-GY.

In the block diagram components the output ports should be connected to the input ports of the other components. In addition the interconnections are permitted only between the same kinds of components, i.e. either between the continuous-time, or discrete-time components. Only the components where a mix of the signals is permitted are A/D and D/A converters.

Components	Permitted connection
I, C, R, SE, SF, Sw	1, 0, TF, GY
1, 0	All
TF, GY	All but not of the same type

Table 5.1 Permitted interconnection of the elementary components

5.2.2   Generation of the Equations

Once the system variables are defined and their correspondence to port variables names established, the next step is to generate equations that describe the constitutive relations of the elements. There are no built-in relations; every relation is either created when the component port is created, or edited later by the user. These relations are described in a manner similar to that of the usual mathematical expressions, as explained in Sect. 3.5.

Relations are not translated to subroutines or functions, but are instead, translated to another string that are called *byte form* and is based on *prefix (Polish)* notation. This notation was selected because, as is well known, it enables writing an expression as a brackets free expression. The constitutive relations are translated to symbolic form that can easily be decompiled to a more familiar mathematical expression, if required (Sect. 5.4). This form of the constitutive relations is directly processed symbolically to generate the system Jacobian matrices in the analytical form. These matrices are needed for the numerical solution of the system equation during the simulation (Sect. 5.3). The byte form of the system equations and Jacobians enables their transformation into stack based expressions for their efficient evaluation during a simulation run (Sect. 5.7).

Without going into too much detail, we now explain how the mathematical model equations are created. During the model building the constitutive relations, which are held in the elementary component ports, are analysed, translated to byte forms, and are stored in an array that is accessible for further processing. The method used is an extension of the approach of [1]. The byte string form is based on the prefix operator form in which the operators are written to the left of its operands, i.e. as

$$Operator \quad Operand1 \quad Operand2 \tag{5.1}$$

Note, there is no spaces in the byte expressions; they are used only for readability.

During the constitutive relation parsing the literal constants found are encoded using operator C and the value of the parameter represented as a real (double) number, i.e.

$$C \ constant\_value \tag{5.2}$$

Thus, e.g. constant 205.452 can be written as $C205.452$.

Similarly the named constants are evaluated and put into a parameter array held at the system model level. They are encoded using operator P and value of its index in this array, i.e.

$$P \ parameter\_index \tag{5.3}$$

Note that parameter evaluation means that the complete hierarchy of parameters is searched, starting with parameters in the particular component, until the complete parameter expression is evaluated (Sect. 4.8.2).

Every occurrence of a port variable name, or time, is encoded using operator V and the integer value of its system variable index

$$V \; var\_index \tag{5.4}$$

To find the corresponding index the global variable, which corresponds to the variable name, is used as the key into the symbol table. The system variables indices, as already noted, run from 1 to n, and where n is the total number of the variables in the model. The index zero (0) is retained for *time*.

In the same vein the time derivative of a variable is encoded using operator D, i.e.

$$D \; var\_index \tag{5.5}$$

Discrete variables are defined similarly using operator Z and the discrete system variable index using the discrete symbol table, i.e.

$$Z \; var\_index \tag{5.6}$$

Various operators are used to describe operations between variables and parameters in the constitutive expressions. These include unary, binary, ternary, and some special operators.

Unary operators are encoded by U and a symbol that describes the operation in question, e.g. U+ for unary plus (not used often), or U− for negation of a variable. Elementary mathematical functions are also treated as unary operations with the second byte used to designate the function, and the operand corresponding to function argument.

Binary operations are encoded similarly. These operators are denoted by B and are used to describe addition (+) and subtraction (−), multiplication (*), division (/) and the modulus operation (%), as well as for relational operators >, <, >=, <=, and logical operators AND and OR. Thus e.g., 2.8*v, where v is a port variable, is encoded as

$$B * C2.8 \; V \; var\_index$$

where *var_index* is the value of the system variable index corresponding to the variable v.

The ternary operator ?: is coded as

$$T? \; Conditional\_part \; Expression1 \; Expression2 \tag{5.7}$$

The operator ?: is borrowed from the C language and is used for *if then else* constructs. The constitutive relationship describing some physical process often cannot be described by a single expression valid throughout the range of the variables, but by two or more expressions valid in different parts of the range. This is the case, for example, with dry friction in mechanics, and in semiconductor models in electronics. To describe such relationships we use ternary ?: operators. The conditional part is described using relational or logical operators, and the expressional parts are constructed in the usual way. The expression is encoded as given in (5.7). Thus, for example, expression

$$c > 0 \, ? \, e \, : \, e + 1$$

which reads as e, if c is greater than zero, and $e + 1$ otherwise, is encoded as

$$T?B > V12C0V43B + V43C1$$

where 12 is used as the index of variable c and 43 of variable e.

We also introduce single- and two-variable operators for encoding the functions defined by tables of values (see Sect. 7.5 for an example), and functions defined symbolically.

The element constitutive relations are defined as

$$var = \text{expression} \tag{5.8}$$

where *var* stands for the port variable and the *expression* is the constitutive relation of the element. Most processes are described by a single relationship of this form (Sects. 2.6 and 2.7). This is the case for processes at *Resistor* or *Transformer* ports. Such relations are written in implicit form as

$$var - expression = 0 \tag{5.9}$$

and encoded as

$$B - V \, var\_index \, expression \tag{5.10}$$

with the understanding that the last expression is equal to zero.

In *Capacitive and Inertial* components, there is also a derivative relationship of the form

$$\frac{dx}{dt} = y \tag{5.11}$$

where x is the internal state variable displacement or momentum of the process and y is a port effort or flow. The *Integrator* input-output relationship is described in a similar form. Such equations are also written in the implicit form as

$$\frac{dx}{dt} - y = 0 \tag{5.12}$$

and encoded by

$$B - D \; var\_x\_index \; \text{V} \; var\_y\_index \tag{5.13}$$

and assuming that this expression evaluates to zero.

Now that we have explained how the constitutive equations are encoded, we can describe the procedure for generating the mathematical model of the system based on the information held in the component ports. Generation of the mathematical model is done using several passes through the corresponding Bond Graphs or block diagrams. During these passes only components of specific types are visited and their constitutive relations are translated. If during this process syntactical errors are found, or some parameters are not defined, the translation process is stopped and a message explaining the cause is displayed. This, however, normally should not happen because when the constitutive expressions are edited the syntactical analysis has already been executed and the expressions are accepted and stored only if the checks have been passed.

We now summarize the forms of the model equations generated. The model consists of three groups of equations. We introduce notation that permits their description in a compact way. All variables appearing in a time-derived form, as in (5.12), are called *differentiated* variables and are described by a vector $\mathbf{x}$ of dimension n_d. These variables represent displacements of *Capacitive* elements, momenta of *Inertial* elements, outputs of *Integrators*, and inputs of *Differentiators*. All other variables are *algebraic* variables and are denoted by a vector $\mathbf{y}$ of dimension n.

In the first pass the differential equation part of the model are generated. These equations can be compactly described in vector form as

$$\frac{d\mathbf{x}}{dt} = \mathbf{B}_1 \mathbf{y} \tag{5.14}$$

were $\mathbf{B}_1$ is $n_d \times n$ dimensional incidence matrix having in each row a 1 in the column of an algebraic variable and a 0 elsewhere (see (5.11)).

In the second pass the constitutive relations corresponding to effort or flow junctions are generated, as explained in Sect. 2.5.7, based on the sense of power of the ports. These are translated to the byte form. This also holds for summators, where the sign of the input ports define the output relationships. The signal nodes need no constitutive relations at all, as their roles are taken care of during the generation of the model variables (Sect. 5.2.1). These equations can be written compactly as

$$\mathbf{B}_2\mathbf{y} = \mathbf{0} \tag{5.15}$$

where $\mathbf{B}_2$ is a rectangular matrix of corresponding dimension and is of full row rank.

The last group represents the constitutive relations of elementary components. They are in general nonlinear algebraic relationships which can be written implicitly as

$$\mathbf{f}(\mathbf{x}, \mathbf{y}, t) = \mathbf{0} \tag{5.16}$$

where $\mathbf{f}$ is a vector function of the corresponding dimension.

Equations (5.14)–(5.16) constitute a system of *differential-algebraic* equations (DAEs). Because of (5.14), such a form is usually called *extended*. Thus, direct application of the element constitutive relation, as described in Sects. 2.5 and 2.6, leads to a mathematical model of the system in the form of a system of differential-algebraic equations in extended form. In the next subsections we discuss some characteristics of such models and their solution.

In mechatronic applications, in addition to the continuous-time model described above, there is also a discrete part. This is generated in the form of explicit equations of general form

$$\mathbf{z}[k+1] = \mathbf{h}(\mathbf{z}[k], \mathbf{u}[k]) \tag{5.17}$$

where $\mathbf{z}[k]$ is value of the discrete variables at the current sampling interval k, and $u[k]$ is the current value of the inputs. The equation predicts the value of discrete variables z at the next sampling interval in terms of the current values of the variables and the inputs. The inputs are generally generated by A/D converters and thus dependent on the value of the continuous-time variable at the sampling time $k \cdot T_s$, where T_s is the sampling interval. Therefore evaluation of the discrete equation should be coordinated with solution of the continuous-time model.

5.2.3   *The Characteristics of the Model*

Systems composed of both differential and algebraic equations generally differ significantly from ordinary differential equations and are much harder to solve [2]. DAEs may be characterised by indices that measure the difficulties encountered when solving them. Several types of indexes are defined. We do not give here their precise definition; this is found in [3–6].

The *differentiation index* is defined as the number of times the DAEs should be differentiated with respect to time in order to reduce them to a system of ordinary differential equations [3, 4]. We illustrate this on an example of semi-explicit equations

$$\left.\begin{array}{l} \dot{\mathbf{x}} = \mathbf{f}(\mathbf{x}, \mathbf{y}, t) \\ \mathbf{g}(\mathbf{x}, \mathbf{y}, t) = \mathbf{0} \end{array}\right\} \tag{5.18}$$

Differentiating the second equation with respect to time yields

$$\left.\begin{array}{l} \dot{\mathbf{x}} = \mathbf{f}(\mathbf{x}, \mathbf{y}, t) \\ \dfrac{\partial \mathbf{g}}{\partial \mathbf{x}}\dot{\mathbf{x}} + \dfrac{\partial \mathbf{g}}{\partial \mathbf{y}}\dot{\mathbf{y}} = -\dfrac{\partial \mathbf{g}}{\partial t} \end{array}\right\} \tag{5.19}$$

Thus, if the partial derivative matrix $\partial \mathbf{g}/\partial \mathbf{y}$ is not singular, this is a system of ordinary differential equations and, hence, (5.18) has a differential index one. If, on the other hand, the partial derivative matrix is singular, we proceed as described in [4]. Supposing that $\partial \mathbf{g}/\partial \mathbf{y}$ is of constant rank, we can transform (5.19) at least locally to the form

$$\left.\begin{array}{l} \dot{\mathbf{x}} = \mathbf{f}_1(\mathbf{x}, \mathbf{y}, t) \\ \mathbf{g}_1(\mathbf{x}, t) = \mathbf{0} \end{array}\right\} \tag{5.20}$$

This can be achieved by expressing the y-variables that appear in some of the algebraic equations of (5.18) as functions of the x- and other y-variables, eliminating them from the system. As a result, the new algebraic equation is independent of the y-variables.

As an illustration, we return to the see-saw problem of Sect. 2.7 and show that (2.72)–(2.75) and (2.98)–(2.102), which describe the system, can easily be converted to the form of (5.20). Thus, forces F_{b1x} and F_{b1y} can be evaluated from the last two equations in (2.72) of Body 1 motion and substituted into the first two. We can also eliminate the gravity force G_{1y}. In a similar way, we eliminate the corresponding variables in (2.73) of Body 2 motion. In the same way we can eliminate components F_{Cx}, F_{Cy} and G_y from (2.101) of the platform mass center motion, and the moment M_C in (2.102) of the platform rotation. Next, we can eliminate moments in (2.102) by use of the relations in (2.98)–(2.100), and (2.75). The angular velocity in (2.75) is eliminated using (2.74), (2.75) and (2.102). In this way, we get the first part of the system equations, which involves derivatives of variables with respect to time, as shown in (5.21).

We can now eliminate the velocity components from the system. Thus, the velocity components in (2.72) can be eliminated using (2.98), (2.100), (2.74), (2.75) and (2.102). We handle the velocity components in (2.73) and (2.101) in a similar way. Finally, we obtain the system of algebraic equations in (5.22), which contains only differentiated variables: momenta and the rotation angle.

$$\dot{p}_{b1x} = -F_{1x}$$
$$\dot{p}_{b1y} = -F_{1y} - m_1 g$$
$$\dot{p}_{b2x} = -F_{2x}$$
$$\dot{p}_{b2y} = -F_{2y} - m_2 g$$
$$\dot{p}_{Cx} = F_{1x} + F_{2x} - F_{Px}$$
$$\dot{p}_{Cy} = F_{1y} + F_{2y} - F_{Py} - mg \tag{5.21}$$
$$\dot{K}_C = (a \sin \phi + c \cos \phi)F_{1x} + (-a \cos \phi + c \sin \phi)F_{1y}$$
$$\quad + (-a \sin \phi + c \cos \phi)F_{2x} + (a \cos \phi + c \sin \phi)F_{2y}$$
$$\quad + (b \cos \phi)F_{Px} + (b \sin \phi)F_{Px}$$
$$\dot{\phi} = K_C / I_C$$

$$p_{b1x} = (m_1/I_C)(a \sin \phi + (b+c) \cos \phi)K_C$$
$$p_{b1y} = (m_1/I_C)(-a \cos \phi + (b+c) \sin \phi)K_C$$
$$p_{b2x} = (m_2/I_C)(-a \sin \phi + (b+c) \cos \phi)K_C$$
$$p_{b2y} = (m_2/I_C)(a \cos \phi + (b+c) \sin \phi)K_C \tag{5.22}$$
$$p_{Cx} = (m/I_C)(b \cos \phi)K_C$$
$$p_{Cy} = (m/I_C)(b \sin \phi)K_C$$

In this way, the equations of the see-saw motion can be reduced to eight equations in differential form and six algebraic equations. These contain eight differentiated and six algebraic variables. The differentiated variables are constrained by the algebraic equations, and hence only two of them are independent. This is to be expected, for the see-saw is a single-degree-of-freedom system and, as such, its dynamics can be described by only two variables, the angle of rotation and its angular momentum (or angular velocity).

We now return to (5.20) and differentiate the last equation with respect to t,

$$\frac{\partial \mathbf{g}_1}{\partial \mathbf{x}} \dot{\mathbf{x}} = -\frac{\partial \mathbf{g}_1}{\partial t}$$

Substituting from the first (5.20), we obtain the expression

$$\frac{\partial \mathbf{g}_1}{\partial \mathbf{x}} \mathbf{f}_1(\mathbf{x}, \mathbf{y}, t) = -\frac{\partial \mathbf{g}_1}{\partial t} \tag{5.23}$$

This last equation constitutes a *hidden constraint* that the solution of the system must satisfy. Thus, if the matrix

$$\frac{\partial \mathbf{g}_1}{\partial \mathbf{x}} \frac{\partial \mathbf{f}_1}{\partial \mathbf{y}} \tag{5.24}$$

is invertible, then the first (5.20) and (5.23) constitute an index 1 problem. Differentiating with respect to time, (5.23) gives a differential equation with respect to variable $\mathbf{y}$. Hence, the system of (5.20) is of differential index 2. The algebraic variables that need two differentiations in order to express them as differential equations often are called index-2 variables.

Returning to (5.21) and (5.22), we see that they constitute a system of DAEs of index-2. The hidden constraints can be found by differentiating the momentum relations of (5.22) and by substituting from (5.21). We do not give them here because of their length. It should be noted that reaction forces between bodies and the platform, as well as between the frame and the pivot (Fig. 2.15), are index-2 variables. These variables correspond to the Lagrangian multipliers of the constrained body mechanics. In the original formulation of the see-saw problem all the algebraic variables are not of index-2 type. Thus, for example, the velocity components v_{lx} and v_{ly} of (2.72) are of index-1 because we only need one differentiation to get the corresponding differential equations, e.g.

$$\dot{v}_{1x} = \dot{p}_{b1x}/m_1 = F_{b1x}/m_1$$

Equations (5.14)–(5.16) are semi-explicit differential-algebraic equations. To simplify the notation we define a variable

$$\mathbf{X} = \begin{pmatrix} \mathbf{x} \\ \mathbf{y} \end{pmatrix} \tag{5.25}$$

The equations can now be represented as

$$\mathbf{g}(\dot{\mathbf{X}}, \mathbf{X}, t) = \mathbf{0} \tag{5.26}$$

We define the *leading coefficient* matrix by

$$\mathbf{A} = \frac{\partial \mathbf{g}}{\partial \dot{\mathbf{X}}} \tag{5.27}$$

In our case this matrix is extremely simple, i.e.

$$\mathbf{A} = \begin{pmatrix} \mathbf{I} & \mathbf{0} \\ \mathbf{0} & \mathbf{0} \end{pmatrix} \tag{5.28}$$

where $\mathbf{I}$ is the identity matrix of dimension equal to the number of differentiated variables.

Another type of index introduced for the detailed study of DAEs is the *tractability index* [5, 6]. It is not based on differentiation, but uses the underlying vector space. The vector space of a column vectors is denoted as $\mathbf{R}^m$, where m is the (constant) dimension. Important roles in the *tractability* approach to DAEs are played by two subspaces. The *image* of a matrix **A**, denoted as im(**A**), is a space consisting of vectors **X** such that **X** = **Au** for some **u** ∈ **R**^{m}. The *null-space* of **A** is denoted as ker(**A**) and is a space of vectors **u** such that **Au** = **0**. Using projectors onto the ker(**A**) it has been shown that DAEs of tractability index 1 can be transformed to ordinary differential equations in state-space form. Such systems are called transferable (to state-space form). The others—those of higher index—are not transferable. Unfortunately, many DAEs of engineering interest are not transferable of index ≥2. Characterization of an important class of index-2 systems has been developed. Particular attention has been paid to DAEs arising in the modelling of electrical systems [7, 8].

There is also an important DAEs index—the *perturbation index* [3]. It is based on the behaviour of the solution under the influence of perturbations. This index can be used to explain why higher index DAEs are very sensitive to perturbations caused by, for example, numerical inaccuracies, discontinuities, and even changes in step size.

After this short introduction to some of the important concepts of DAEs, we now return to our problem, the numerical solution of the system represented in Eqs. (5.14)–(5.16). A basic characteristic of such a system is that the equations are in semi-explicit form. They could be converted to a fully implicit form by eliminating variables on the right hand side of (5.14). The index of the resulting system will be one degree lower.

The important characteristic of the original equations is that the leading coefficient matrix of (5.26) has an extremely simple form. Thus, its null-space is constant. On the other hand, the reduced system can, in general, contain non-linear functions of the derivatives of variables with respect to time. As a result, the leading coefficient matrix depends generally on the system variables. The corresponding null-space is not constant, but changes during the solution. It has been shown that a system of tractability index 1, in which the leading coefficient matrix null-space changes with the solution, behaves analytically and numerically as index 2 tractable DAEs with a perturbation index of 2 [12]. This confirms the assertion in [2] that semi-explicit DAEs behave like fully implicit ones of one index lower.

To show the similarity of (5.14)–(5.16) to some other formalisms, we consider models from two fields important to mechatronics. These are the Lagrangian formulation of equations of constrained rigid bodies and the charge/flux formulation of modified nodal equations (MNA) from electrical circuit analysis.

The classical approach based on the Lagrangian formulation leads to equations of motion of constrained bodies in terms of the generalized coordinate vector **q** of the form [9]

$$\left.\begin{aligned}
\dot{\mathbf{q}} &= \mathbf{v} \\
\mathbf{M}(\mathbf{q})\dot{\mathbf{v}} &= \mathbf{Q}(\mathbf{q}, \mathbf{v}) - \mathbf{G}^T\lambda \\
\mathbf{g}(\mathbf{q}) &= \mathbf{0}
\end{aligned}\right\} \tag{5.29}$$

where is $\mathbf{M}$ is the mass-inertia matrix, $\mathbf{Q}$ is a vector of generalized forces, and λ is a vector of Lagrange multipliers. The last equations represent the position constraint that the coordinates must satisfy. Note that $\mathbf{G} = \partial\mathbf{g}/\partial\mathbf{q}$. (The superscript T denotes matrix transposition.) The first equation is put in the form of (5.14) because in the original formulation (5.29) it is of second order with respect to the generalized coordinates. This normally leads to equations in extended form with respect to the position coordinates. Such equations are known to be of index 3. Different approaches have been used to lower the index of such equations [2, 3, 9].

In Chap. 9 we solve a problem of this type using the bond graph modelling approach that naturally leads to equations with velocity constraints, i.e. of the form

$$\left.\begin{array}{l} \dot{\mathbf{q}} = \mathbf{v} \\ \mathbf{M}(\mathbf{q})\dot{\mathbf{v}} = \mathbf{Q}(\mathbf{q},\mathbf{v}) - \mathbf{G}^T\lambda \\ \mathbf{G}(\mathbf{q})\mathbf{v} = \mathbf{0} \end{array}\right\} \qquad (5.30)$$

Such equations are of differentiation index 2. The general form of the equations thus correspond to the Lagrange formulation with the velocity constraint replacing the positional constraints.

Analysis of electrical circuits is usually based on the classical modified nodal analysis (MNA). There is only one type of junctions in circuits, i.e. the nodes. The governing equations are developed by applying the Kirchhoff current law to every node in the circuit. The variables consist of nodal potentials and currents in the voltage-controlled elements (inductors and sources). The constitutive relations of voltage-controlled elements are appended to the system equations. The charge/flux oriented MNA introduces charges and fluxes also as system variables. Equations describing inductors are used in the same way as those defined by inertial elements of bond graphs, i.e. using the extended form of equations

$$\left.\begin{array}{l} \dot{\Phi} - u_2 + u_1 = 0 \\ \Phi - L \cdot i = 0 \end{array}\right\} \qquad (5.31)$$

Here $u_2 - u_1$ is the voltage across the inductor, Φ is the flux, i is the current through the inductor, and L is the inductance parameter (for the linear inductors). The equations are added along with the constitutive relations for capacitors to the nodal equations. The constitutive relations of current-controlled elements are set directly into the nodal equations.

As an illustration of equations formed using the charge/flux based MNA approach, consider the simple of circuit in Fig. 5.2. The equations read

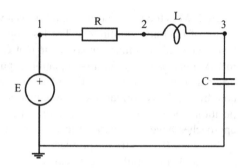

$$
\left.
\begin{aligned}
&-i_1 + (u_1 - u_2)/R = 0 \\
&-(u_1 - u_2)/R + i_2 = 0 \\
&-i_2 + \dot{q} = 0 \\
&u_1 = V \\
&\dot{\Phi} - (u_2 - u_3) = 0 \\
&\Phi = L \cdot i_2 \\
&q = C \cdot u_3
\end{aligned}
\right\}
\qquad (5.32)
$$

The first three equations are written by applying the Kirchhoff current law to the three nodes. The fourth equation gives the constitutive relation of the voltage source, the next two describe the inductor, and the last is the constitutive relation of the capacitor.

Charge/flux based MNA is used to describe charge and flux accumulations in circuits in a better way. It has also been shown that circuit equations based on this formalism are better suited to simulation [7, 8]. The explanation for this is found in the leading coefficient matrix null-space that, for the charge/flux based formulation, does not change during the solution.

There are few papers in the bond graph literature dealing with the analysis of the DAEs formulation of bond graph models. Important papers in this respect are [10, 11]. These show that the index of the system model can be larger than 2. A similar analysis is made for electrical circuits in [12]. We believe that most models of engineering systems have indices not greater than two. Higher index systems at the current state of solver techniques are difficult to solve by a general-purpose modelling and simulation system. Model reformulation and some specialized approaches are then necessary.

In [13] an approach based on tearing variables for dependent storage elements has been proposed. It suggests changing the model by introducing Langrange multipliers. Structural type analysis, such as the causality analysis already discussed in Sect. 2.10, is generally of limited value, as the structure of real problem models generally change during solution. The problems can be treated more generally using the projectors of [5, 6]. But, unfortunately, this is not an easy task.

The conclusion that can be drawn from these discussions is that the form of the model generated by the program is suited to simulation because of the constant leading coefficient matrix. We believe it serves as an acceptable frame for the successful modelling and simulation of the mechatronic systems in Part 2.

5.3   Numerical Solution Using BDF Methods

This section describes the methods and strategy for the solution of the generated differential-algebraic equations. Among the possible candidates for solving DAEs, two methods have attracted most attention: those based on backward difference formula methods, known as BDF methods; and various kinds of the Runge-Kutta method [3, 4]. Among the implementations of BDF methods, perhaps the best-known software is DASSL, of which a detailed description can be found in [3]. It is freely available from the NETLIB web repository. Similarly, among the implementations of the Runge-Kutta methods, possibly the most useful is Radau5 of [4], which also is freely available from those authors.

We use an implementation of the BDF method for the solution of the generated DAEs. One of the reasons for this choice is that we have had experience with BDF methods from the time of the famous DIFSUB program [14]. BDF codes are widely used in electronic circuit simulators and continue to attract attention as capable, general-purpose methods for solving DAEs. Part 2 shows that it also is a method with which it is possible to solve mechatronics problems based on the bond graph modelling approach.

5.3.1   The Implementation of the BDF Method

The solver used for solving DAEs of the model system is based on the *variable coefficient* version of BDF [3, 15]. In comparison, DASSL uses a fixed coefficient implementation of BDF [3]. The variable coefficient form is perhaps the most stable implementation of the BDF methods, though it is less efficient because it requires frequent re-evaluation of the partial-derivative matrix. The reason for using it here is that the solution of DAEs is not an easy problem, and the higher level of stability of this method is welcome. Unfortunately, the authors are aware of one study only that compares these two BDF implementations [15]. The frequent re-evaluation of the partial-derivative matrix is less expensive in our approach, as we use it in an analytical form and not by a numerical approximation. This improves the stability of the BDF code. In the following section we describe in some detail the method we use. We use a similar notation as in [3].

The system we solve numerically is described by (5.26). The numerical solver generates approximations X_i to the true solution $X(t_i)$ (Fig. 5.3). The BDF method is a variable step and variable order predictor-corrector method. Because of the

Fig. 5.3 The approximation
of the solution at discrete
times

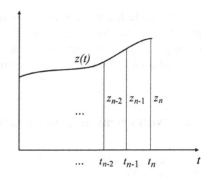

stability requirement, the order k of the method is limited to 5. The method predicts values of the solution at the next instant of time t_{n+1} by evaluating a polynomial that interpolates the last $k + 1$ values, i.e.

$$\mathbf{X}_{n+1}^0 = \mathbf{P}(t_{n+1}) \tag{5.33}$$

where $\mathbf{P}$ are the prediction polynomials defined by

$$\mathbf{P}(t_{n-i}) = \mathbf{X}_{n-i}, i = 0, 1, \ldots, k \tag{5.34}$$

The predicted value of the time derivative at time t_{n+1} is similarly found by evaluating the time derivative of the interpolating polynomial, that is

$$\dot{\mathbf{X}}_{n+1}^0 = \dot{\mathbf{P}}(t_{n+1}) \tag{5.35}$$

In the DIFSUB code of [14], the interpolating polynomials were originally represented by Taylor series. The past history is held in the form of Nordsieck vectors of the current values of the variables and their scaled derivatives up to the order k. It was found that it is much more efficient to use interpolating polynomials based on the Newton interpolating polynomial with modified divided differences [16, 17]. DASSL follows this approach and it is also used in our code. A detailed description of the implementation details can be found in [3, 17].

The approximation to the solution value is determined using corrector polynomials $\mathbf{Q}$,

$$\mathbf{X}_{n+1} = \mathbf{Q}(t_{n+1}) \tag{5.36}$$

such that

$$\mathbf{g}\left(\dot{\mathbf{Q}}(t_{n+1})\right), \mathbf{Q}(t_{n+1}), t_{n+1}\right) = \mathbf{0} \tag{5.37}$$

There are several ways correction polynomials can be defined [14]. The method we use defines it as a polynomial that interpolates through the same last k points as the predictor, i.e.

$$Q(t_{n+1-i}) = \mathbf{X}_{n+1-i}, \ i = 1,\ldots,k \tag{5.38}$$

Continuing in the same way as in [3, 15], we arrive at expressions for approximating the derivative, known as the *differentiation formula*

$$\dot{\mathbf{X}}_{n+1} = \dot{\mathbf{X}}^0_{n+1} - \frac{\alpha^0(n+1)}{h_{n+1}} \left(\mathbf{X}_{n+1} - \mathbf{X}^0_{n+1}\right) \tag{5.39}$$

Here

$$h_{n+1} = t_{n+1} - t_n \tag{5.40}$$

is the attempted step size and

$$\alpha^0(n+1) = -\sum_{i=1}^{k} \alpha_i(n+1) \tag{5.41}$$

where

$$\alpha_i(n+1) = \frac{h_{n+1}}{t_{n+1} - t_{n+1-i}}, \ (i \geq 1) \tag{5.42}$$

The coefficient α_i in (5.42) depends on the ratio of the current and accumulated previous step sizes and changes each time the step size or order changes. Such a formula is known as a *variable coefficient* formula. In DASSL, a simplified form of this formula is used in which the coefficient is independent of the step-size (denoted in [3] as α_s). The formula was developed using corrector polynomials that interpolate at equidistant time points.

After substituting in (5.37), we get

$$\mathbf{g}\left(\dot{\mathbf{X}}^0_{n+1} - \frac{\alpha^0(n+1)}{h_{n+1}} \left(\mathbf{X}_{n+1} - \mathbf{X}^0_{n+1}\right), \mathbf{X}_{n+1}, t_{n+1}\right) = \mathbf{0} \tag{5.43}$$

This is an implicit vector equation that can be solved iteratively using the predicted values as the starting point. The modified Newton method is used. If we denote by $\mathbf{d}_m$ the corrections at the m-th iteration step, the corrections can be found by solving the linear equation

$$\mathbf{J}\mathbf{d}^m_{n+1} = -\mathbf{g}\left(\dot{\mathbf{X}}^m_{n+1}, \mathbf{X}^m_{n+1}, t_{n+1}\right) \tag{5.44}$$

where $\mathbf{J}$ is the partial derivative (Jacobian) matrix

$$\mathbf{J} = \frac{\partial \mathbf{g}}{\partial \mathbf{X}} + \alpha \frac{\partial \mathbf{g}}{\partial \dot{\mathbf{X}}} \tag{5.45}$$

with

$$\alpha = -\frac{\alpha^0 (n+1)}{h_{n+1}} \tag{5.46}$$

The values of the variables and the derivatives are updated for the next iteration by using the formulas

$$\left. \begin{array}{l} \mathbf{X}_{n+1}^{m+1} = \mathbf{X}_{n+1}^{m} + \mathbf{d}_{n+1}^{m} \\ \dot{\mathbf{X}}_{n+1}^{m+1} = \dot{\mathbf{X}}_{n+1}^{m} + \alpha \mathbf{d}_{n+1}^{m} \end{array} \right\} \tag{5.47}$$

The solution (5.44) is accomplished by the method of LU decomposition and back substitution. Because of the sparse structure of the matrix, we use a sparse package for the LU decomposition of the matrix and the solution of the corresponding systems of triangular equations. As a practical implementation of the method, we use the Sparse Linear Equation Solver (version 1.4) of the University of California, Berkeley.[1] The solver was originally developed for circuit simulators. We used its basic functionality only and, in particular, the possibility to decompose the matrix at a fixed pivotal order once the complete decomposition is done. This way, if the matrix has not changed too much, then decomposition can be performed quite efficiently. The strategy used for solving (5.44)–(5.47), including the stopping criterion, is the same as described in [3] for DASSL.

We use a similar strategy for accepting an integration step, the step size selection, and change of order, as in [3]. This has its roots in [17]. The basic difference is a slightly different expression for the estimated principal part of the truncation error. Other differences are described later. The estimation used for the truncation error is of the form

$$\theta_{n+1} = \alpha_{k+1}(n+1) \left\| \mathbf{X}_{n+1} - \mathbf{X}_{n+1}^0 \right\| \tag{5.48}$$

where $\|\ldots\|$ is a weighted mean square root norm. The expressions are slightly different from DASSL because of the variable coefficient form of the differentiation formula in (5.39).

---

[1]The code and documentation are freely available through the NETLIB repository. We express our thanks to the University of California, Berkeley, Department of Electrical Engineering, and to the authors of this really sophisticated software.

Fig. 5.4 Coordinates of the nonzero partial derivative matrix entry

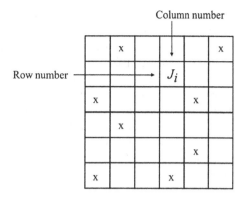

5.3.2   The Generation of the Partial Derivative Matrix

Solving (5.44) requires evaluating the partial derivative matrix of (5.45). This matrix is often approximated numerically. We generate analytical expressions for this matrix by symbolic differentiation of the model equations.

The matrix is of sparse structure, having a small number of nonzero elements per row and column. We generate and store only the expressions of nonzero elements. One of the often-used methods for storing sparse matrices is based on the coordinate scheme. It is used here because it is very easy to implement (Fig. 5.4).

Thus, for every nonzero element of the partial derivative matrix we generate an analytical expression and record its row and column number. The procedure is straightforward and is as follows

1. Take the left side of the expression of the first system model equation.
2. Find the list of variables and the derivatives of variables that appear in the expression.
3. For every variable evaluate, in turn, the partial derivative of the expression by symbolic differentiation. The evaluated expressions are stored successively in the corresponding string array. Store, in the separate arrays, the index of the equation (row index) and of the variable (column index).
4. Repeat step 2 with the next expression until all equations have been processed.

The values of the partial derivative matrices are evaluated during the simulation using stack oriented routines, as described in Sect. 5.7.

5.3.3   The Error Control Strategy

We now describe the strategies used for error control. We do not apply any projections on the model equations, but try to solve them directly. The BDF code, as used here, can handle DAEs of index 1 and 2, if some precautions are taken. Putting

aside the problem of initialisation that we consider in Sect. 5.5, one of the most
sensitive points is error control. This concerns the strategy for step acceptance and
how the next step and/or the order of the method is chosen.

Error control in BDF codes is based on testing the truncation error estimates
against predefined error tolerances. The truncation error is estimated by (5.48),
which is similar to that in DASSL, with some difference in the leading coefficient.
The error is thus proportional to the corrections made in the Newton iterative
loop. If the error test is not satisfied a new, smaller step size is chosen until the test
is passed. This works well for index-1 DAEs. In index-2 equations there is a
repeatable failure of the error tests and the procedure is finally stopped. The
problem is that errors in some algebraic variables do not approach zero as the step
size approaches zero. As a result, these cannot be made less than the tolerance (in
some norm) with decreasing step size. Thus, solving higher-index DAEs requires
changing the way the error is tested.

The simplest way to do this is suggested in [3]: All the algebraic variables are
removed from the error tests, the next-step-size test, and the order-selection tests,
and these tests are applied to the differentiated variables only. This has the draw-
back that the algebraic variables are partially out of control. They, however, satisfy
the corrector equation solving loop and are accepted if the differentiated variables
pass the integration error test. The reasoning behind such a strategy is that errors in
the algebraic variables do not influence directly the future state of the system. In this
way, integration can be executed successfully, although with the price of lower
accuracy in the algebraic variables. In many cases this is acceptable, as long as high
accuracy in such variables is not required.

We also scale the algebraic equations by the factor $1/h$, where h is the current
step-size. This scaling, as discussed in [3], improves the ill conditioning of the
iteration matrix at very small step sizes. This, in turn, improves the accuracy of the
differentiated variables and, to a lesser extent, of the algebraic variables. In any
case, scaling is recommended. We have not applied scaling directly in the model
equations because these become quite difficult for the user to understand (Sect. 5.4).
Instead, we simply modify the values of the left side of the equations and of the
partial matrix elements when they are evaluated during iteration.

We have also implemented another strategy based on *local error* control. This
has its roots in a study of the local error control of DAEs of index-1 and 2 in [18].
Using the notation of Sect. 5.3.1, the proposed estimation of the local error is given
by

$$\mathbf{S}_{n+1} = \mathbf{J}^{-1}\mathbf{A}\left(\kappa\mathbf{I} - \frac{(\alpha^0)^2}{\alpha^0(n+1)\cdot h^2}\mathbf{J}^{-1}\mathbf{A}\right)\boldsymbol{\theta}_{n+1} \qquad (5.49)$$

Here matrix $\mathbf{J}$ is the last evaluated and decomposed partial derivative matrix
defined in (5.45), $\mathbf{A}$ the leading coefficient matrix of (5.27), and $\boldsymbol{\theta}_{n+1}$ the truncation
error estimate of (5.48). The α^0 is the value of the BDF parameter in (5.41) used in
the last evaluated partial derivative matrix, and h is the value of the step size used at

that time. Parameter κ is a user-supplied factor used to weight the inherent differentiation in index-2 DAEs. This factor can be set from 1 to a value of the order of $1/h$.

In the implementation of the formula the inverse of the Jacobian is not required because the matrix is already evaluated and decomposed into LU factors. Corresponding terms in (5.49) can be evaluated by back substitutions from the corresponding linear system. The cost of the formula is, in essence, these two back substitutions. We also use this formula for step size and order selection, as suggested in [18]. The basic strategy, as described in [3], is retained using the local errors estimates evaluated by (5.49) and used for the next step size and order selection. In [18] it was shown that the error control based on the local error is satisfactory for index-1 and index-2 DAEs. It is expected that this control strategy is an improvement over the error control of using the first strategy. We have found this to be true at least for some of the problems treated in this book (Chap. 9). We also found that the local error control strategy works much better if the algebraic equations are scaled by the step-size.

To summarize the integration control used for solving the model equations during simulation, we describe the parameters and options at the user's disposal. First of all, after the mathematical model is built and suitable data are created, the user has to select the

- Simulation interval
- Output interval
- Maximum step size

After the simulation is finished, it is possible to continue or to restart the simulation. Output is generated at every output interval value. The maximum step size that the integrator selects is limited. By default, both the output interval and maximum step size are set to 1/100 of the simulation interval. For many problems, however, quite a short maximum step size is often necessary.

The BDF method "likes" relatively tight error tolerances. Thus, the default values of the absolute and relative error tolerance are set to 10^{-6}. These can be changed, up or down. Lower values are, in general, not advised. On the other hand, some problems are not easy to solve with too tight a tolerance, i.e. 10^{-12} or 10^{-15}.

Currently only one integration method is implemented, the variable coefficient BDF method. It is possible, however, to scale the algebraic equations (this is the default behaviour), or to disable scaling. Both error control strategies described above are implemented:

- Control of errors in differentiated variables only (default)
- Local error control

In the last option it is possible to input the differentiation weight parameter value, too. It is set to 1 by default.

5.4   Decompiling the Model Equations

The system model equations, including the partial derivative matrices, are generated automatically by the *BondSim* software. They are generated in the byte form, as explained in Sects. 5.2 and 5.3, and hence are not easily readable. We can, however, *decompile* them into a more readable form. The system variables used in the decompiled form are not the original symbols, but are of the form $Y(1)$, $Y(2)$, etc., and t for time. We will show how they can be interpreted.

As an example consider simple problem of Sect. 2.7.1. We start the *BondSim* program first and then open the project *Body Spring Damper Problem*. To build the corresponding mathematical model we chose the command *Build* from *Simulation* menu, or click the corresponding toolbar buttons. The process can be observed in *Build* tab of *Output* window at the bottom of the screen. To show the model created we open the *Simulation* menu, submenu *Show Model*, and then the command *Write Equations*, or the corresponding toolbar button (in the form of a pencil). The model appears in the *Output* window. The decompiled model can be copied into a text file by clicking on the equations with the right mouse button and selecting *Copy*. After the *Save As* dialogue opens we define the name of the text file and the destination of the decompiled data. The result is shown in Fig. 5.5.

```
Equations of the model:
    EQ(1) = Y'(8)-Y(7) = 0
    EQ(2) = Y'(9)-Y(3) = 0
    EQ(3) = Y'(12)-Y(11) = 0
    EQ(4) = (Y(3)-Y(2))-Y(5) = 0
    EQ(5) = ((-Y(6))+Y(4))+Y(1) = 0
    EQ(6) = (((-Y(4))-Y(1))+Y(10))-Y(7) = 0
    EQ(7) = ((-Y(2))+Y(3))-Y(11) = 0
    EQ(8) = Y(1)-(P(1)*Y(5)) = 0
    EQ(9) = Y(2) = 0
    EQ(10) = Y(8)-(P(2)*Y(3)) = 0
    EQ(11) = Y(10)-P(3) = 0
    EQ(12) = Y(4)-(P(4)*Y(12)) = 0
List of differential variables:
    Y(8), Y(9), Y(12)
Indexes of the algebraic equations:
    EQ(4), EQ(5), EQ(6), EQ(7), EQ(8), EQ(9), EQ(10), EQ(11), EQ(12)
The initial value expressions :
    Y(8)= 0
    Y(9)= 0
    Y(12)= 0
    Y(10)= P(3)
Parameters :
    P(1) = 150, P(2) = 5, P(3) = 500, P(4) = 112500
```

Fig. 5.5 The decompiled form of the model equations

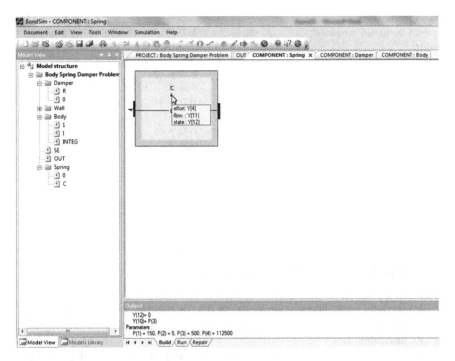

Fig. 5.6 The project screen with port variable tooltip window

In addition to the model equations the other information is also generated as well, such as a list of differential variables, indices of algebraic equations, the initial values expressions, and the parameters.

In order to interpret the system variables we use the symbol tables generated during model build up. To inspect the variables of a component, e.g. the Spring in Fig. 5.6, we need to open it and move the mouse cursor over the ports of the elementary components, e.g. over the port of C component. A tool tip window appears showing the system variables representing the port effort, flow and state variables, respectively. (Note the compiled equations at the bottom.)

In a similar way data for the partial derivative matrix can be decompiled using the command *Matrices of Partials* (Fig. 5.7). For every nonzero element of the matrix the expression, accompanied by row and column index, is given. Data for the leading coefficient matrix of (5.27) also is given.

Fig. 5.7 The decompiled
form of the partial derivative
matrix

```
      Matrix of partial derivatives:
        J(1) = -1
        IRow(1) = 1      JCol(1) = 7
        J(2) = cj
        IRow(2) = 1      JCol(2) = 8
        J(3) = -1
        IRow(3) = 2      JCol(3) = 3
        J(4) = cj
        IRow(4) = 2      JCol(4) = 9
        J(5) = -1
        IRow(5) = 3      JCol(5) = 11
        J(6) = cj
    -----------------------------------------
```

5.5   The Problem of Starting Values

In this and the next section we describe the approaches used for solving two
important issues of simulation: the generation of starting values and integrating
across discontinuities.

The problem of values from which the simulation starts is one of the difficult
problems of DAEs that has attracted much attention. In the equations in state-space
form, such as

$$\left.\begin{array}{l} \dot{\mathbf{x}} = \mathbf{f}(\mathbf{x}, t) \\ \mathbf{y} = \mathbf{g}(\mathbf{x}, t) \end{array}\right\} \tag{5.50}$$

this is not much of a problem. To start solving the equations it is enough to specify
the initial values of the state variables, e.g. by specifying the values at initial time
$t = 0$,

$$\mathbf{x}(0) = \mathbf{x}_0 \tag{5.51}$$

In this way, the time derivatives of the variables needed by the integration
routine can be evaluated directly from the first equation

$$\dot{\mathbf{x}}(0) = \mathbf{f}(\mathbf{x}_0, 0) \tag{5.52}$$

and the integration can start. The output values are generated using the second
equation. The problem, thus, is decoupled into solving the differential equation and
generating the outputs. This is well known and it is why much attention in bond
graphs, and in other approaches, has been given to setting the model equations into
this form. In modelling real systems, however, models in the state space form
generally are too restrictive. This makes it necessary to deal with DAEs.

Because there is no decoupling of equations and variables in DAEs, we have to cope with the complete system simultaneously. In DAEs of index 1, such as of (5.18), some of the similarity with the state-space form is retained. Because the Jacobian matrix $\partial \mathbf{g}/\partial \mathbf{y}$ is not singular, it is possible to find values of the algebraic variables $\mathbf{y}$, given values of the differentiated variables $\mathbf{x}$. Hence, if values of the latter are known at the initial time, values of the former can be found by solving the algebraic equations

$$\mathbf{g}(\mathbf{x}_0, \mathbf{y}, 0) = \mathbf{0} \tag{5.53}$$

This can be accomplished using iterative methods, typically Newton type methods, for which we need only an initial guess of their values to start the iteration. Once values are found to the prescribed accuracy, the starting values of variables are known. Hence, in index-1 DAEs, the differentiated variables play the role of state variables, in the sense that they are independent and completely define the values of all other variables.

Index-1 DAEs are typically solved using BDF methods. To start the solution we need also the time derivatives of the variables. For the semi-explicit DAEs of (5.18), derivatives of the differentiated variables with respect to time can be found easily by evaluating the right-hand side of the first equation at initial time. For algebraic variables, on other hand, we first need to find the time derivative of the algebraic equations. We can do this by symbolic differentiation. The resulting equations are linear in the time derivative of the algebraic variables, as can be seen from the second equation in (5.19) and, hence, are solved readily for the time derivative of the variables.

Values of the variables and their time derivatives constitute a set of *consistent* starting values for the system. If we find such values with sufficient accuracy, integration continues smoothly. Otherwise, we can expect wide fluctuations, which hopefully converge to the solution.

For higher index DAEs the problem is more complicated. Looking at (5.20), which is often used as a prototype of semi-explicit index-2 DAEs, we see that the differentiated variables are constrained by the algebraic equations. Hence, all of their values cannot be set independently. The independent part plays the role of system state variables. These can be set initially to appropriate values. The others then are found by solving the second equation. Even for these we need starting values that are used to initialise the iteration.

An independent set of differentiated variables traditionally is found in bond graphs by the causality assignment procedure (Sect. 2.10). More generally, they are found using the subspace structure of the underlying equations [5, 6]. This way, only part of the first equation in (5.20) is an ordinary equation that involves integration. The other constitutes algebraic equations that involve differentiation of the variables. But this is only part of the story and, perhaps, the easier part. What about the algebraic equations? To find them we need to differentiate the second equation in (5.20). After substituting from the first equation, we get the hidden constraint of (5.23). If the corresponding Jacobian given in (5.24) is not singular, we solve the hidden constraint for the algebraic equations. For this, however, we again need starting values.

Determination of the initial values of the time derivatives of variables is accomplished in a similar way as that used for index-1 DAEs. Initial values of the time derivatives of the differentiated variables are found from the first equation in (5.20). But for the algebraic variables we again need to differentiate the hidden constraint with respect to time and find the corresponding values of the variables from the resulting linear equations.

In DAEs of index higher than 2, the problems are even more difficult, as more differentiation is necessary and the subsequent equations are more difficult to solve. Different approaches have been reported for solving the problem of starting values [19–23]. Many are concerned with index-1 DAEs, but some also treat index-2 equations. In general, they resort in some way or other to differentiation. A good survey of these is found in [3, 23]. One of the problems is how to determine the index of the system. Approaches based only on structural information, as in [20], are often not feasible, as the system structure can change because, for example, some of the constitutive relations are defined conditionally, depending on the values of variables. As a simple example, we return to the see-saw problem (Sects. 2.7 and 5.2.3). The model generated for this system is of index-2. If we assume, however, that there is dry friction—as there is!—between the see-saw seat and the children, when the children stick to the seat this becomes an index-2 problem. When they slip even for a short time, however, the model changes to index-1, because the differentiated variables become independent.

The problem of consistent starting values is not only important at the start of the simulation, but also after every discontinuity. In addition, any approach to the problem of starting values assumes that the user supplies the necessary data. Even under the best conditions this is often too much to ask. The modelling approach in this book has been developed to help the user design or analyse mechatronic equipment. For this we need a simpler approach to the starting of the simulation.

The approach we use is close to that of [19, 21]. It was prompted by how we start real equipment. Typically, we switch the power on and, after some time, the system settles down to the appropriate operating state and we start using it. We thus assume that the system starts off un-energized, i.e. with efforts and flow equal to zero. We also assume that the time derivatives of all variables are equal to zero. By default, the starting values of all differentiated variables defined in the corresponding elements (capacitive, inertial, and integrator elements) are also set to zero, but can be changed to some other value if required. These are taken as an estimation of consistent starting values.

To find a starting value for the simulation we use a version of the BDF solver that employs a first-order method (implicit Euler) and a fairly small and constant step size. To simulate the transients that can be quite intensive until the system settles down, we simulate the system for several steps without error control. We then advance the simulation time for one step more and check the error. If the error test is passed, we integrate back to the initial time and use the state reached as the starting state for the simulation. If, on the other hand, the test is not passed, we repeat the starting procedure with a smaller step. The starting procedure is stopped and failure reported if either the step was reduced to the minimum without success, or if too many initialisations have been attempted. During initialisation the corrector equations are solved at every step to the prescribed accuracy. The partial derivative matrix is updated and decomposed at every simulation or iteration step. If, for any reason, it is not possible to solve these equations, initialisation is stopped with the appropriate message. In Part 2, in which different mechatronic problems are solved by bond graph modelling, all initialisations are done using this approach quite successfully. The reader can test this for herself.

5.6   The Treatment of Discontinuities

Discontinuities give reality to the models. There are numerous examples of engineering systems where such features are necessary. This is the case, for example, when dealing with dry friction and impact in mechanical systems. In electronics, processes are often described by different expressions, depending on the range of variable values. This, of course, complicates the problem of equation solving.

We analysed several schemas for dealing with discontinuities. In one of these during the model generation the sub-expressions contained in the constitutive equations and which define the discontinuities in the model are extracted and stored in a separate array. These sub-expressions are decompiled and shown below (Fig. 5.8). We may use these functions to generate the state transition tables. The location time instant of the switching was found by a binary search using interpolating polynomials that the BDF solver maintains or by repeating the time step. After the discontinuity, the integration was reinitialised using a method similar to that used for simulation initialisation (Sect. 5.5). We found, however, that this approach did not work as expected: Many times there were conflicts during the Newton and the corresponding step-advancing steps. A possible explanation for such behaviour is that there is no guarantee that when the system crosses the

```
The model switch functions :
  Switch(1) = ((t<1E-009)?0:1)
  Switch(2) = (((Y(7)/0.75)<0.5)?0:1)
  Switch(3) = (((Y(7)/(1*0.0258))>50)?0:1)
  Switch(4) = (((Y(9)/0.75)<0.5)?0:1)
  Switch(5) = (((Y(9)/(1*0.0258))>50)?0:1)
  Switch(6) = ((Y(12)<=(1+(0*(SQRT (0.6-Y(13))-
              SQRT(0.6)))))?0:((Y(42)>=0)?((Y(42)<((Y(12)-1)-(0*(SQRT (0.6-Y(13))-
              SQRT (0.6)))))?1:2):((-Y(42)<((Y(12)-1)-(0*(SQRT (0.6-Y(13))-SQRT
              (0.6)))))?3:4)))
```

Fig. 5.8 Generation of state transition functions

discontinuity it stays on the other side for some finite time and continue its motion. It is quite possible that the system returns back, and under actions of the driving efforts tries again to go over the discontinuity. Eventually it goes over the discontinuity. Thus, mechanism often is not smooth

We thus decided not to design a state transition mechanism to control dealing with the discontinuities, but to try to go straight over the discontinuity; we simply let the solver handle the discontinuities. The conditional expressions are retained in the model expressions and those of the partial derivative matrix elements. When the discontinuity is encountered, the solver tries to go over it. The behaviour of the solver is similar to that at the initialization. During first few steps the error is not controlled. After that if the error control requirements are satisfied integration continues. Otherwise, the step-size is reduced and the procedure is repeated.

This perhaps is neither the most efficient nor the best technique that might be used, but it works. It is capable of dealing with impulsive forces appearing e.g. at the impacts. It works well in all the examples treated in this book, as well as with both error control strategies discussed in Sect. 5.3.3.

To help the solver deal with discontinuities it is advisable to ensure continuity of sub-expressions at their boundaries, if possible. In many cases discontinuities appear because of the simplification of the model. Thus, a complicated non-linearity presents fewer problems than an on-off discontinuity. This is particularly true at the start of a simulation.

5.7   A Stack Based Approach to Function Evaluation

The efficient simulation of models described in terms of Bond Graphs and Block Diagrams is based on compiled routines. Thus, most of the support for the operations described in the previous chapters is coded in C++ language and compiled using Visual C++ compiler in MS Visual Studio 2012 environment. However, there are parts that cannot be precompiled because they depend on data defined during a project model development. These parts are connected to the constitutive relations and parameters of different elementary and block diagram components, which depend on the problem under study.

The solver, in our case BDF, during the simulation repetitively evaluates several groups of the functions that are based on the constitutive relations. Thus, it evaluates the left side of the mathematical model equations (see Fig. 5.5) in the attempt to make them approximately equal to zero. These left sides are really the constitutive relations of the elementary components of which the model consists.

BDF solver as described in Sect. 5.3.1 needs the values of the partial derivative matrix (Jacobian) (5.45) and rate derivative (leading coefficient) matrix (5.27). In the current implementation these matrices are generated analytically by applying the symbolic differentiation on the constitutive relations. It is necessary to find an efficient method for evaluation of these matrices, in particular the first one.

Finally, if the model contains discrete parts it is necessary to evaluate (5.17) at every sampling instance. These equations are normally executed after a completed BDF step.

As discussed in Sect. 5.2.2 all these expressions are written as byte strings based on the *Polish* (*prefix*) notation. Because the corresponding expressions do not contain parentheses they could be evaluated by one pass through the corresponding string from the left to the right. This method was used in the *BondSim* program packed with the first edition of this book [24]. Later the evaluating method was upgraded by using the *CIL* (*Common Intermediate Language*),[2] which is an object-oriented assembly language entirely based on stacks. As a result the efficiency of the functions evaluation increased at least twofold in comparison to the interpreter approach. Here we will give an overview of the approach used trying not to be too technical.

The prefix strings can be used to generate the stack based evaluation. Thus, addition of two numbers written in the prefix notation as

$$B + \ value1 \ value2 \tag{5.54}$$

is evaluated using the stack operations as:

1. *value1* is pushed onto the stack.
2. *value2* is pushed onto the stack.
3. *value2*, and then *value1* are popped from the stack.
4. *value1* is added to *value2* and the result is pushed onto the stack.

Note that reading of the byte expression is from left to right, and thus the first operand is pushed first onto the stack, and then the second. The popping of the operands was made in the opposite order. The stacks support *LIFO* (Last In First Out) order. However, operations defined by prefix notation apply the operator, addition in this case, starting at the first operand and then the second. These operations are in accordance with the stack transitional behaviour used in *CIL*.[3] Stack operations can be defined for any byte strings, which include the operations and function calls discussed in Sect. 5.2.2.

The fundamental question is how to implement these operations. Owing to NET technology that Windows supports through the NET Framework[4] use was made of the *Dynamic Assembly*, a powerful concept that allows extending the compiled code dynamically at the run time. The implementation of this concept was eased due to Microsoft extension of C++ language known as C++/CLI, which is supported by Visual C++.

---

[2]Formerly known as Microsoft Intermediate Language or MSIL.

[3]It is also possible to interpret prefix byte strings by reading from right to left.

[4]Currently NET Framework 4.5 is used.

Fig. 5.9 Structure of
dynamical class ModelClass

Attributes:
    systemVariables

Methods:
    Constructor
    ModelEquations
    ParcDerivativeMatrix
    RateDerivativeMatrix
    ZEquations

We create the assembly builder first, which is used to create the *Math Model Module* builder object. This object is used to create *ModelClass* type (class) builder object. This class plays the fundamental role in the model evaluation because it defines dynamically all necessary operations mentioned in the beginning. The structure of the class is shown in Fig. 5.9. Note that in NET every class has *Object* class as a top superclass (similarly as MFC has *CObject* as its top superclass).

Using type builder we can define the class methods. We define first the class constructor. The procedure is

- Define constructor builder. The constructor is with one argument.
- Get CIL generator for the constructor.
- Using the generator define the constructor by a stream of CIL commands. The base class constructor is called first, and then the pointer (tracking handle) of the Object passed as the constructor argument is assigned to *systemVariable* attribute field.

We define the class method functions in a similar way, but this time we take into account that these functions are defined by the byte strings in the *Polish* notation. Thus for every type of member function there is a function that reads the corresponding byte strings and translate them into CIL streams of the stack operations. Thus, the procedure is

- Define method builder
- Get CIL generator for the method
- Call a function which reads the function byte strings and emit CIL stack operations instructions using the method CIL generator.

In this way we define *ModelEquations* method for evaluating left sides of the mathematical model equations, *ParcDerivativeMatrix* and *RateDeivativeMatrix* nonzero elements, and *ZEquetions* method for evaluating the discrete model relationships.

These methods are CIL equivalent of the byte strings expressions. They need the other data as well as are the system variables, parameters and functions discussed in Sect. 3.5. To that purpose we define a *SystemVariables* class that warps all system variables and define the basic function that the constitutive relations may use. Now we may create a *SystemVariables* object containing all necessary information, and create a *ModelClass* object using the constructor defined above and pass to its field the pointer (tracking handle) to just created *SystemVariables* object.[5]

Once the *ModelClass* object is constructed it can be used to call its member functions in Fig. 5.9 to evaluate quantities that BDF solver needs. The first time these functions are called, they are compiled into machine code of the current processor using JIT (Just In Time) compiler of Windows NET Framework. This normally happens at beginning of the simulation during the initialization step. After this step the evaluation of these functions is really very quick.

References

1. Reverchon A, Ducamp M (1993) Mathematical software tools in C++. Wiley, Chichester
2. Petzold L (1982) Differential/algebraic equations are not ODEs, SIAM. J Sci Statist Comput 3:367–384
3. Brenan KE, Cambell SL, Petzold LR (1996) Numerical Solution of initial-value problems in differential-algebraic equations. Classics in Applied Mathematics, SIAM, Philadelphia
4. Hairer E, Wanner G (1996) Solving ordinary differential equations II, Stiff and differential-algebraic problems, 2nd revisited. Springer, Berlin, Heidelberg, New York
5. Griepentrog E, März R (1986) Differential-Algebraic equations and their numerical treatment. BSB Teubner, Leipzig
6. März R (1992) Numerical methods for differential-algebraic equations. Acta Numerica 141–198
7. Tischendorf C (1995) Solution of Index-2 differential algebraic equations and its applications in circuit simulation, Ph.D. thesis Humboldt University Berlin. Logos Verlag, Berlin
8. März R, Tischendorf C (1997) Recent Results in solving index-2 differential-algebraic equations in circuit simulations, SIAM J Sci Comput 18:139–159
9. Haug EJ (1989) Computer-aided kinematics and dynamics of mechanical systems, vol. I: Basic Methods. Allyn and Bacon, Needham Heights, Massachusetts
10. Van Dijk J, Breedveld PC (1991) Simulation of system models containing zero-order causal paths—I. Classification of zero-order causal paths. J Franklin Inst 328:959–979
11. Van Dijk J, Breedveld PC (1991) Simulation of system models containing zero-order causal paths—II. Numerical implications of class 1 zero-order causal paths. J Franklin Inst 328:981–1004
12. Schwartz DE, Tischendorf C (2000) Structural analysis of electrical circuits and consequences for MNA. Int J Circ Theor Appl 28:131–162
13. Borutzky W, Cellier F (1996) Tearing in bond graphs with dependent storage elements. In: Proceedings of symposium on modelling, analysis and simulation, CESA'96, IMACS multi conference on computational engineering in systems applications, vol 2. Lille, France, pp 1113–1119
14. Gear CW (1971) Numerical initial-value problems in ordinary differential equations. Prenice Hall, Englewood Cliffs
15. Jackson KR, Sacks-Davis R (1980) An alternative implementation of variable step-size multistep formulas for stiff ODEs. ACM Trans Math Softw 6:295–318

---

[5]In CLI because the objects are not fixed in memory an analogue to C/C++ pointers known as tracking handles are used.

16. Krogh FT (1974) Changing step size in integrations of differential equations using modified divided differences. In: Proceedings of conference numerical solution of ODEs, lecture notes in mathematics 362, Springer, New York
17. Shampine LF, Gordon MK (1975) Computer simulation of ordinary differential equations. FH Friedman and Co., San Francisco
18. Sieber J (1997) Local error control for general index-1 and index-2 differential-algebraic equations. Humboldt University Berlin, Preprint
19. Sinovec RF, Erisman AM, Yip EL, Epton MA (1981) Analysis of descriptor systems using numerical algorithms. IEEE Trans Aut Cont 26:139–147
20. Pantelides CC (1988) The consistent initialization of differential-algebraic systems. SIAM J Sci Comput 9:213–232
21. Kröner A, Marquardt W, Giles ED (1992) Computing consistent initial conditions for differential-algebraic equations. Comput Chem Eng 16:S131–S138
22. Brown PN, Hindmarsh AC, Petzold LR (1998) Consistent initial conditions calculations for differential-algebraic systems. SIAM J Sci Comput 19:1495–1512
23. Schwarz DE (2000) Consistent initialization for index-2 differential algebraic equations and its applications to circuit simulation. Ph.D. thesis Humboldt University, Berlin
24. Damic V, Montgomery J (2003) Mechatronics by bond graphs: an object-oriented approach tot modelling and simulation. Springer, Berlin, Heidelberg

Part II
Applications

Chapter 6
Mechanical Systems

6.1 Introduction

In this chapter we start the study of problems from Mechatronics using the bond graph modelling approach developed in Part 1. The general procedure consists of

1. Analysis of the problem under study
2. Development of the corresponding model in terms of bond graphs, and
3. Analysis of the behaviour by the simulation

Bond graph modelling and simulation will be undertaken using the *BondSim* program, which is freely available from the authors web page (see also Appendix). Readers are advised to use it when reading material given in this and subsequent chapters.

In this chapter we study simple, mostly one-dimensional, mechanical problems. One reason for this is to familiarize the reader with using the *BondSim* software on relatively simple problems, though these problems are of interest in their own right. We start with the well-known Body Mass Damper problem already discussed in Sect. 2.7.1. After that we continue with the study of the influence of dry friction on the system, which introduces discontinuities in the model equations. We then study the Bouncing Ball problem. This also involves discontinuities, but is better known for its chaotic behaviour. The section concludes with a discussion of higher index problems. It is not the intention to show that it is possible to solve all such problems using the methodology developed; we simply show that a class of such problems of interest in Mechatronics can be solved in an acceptable way from an engineering point of view. This is particularly true for some problems in multibody dynamics. In this chapter we will study only the simple pendulum problem. Further discussion of solving problems in multibody dynamics is left to Chap. 9.

© Springer-Verlag Berlin Heidelberg 2015
V. Damić and J. Montgomery, *Mechatronics by Bond Graphs*,
DOI 10.1007/978-3-662-49004-4_6

6.2   The Body Spring Damper Problem

6.2.1   The Problem

We start by analysing a well-known problem from engineering mechanics, the simple Body Spring Damper system (Fig. 6.1), which has already been discussed in Sect. 2.7.1. The system consists of a body of mass m that can translate along the ground, and is connected to a wall by a spring of stiffness k and a damper having the linear friction velocity constant b. An external force F acts on the body parallel to the ground. The effect of dry friction is not included in the model. The modelling of dry friction is a topic of Sect. 6.3.

We analyze the transient behavior of this simple system, and also its behaviour under sinusoidal forcing. We first develop a bond graph model using the *BondSim* program. After that we analyze the dynamical behaviour of the system by simulation. The explanations of the procedures used are somewhat detailed and serve as an introduction to using the *BondSim* program.

6.2.2   The Bond Graph Model

6.2.2.1   System Level Model

Before we start with model development, we must launch the *BondSim* application. This can be done in the usual way for Windows environment, e.g. by double clicking the application's shortcut *BondSim 2014* on the computer desktop; or using the Windows *Start* button at the left corner of the computer screen, then choosing the application from the list of the recently used applications, or choosing the *All Programs* command, and then the folder *BondSim* and finally the shortcut *BondSim2014*.[1] The main program screen then appears having the title *BondSim Application* (see Fig. 4.2). There is a menu bar just below the program title showing the main menu commands—*Project, View, Tools* and *Help*—from which the program commands are invoked (Sect. 4.2). There is also a row of the toolbar buttons used to invoke some of the more often-used commands. Some of these buttons are familiar to Windows users, but others are specifically designed for *BondSim*. By moving the mouse cursor over the buttons, a tool tip appears that contains a short description of the corresponding command. Some of these buttons are currently disabled.

To begin with the model development for the *Body Spring* Damper problem, we define a new project first. This can be done using the *New* command in the *Projects* menu, or by using the *New Project* toolbar button. In the dialogue that appears

---

[1]The current program version is BondSim 2014, but similar operation are expected to be valid for the future versions of the program as well.

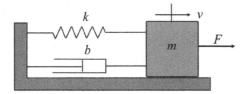

Fig. 6.1 The body spring damper problem

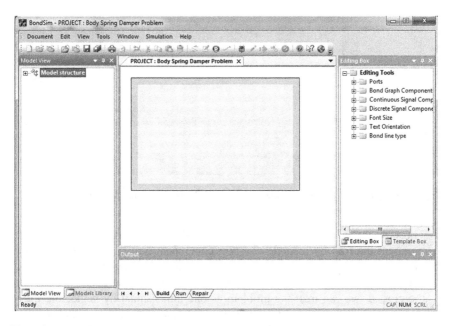

Fig. 6.2 The project document window

(Fig. 4.3), we type a suitable project name, e.g. "*Body Spring Damper Problem*", and then click the *OK* button. The command is accepted if there is no existing project with the same name. The dialogue closes, and a new empty document appears having the title of the new project as shown in Fig. 6.2. The menu also changes to show the commands that can be used at the document level (Sects. 4.6 and 4.7).

If we expand the *Model Structure* at the left it shows only one entry: *Body Spring Damper Problem* (Fig. 6.3). The central document window serves to define the structure of the system model. The basic tools for bond graph model development are contained in the *Editing box* (Fig. 6.2 at the right). We can expand the first two branches: *Ports* and *Bond Graph Components*.

The model development starts by decomposing the system into components. Looking at the problem schematics of Fig. 6.1, we identify the following components:

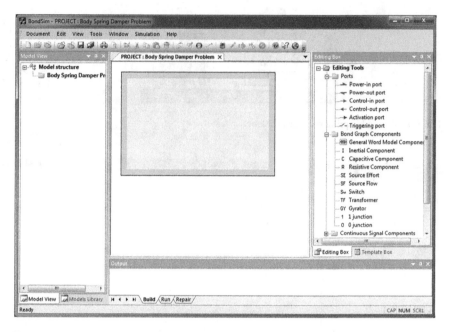

Fig. 6.3 The project document window with expanded model structure and editing tools

- The wall at the left is used to fix the left ends of the spring and damper. We represent the wall by a single two-port component Wall; the ports being the places where the spring and the damper are connected. There is also the ground over which the body moves. But, if we neglect the friction between the body and the ground, we can treat the problem as one dimensional by neglecting inter-actions in the vertical direction (which simply means that the ground reaction force is equal to the body weight).
- The spring of stiffness k, which can be represented by a two-port component Spring, which ports represent interactions at the spring connection ends.
- The damper, of linear viscosity coefficient b, is represented by a two-port component Damper. The ports again represent the interactions at its connection ends.
- The body itself, which is connected to the spring and the damper on the left side, and on the right side acts a force. The body can be modelled by a three-port component Body.
- The force is acting on the body in the direction parallel to the ground. Its effect can be represented by a source effort component.

To create these components we will use the drag and drop technique described in Sect. 4.6.1 (Fig. 6.4). We start with Wall component. However, because we wish to write its name in the vertical direction, we will first expand the *Text Orientation* branch in the *Editing Box*, and select *Vertical*. Now we can drag the *General Word Model Component* and drop it near the left working area border, but not too close

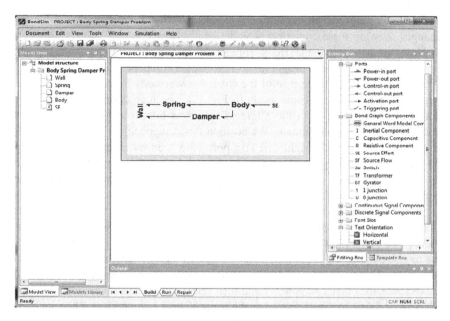

Fig. 6.4 General structure of the model

(see the box below). When the caret appears, we start typing the word "*Wall*" and text runs vertically. To end the editing we may click somewhere outside of the text entered. We will change *Text Orientation* again to *Horizontal* to return to the default text orientation mode.

The created component has two power ports. We will drag them around the component border and position them on the right side. We assume that power is delivered by the spring and damper to the wall. The upper ports half-arrow is already directed into the wall. But, the lower one has the opposite power sense. To change this we will select the port and then click *Change Port* toolbar button (with icon in the form of two opposite power ports).

Next, we will drag again a word model component and drop at the right of the Wall, and type-in "*Spring*". When we click outside the component it is created with two power ports directed from left to right. We change the power sense of the both by clicking the toolbar button *Change Ports* of the selected component (when a component is created it is already in the selected state as indicated by the red bounding box). We drag the component so that its left port is at the same height as the upper Wall port. Now we may connect the spring to wall by a bond by clicking Wall port and moving the mouse to the left Spring port and clicking again (Fig. 6.4).

In a similar way we may create the Damper. This time we will place it somewhere to the right not to overlap the Spring. We again change power flow direction of its ports and connect the left port to the bottom wall port by bond. Note as we add components in the central windows the *Model structure* tree at the left changes showing these components as the empty leaves.

Now we will create the Body component to the right of the Spring and Damper by dragging and dropping the word model component and typing *"Body"*. We will again change the sense of the created power ports by clicking the toolbar button *Change Ports*. Because *Body* in Fig. 6.1 has three connection points, we drag a power-out port from the *Editing Tools, Port* branch and drop it at the left bottom of the Body component. We position this component so that its left port is at the same height as the right Spring port. Now we may connect the left Body port to the Spring right port, and the bottom Body port to the right Damper port by the bonds. (When drawing the last bond we may need perhaps to modify the bond line to form e.g. a right angle as in Fig. 6.4).

The final component that we create is a *Source Effort*, which describes the effect of a force applied to the body by the environment. To place it we need, however, more place to the right of Body. To enlarge the working area to the right, we place the mouse over the right edge of the working area until the cursor changes its form to a two-side arrow. Then click the cursor and drag the right edge to the right to make space for other components and then release the mouse. Now we may drag the *Source Effort* tool and drop it at the right of Body component. Drag its ports from the right to side facing the Body. Also move the SE component so that its port is at the same level as the right Body port, and interconnect them by a bond. Now we obtain the Bond Graph shown in Fig. 6.4.

Figure 6.4 shows the system model that defines the components, which are seen at the system level and are represented by the word models. To complete the model it is necessary to develop a model of every one of its components. However before we go into this, it is important to define what information we need from the model. For the system we are studying we may be interested in the input force acting on the body, the body position during its motion and the total wall reaction to the spring and damper.

To obtain information on the current value of the input force we can break the bond connecting SE and Body, and drag the SE to the right to make place for a 0-junction. Next we drag the 0 *Junction* from the *Editing Box, Bond Graph Components*, and drop it between the SE and Body (Fig. 6.5).[2] The created junction has three power ports. We remove the bottom one and replace it by a Control-out port from the *Port* branch in the *Editing Box*. It serves to extract the effort signal, i.e. the input force, transferred by the junction. We also change the power port senses of the other two ports so that power flows from the right to the left. Finally we connect the junction power ports to the corresponding ports of the SE and Body (Fig. 6.5).

---

[2]Note that while the previous figures are generated by capturing the computer screens, this figure is generated using Print to File command described in Sect. 4.10.2.

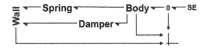

Fig. 6.5 System level model for body spring damper problem

To obtain information on the other two quantities (the body position and the wall force) we drag *Control-out* ports and drop them at the peripheries of Body and Wall. We may now to add a display component to the Bond Graph, by dragging and dropping the *X − Y Display* from *Continuous Signal Components* (Fig. 4.18 left). We add an additional Control-in port to this component, and connect the all its ports to 0-junction, Body and Wall, respectively as shown in Fig. 6.5.

We can assign labels to the plot axes by double clicking the corresponding display ports. Thus, by double clicking the port connected to the Body, a dialog opens as shown in Fig. 6.6. We may now type '*x*' (position) as label of the axis. By default it refers to *Y axis*, but we may select also *X axis*, or *None*.

In a similar way we may assign '*F*' (the input Force) to the top port, and '*Fw*' (the wall force) to the bottom display port. Note we may assign all variables to the *Y axis*, in which case we have *Y − t* plot with all variables plotted to the Y axis, and *time* along the *X* axis. We also can create *X − Y* plots, where one variable is assigned and plotted along the *X* axis and all others to *Y* axis. Finally, if variable is

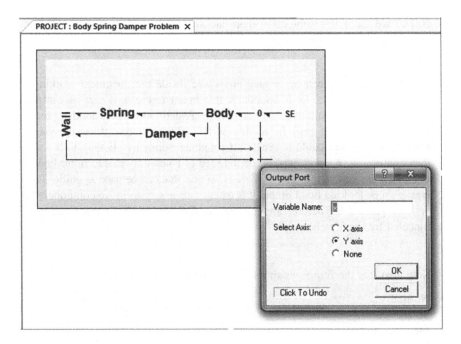

Fig. 6.6 Assigning port names as axis label

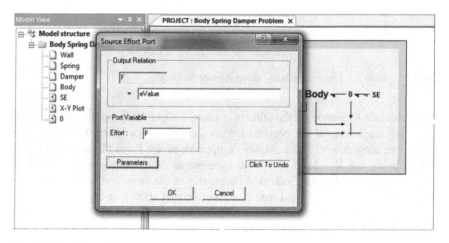

Fig. 6.7 Specifying the input force as a constant force of 500 N

set to *None* (axis), it means that its plot is not currently displayed, but we can change this any time.

To complete the model at the system level we need to specify the force generated by SE (Fig. 6.5). We open the corresponding port by double clicking. A corresponding dialog opens as shown in Fig. 6.7. We edit the port variable effort denoting it as '*F*'. Using the *Parameter* button we change *eValue* to 500 N (for editing the parameters see Sect. 4.8.2).

Before we close this subsection there are some suggestions regarding editing Bond Graphs (see also Sect. 4.6).

Note A component can be created anywhere inside the document working area, but not too close to its boundary. It is better to create it near its center and then move it. Similarly, a port cannot be created too close to another port. It is better to create it away from other ports, and then drag it. When drawing a bond, it is also possible to create intermediate points by clicking to it and then continuing. A bond line can be dragged to change its shape. Bond lines often have a zigzag appearance. There is no grid that can be used as guide. To create a nice looking bond graph, it is usually necessary to move components or their ports a bit before drawing the bonds. Drawing of a bond can be canceled by double-clicking!

Note also that the font sizes that the program supports can be

- Normal
- Small
- Medium, and
- Large

By default, the *Normal* size is pre-set for elementary components titles. Similarly, the *Medium* size is used for general word model components. In a similar vein, *Large* fonts can be used for some of the main word model components, and *Small* for some other components. This is not, of course obligatory; the user can choose the most appropriate size. To change the component font size it must be disconnected from the other components, and

- Select the component (by clicking its title);
- Expand the *Font Size* branch in *Editing Tools* and select the appropriate size;
- Click *Edit Text* toolbar button (in the form of a page with pencil);
- Edit the component title.

Note that to end editing a component title and create the component it is necessary to click outside of the component.

6.2.2.2   The Components Models

To complete the model development it is necessary to develop models of all of the word model components, and to modify the default models of the contained elementary components. We will use basically the same models as in Fig. 2.15.

To define the model of the Body component we double-click it. Because its model has not been defined previously a dialogue appears asking if we would like to create a new component model. We select *Yes*, and a new empty document window appears titled *Body* (Fig. 6.8). The document has four ports—three power ports and one control out port—corresponding to the component ports.

We will first insert a 1-junction, which serves to balance all forces acting on the body from the outside, i.e. the spring, damper and input forces. In addition we will insert an inertial component above the junction. We have to change the power sense of the left and right 1-junction ports, and add at the top of the junction a power out port for connecting to the inertial (I) component. Next we may drag the inertial port around the component so as to face the junction port we have just created. Finally, we will connect the junction ports by bonds to the corresponding document ports as shown in Fig. 6.8.

Because we need to output information about the position of the body, we will drag and drop a Control-out port at the corner of the 1-junction component. This

Fig. 6.8 Model of Body

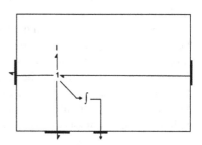

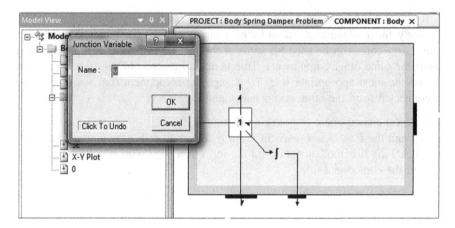

Fig. 6.9 Defining the junction variable

port extracts the body velocity. We may double click the 1-junction and in a dialogue that opens write v (velocity) as name of the junction flow variable (Fig. 6.9).

To obtain the position we need to integrate the velocity. Thus we will drag and drop an integrator from Editing box, *Continuous Signal Components* branch, as we did in Sect. 4.6.3 and Fig. 4.21. Then connect the input integrator port to the output 1-junction port and its output port to the output document port (Fig. 6.9). We may set the input integrator variable to v by double clicking its input port, in the same way as we did with the junction variable in Fig. 6.9. Similarly, we may set the integrator output by double clicking the integrator output port (Fig. 6.10).

We set the output variable to x, and the output relation now defines the output as the time integral of the input. There is also a field that defines the initial value of the output. By default it is set to zero, and we can accept this value.

In order to finish editing the Body model we have to specify the inertial element I. By clicking its port a corresponding dialog opens as shown in Fig. 6.11. We will change the flow variable to v, the state to p (momentum), and we also need to edit the constitutive relation as shown in Fig. 6.11. This relation defines the body momentum p as body mass m times its velocity v. Finally, using the *Parameter* button, we will set the body mass m to 5 (kg). Note also that there is a field which is used to define the initial value of the body momentum. By default it is set to zero. If the body is initially at rest we may accept this value.

In a similar way we define a model of the Spring (Fig. 6.12). The deformation of the spring depends on the relative velocities of its ends. Hence the Spring component can be described by a flow junction 0 whose two ports are connected to the document ports (spring ends); the third one is connected to the port of the capacitive component C that is used to model the process in the spring. To define the spring parameters we will double click the port of the C component, similarly as we did with the inertial component. We change the state variable to the spring deformation

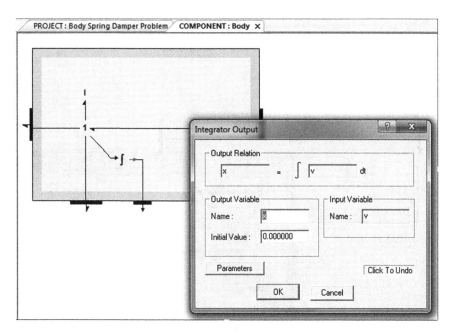

Fig. 6.10 The output integrator port dialog

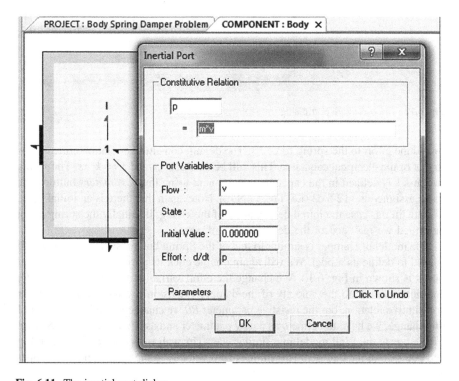

Fig. 6.11 The inertial port dialog

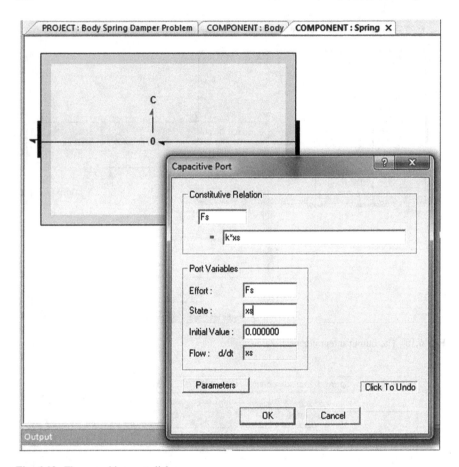

Fig. 6.12 The capacitive port dialog

xs and the effort to the spring force *Fs*. The default constitutive relation corresponds to that of the electrical capacitor. This will be changed to read *Fs = k*xs*. The spring stiffness *k* is defined in the capacitive component using the *Parameters* button. The value assigned is 112.5 e3 (i.e. 112.5 kN/m). Note again that there is an initial value of state, in this case the initial deformation of the spring. If initially the spring is not deformed we may accept the default value of zero.

The model of Damper is similar to that of the Spring but a resistive component R is used to define its model. We will again double click its port and the dialogue that opens is shown in Fig. 6.13. We change the effort variable to the damper force *Fd* and flow variable to the velocity of the damper extension *vd*. We accept the linear constitutive relation, but the resistive parameter *R0* we change to read *b*. Because of this change, we have to define *b* as a new parameter and assign a value 150 (N s/m).

We create the Wall model in a similar way. The wall fixes one side of the spring and one side of the damper. This implies zero velocity at the spring and the damper

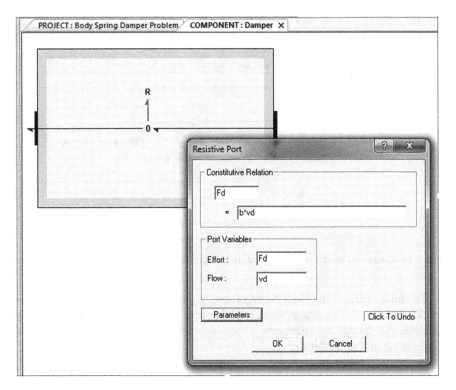

Fig. 6.13 The resistive port dialog

Fig. 6.14 Model of wall

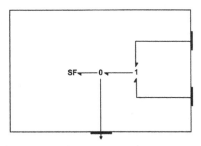

ends. Thus, in the Wall document (Fig. 6.14), we create an effort junction 1 connected to a zero-flow source SF. The other two junction ports are connected to the Wall document power ports and, thus to the spring and the damper ports (Fig. 6.5). The effort at the flow source port is equal to the sum of the forces at the other two 1-junction ports, i.e. it represents the force transmitted to the Wall. To extract this force a 0-junction is inserted between the SF and 1-junction with a control out port, which is connected to the corresponding document output port.

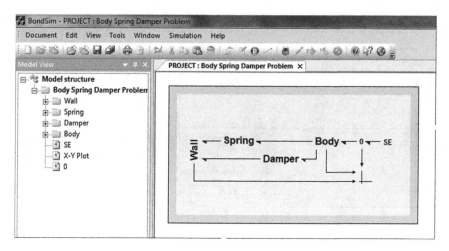

Fig. 6.15 Completed model of body spring damper problem

The final model of the Body Spring Damper problem is shown in Fig. 6.15. The model structure shows that all word model components are defined. We can easily examine the model by alternately expanding and collapsing the corresponding components in the model tree.

6.2.2.3 The Use of Mechanical Components

The Body Spring Damper problem of Fig. 6.1 can also be developed using mechanical components from the *Mechanical Components* library (Sect. 4.9.2). Note the components in this library contain the complete models of the component and are depicted using the mechanical component symbols. We can use these component models as they are or modify them.

Note that at the point of the connection the mechanical components have the common velocity. Hence, it is a 1-junction. Following the common practice in the mechanical circuits for such connections we use the *Connector* components. These are 1-junctions visually represented as dots. To create such a component we drag and drop the *Connector* tool, which we can find in the *Template Box* tab at the right of the main window, and inside the *Mechanical Components* (Fig. 4.23 right).

The procedure parallels that of the previous two subsections (Fig. 6.16):

- Launch *BondSim*, select the command *New* from the *Projects* menu, or by using the *New Project* toolbar button. In the dialogue that opens define a new project name, e.g. "*Body Spring Damper—Mechanical Model*", and then click the *OK* button.
- Select the *Models Library* from the left pane, and then expand the *Mechanical Components* branch and expand the *Wall Left*, pick and drag the *Ground*

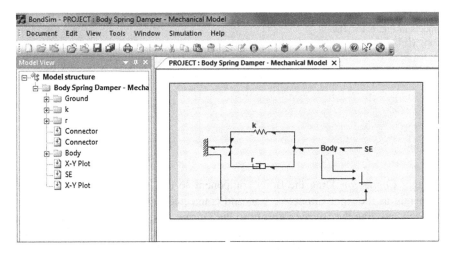

Fig. 6.16 Model of body spring damper problem using the mechanical components

component to the central document window and drop it near the left document working area border.

- Activate the *Template Box* tab, and then expand the *Mechanical Components*. Pick and drag the *Connector* to the main window and drop it to the right of the left wall component. Because we expect that power flows through the *Connector* to the wall, we will select its left port and change its power sense (using the toolbar button *Change Port*). Delete the right connect port (by selecting the port and pressing the *Delete* key). Pick and drag a *Power-in* port from the *Editing Box* tab, the *Port* branch, and drop it at the bottom of the *Connector* component. Move the component so that its left port is at the same height as the *Ground* component and connect the ports by a bond.

- Expand the *Spring* branch in the *Mechanical Components* library. Pick and drag the *k* component to the main window. Drag it to the right and somewhat above the *Connector*. Similarly, expand the *Damper* branch and pick and drag *r* component to the main window, and drag it to right and something bellow of the *Connector*. Arrange the spring and damper component one above the other and connect their left ports to the corresponding ports of the *Connector*.

- Pick and drag the *Connector* from the Template Box, Mechanical Components branch into the main window and drop it to the right of the spring and damper components, at the same height as the left *Connector*. Remove the left port, and change the power sense of the right and top ports. Pick and drag a *Power-out* port from the *Editing Box* tab, the *Port* branch, and drop it at the bottom of the *Connector* component. Connect the top and bottom ports to the corresponding spring and damper ports by bonds.

- Expand the *Mass* branch in the *Mechanical Components* library and pick and drag the *Body* component to the main window, and drag it to the right of the

Fig. 6.17 Component models in Fig. 6.16: The Body (*left*), Ground (*right*)

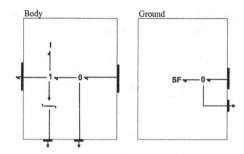

right *Connector*. Drag the Body component so that its left port is at the same height as the right *Connector* port and connected them by a bond.

- Pick and drag the *Source Effort* from the *Editing Box, Bond Graph Components* branch, into the project main window and drop it to the right of the *Body* component. Drag the port around the component so as to face the *Body* component. Position the SE so that its port is at the same height as the right *Body* port and connect them by a bond.

- Pick and drag the *X − Y Display* from the *Editing Box, Continuous Signal Components* branch, into the project main window and drop it at the right lower corner of the working area. Insert also a *Control-in* ports at top left of the component. Connect the left ports to the corresponding *Body* ports. Insert a *Control-out* port at the right boundary of the left wall component and connect it to the bottom port of the *X − Y* display component.

This completes the construction of the project system level model. As the model tree at the left indicates all word model components have predefined models. By expanding the *k* and *r* branches we can see that their models are the same as in Figs. 6.12 and 6.13; we will also double-click C and R ports to define the parameters. The *Body* component differs from the model in Fig. 6.15, because the 0-junction including the *Control-out* port from Fig. 6.15 is included in the component (Fig. 6.17 left). We update the contained elementary components in the same way as we did earlier (see Fig. 6.11). The model of the left wall component (*Ground*) is different also (see Figs. 6.14 and 6.17 right). Because we have already used the *Connector* in Fig. 6.16 the model in Fig. 6.17 right does not contain a 1-junction, only a 0-junction. We need to insert a *Control-out* port at the bottom of this junction and connect it by a signal line to the output document port.

6.2.3 Analysis of the System Behaviour by Simulation

In previous sections the bond graph model of the Body Spring Damper system was developed. The model closely resembles the real system. By running simulations under different conditions, we can observe the time behaviour of different variables in a way that is analogous to observing the behaviour of the real system via the outputs of the instruments. To analyse the dynamical behaviour of the system, we thus change the inputs and parameters, and study the effects.

6.2.3.1 Building the Model

We must first *build* the mathematical model that will be solved during the simulation runs. To build the model, the modelling project must be opened; and the *Build* button on the toolbar clicked, or the *Build Model* command on the *Simulation* menu selected. We analyse the *Body Spring Damper Problem* project developed in Sect. 6.2.2. The problem parameters that were already set during the model development are:

- The body mass $m = 5$ kg
- The spring stiffness $k = 112.5$ kN/m
- Linear damping coefficient $b = 150$ N s/m
- The input force $F = 500$ N

Denoting the natural frequency of the system as

$$\omega_n = \sqrt{\frac{k}{m}} \qquad (6.1)$$

we get $\omega_n = 150$ rad/s $= 23.87$ Hz. Similarly, the damping ratio

$$\zeta = \frac{b}{2\sqrt{m \cdot k}} \qquad (6.2)$$

has value of 0.1.

During the build operation the generated messages appear in the *Output* window at the bottom of the program screen (under *Build* tab). If during this operation an error was found the build operations is stopped and the corresponding error messages appear in the *Output* window. The component document where the error was detected opens and shows the error in the central part of the program screen.

To illustrate a typical error behavior we intently broke the bond in Fig. 6.11 between 1-junction and the left document port. We close the Body document and save it (to the project). Now we click the *Build* button. The program asks if we would like to save the project to the disk first. We may reject this and the build operation continues. However, there is an error, which is indicated by the messages in *Output* window. Also the program opens the Spring model document and selects the port where the error was found (Fig. 6.18).

Thus the message "Improper port connections" refers to this port. To find the error we may click the *Joined Port* toolbar button (or *Show Joined* command in the *View* menu). A new window opens as shown in Fig. 6.19. This document shows the Spring component enclosed in a bounding box, which indicates that the component

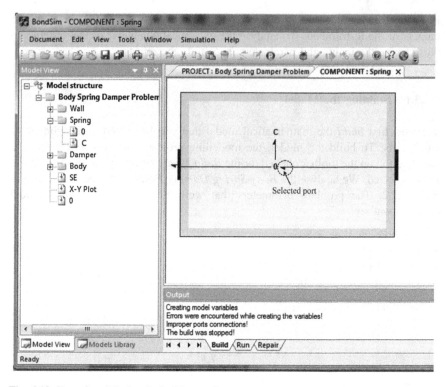

Fig. 6.18 Error found during the build operation

Fig. 6.19 The outlined port
indicating the broken
connection

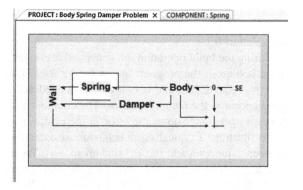

is open, and the bond going from this component to the Body component. It is
shown in blue colour. Note that the connected to port is not filled with the colour
but only its outline is shown. This indicates that the port is not connected inside the
component model. Of course we can correct this by connecting the bond we earlier
had broken!

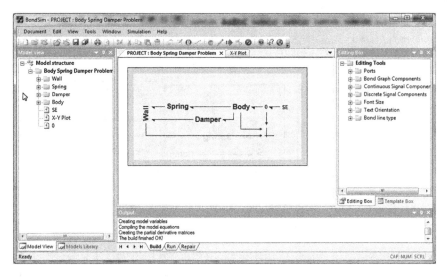

Fig. 6.20   The screen appearance after the successful build operation

Figure 6.20 shows the screen after the successful build operations. We see that all components documents are closed and the main document is now active. In addition there is also a $X - Y$ *Plot* tab, which serves to display the plots generated by the $X - Y$ *Display* component during the simulation.

After the model is built successfully we can continue with simulation. As already discussed in Sect. 5.4 we can examine the complete model built, or print it out. During the simulation, or after it, we can click $X - Y$ plot tab to view the plots generated during the simulation. There are the messages generated at the Output window at the bottom during the simulation.

6.2.3.2   Running Simulations

The model is created with zero initial conditions for all variables. We change only the value of the input force. The simulation of such a model corresponds to simulating the system step response to an applied force. To start, we click the *Run* toolbar button, or choose the *Run* command on the *Simulation* menu. The command displays a dialogue used to define the necessary simulation parameters, such as the duration of the simulation run, the increment at which output values will be generated, error tolerances etc. (Fig. 6.21).

The natural frequency of the system is about 24 Hz, hence the period is about 0.04 s. We can expect the transient to die out after about five periods, or 0.2 s. We use a somewhat greater simulation period, e.g. 0.5 s, to be sure that the transient has settled down. We set the simulation period to 0.5 s, and the output interval and the maximal step-size to one hundredth of this value, i.e. to 0.005 s. The maximum step-size should not be greater than the output interval. It is usually set to the same

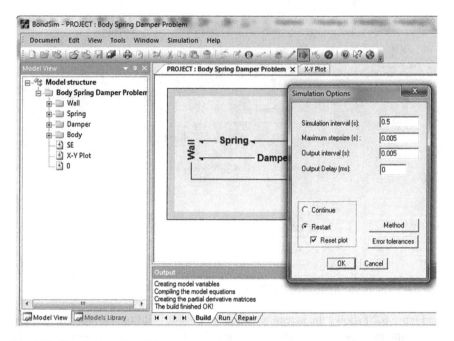

Fig. 6.21 Simulation option dialog

value as the output interval. We can choose a smaller value for the output interval if we wish to get better resolution and smoother diagrams. This means, however, that there are more calculations and the simulation run needs more computer time and memory resources.

Default error tolerances are set to 10^{-6} for both the absolute and the relative errors. The default integration method is the *BDF* method with a sparse matrix linear equation solver (Sect. 5.3). Accepting the simulation data by pressing the OK button, the simulation starts.

When the simulation starts, the messages in the *Output* window, under the *Run* tab, inform the user of the simulation steps completed. Simultaneously, at the right end of the status bar, a progress bar appears. This shows the progress of the simulation. When the simulation finishes, the progress bar disappears and a message informs the user of this fact. If the simulation time is short, as is in this case, the progress bar quickly disappears.

Note that in the central window there is already a tabbed window containing the $X - Y$ plot. We can activate this by clicking the tab. However, to clearly see the change in the body position with time, because of different scales, we need to deactivate the input and wall forces (by double clicking the corresponding display ports and setting the axes to *None*, Fig. 6.6). The resulting plot is shown in Fig. 6.22.

We can examine the values of the variables by clicking inside the plot. Horizontal and vertical lines appear which intersect on the curve. The corresponding values of the variables appear below the plot.

Fig. 6.22 The response of the body position to a step in the applied force

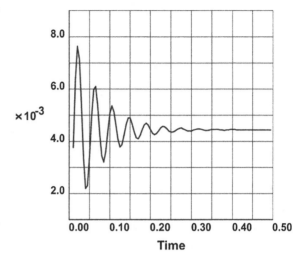

It is also possible to get a list of output values by clicking on the plot with the *right* mouse button. A drop-down menu appears from which we can choose the *Show data* command. This command creates a list box containing values of the variables. By scrolling through the list box we can examine values of variable pairs generated by the simulation. It should be noted that the number of value pairs is limited. The values listed are symmetrical with respect to the current value selected by the mouse. If there is large number of output values, the values throughout the complete range can be examined by a combination of mouse clicking and the use of the list box.

In the present example, we can compare the values obtained by simulation with the exact solution of the problem under study. The equation of motion of the Body Spring Damper system can be written as

$$m\frac{d^2x}{dt^2} + b\frac{dx}{dt} + kx = F \tag{6.3}$$

The response to a step F of the force can be written as (see e.g. [1])

$$x = x_0\left[1 - \frac{e^{-\zeta\omega_n t}}{\sqrt{1-\zeta^2}}\cos\left(\omega_n\sqrt{1-\zeta^2}t - \varphi\right)\right] \tag{6.4}$$

where $x_0 = F/k$ and

$$\tan\varphi = \frac{\zeta}{\sqrt{1-\zeta^2}} \tag{6.5}$$

Table 6.1 Response of the body position to a step in the applied force

Time, s	Simulation	(6.4)
0.01	0.003761887676	0.003761880557
0.05	0.003448204502	0.003448187707
0.1	0.005076850687	0.005076930584
0.2	0.004465699802	0.004465689113
0.3	0.004406268287	0.004406235937
0.4	0.004455428889	0.004455470625
0.495	0.004444573861	0.004444576945

Table 6.2 The simulation statistics

Name	Value
The order of the method at the last step	5
The number of the steps taken	431
The number of the function evaluations	901
The number of the partial derivative matrix evaluations	468
The relative error tolerance	10^{-6}
The absolute error tolerance	10^{-6}
The total elapsed time s	0.03

Table 6.1 compares values obtained by simulation and by direct calculation according to (6.4) and (6.5). The values are rounded to ten figures. The values obtained agree to six digits, which corresponds to the set error tolerances (10^{-6}). A better agreement can be obtained by lowering the error tolerances.

We also look at the simulation statistics using the *Show statistics* command on the *Simulation* menu. The results are summarized in Table 6.2.[3] The simulation was relatively efficient using the maximal order of the method (5) at the end of the simulation.

6.2.3.3   Response to Harmonic Excitation

The study of mechanical vibration pays great attention to the influence of forces that vary harmonically (see e.g. [1]). We can study such influence by defining a force at the power port of the force component (Fig. 6.7) as

$$F = F0^* \sin(omega^* t) \tag{6.6}$$

---

[3]The simulation was conducted on a laptop with an Intel Core i7 CPU 2.0 GHz processor, installed memory (RAM) 16 GB. The elapsed time is for the simulation runing in the backgound.

The amplitude *F0* = 500 N and frequency *omega* of the applied force are parameters, which can be set to any appropriate value. We study the effects of the harmonic force on the system by changing its frequency.

The simulation interval is chosen as 1.0 s and the output interval is 0.001 s for better resolution. Some typical results are given in Figs. 6.23, 6.24, 6.25 and 6.26.

In vibrations as well as in other fields an important role in design is played by amplitude-frequency and phase-frequency response diagrams. Determination of such diagrams is not an easy task particularly for non-linear systems (see [2] for

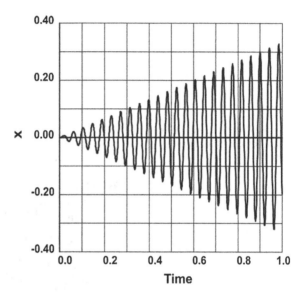

Fig. 6.23 Resonant response of the undamped system (ω_n = 150 rad/s, ζ = 0)

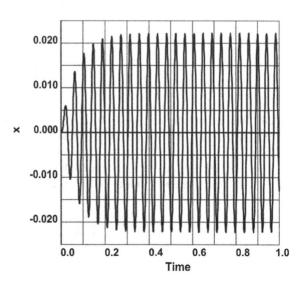

Fig. 6.24 Resonant response of the lightly damped system (ω_n = 150 rad/s, ζ = 0.1)

Fig. 6.25 Force at the wall at frequency of 212.1 rad/s ($\sqrt{2}\omega_n$, $\zeta = 0.1$)

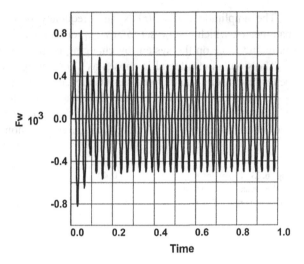

Fig. 6.26 Force at the wall at frequency of 300 rad/s and, $\zeta = 0.1$

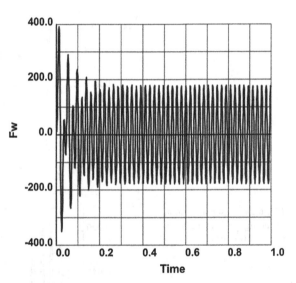

detailed discussion). In the case of linear systems such as the *Body Spring Damper* system these diagrams can be found from the Fourier transform of the impulse response of the system [1–3].

The impulse response of the system can be approximated by the time response to a very short force pulse of unit area (Fig. 6.27). Such a pulse can be described as

$$F = \begin{cases} F_p, & t < T_p \\ 0, & t \geq T_p \end{cases} \qquad (6.7)$$

Fig. 6.27 A short pulse of unit area as an approximation of the unit impulse

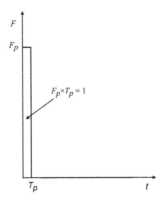

Using the question operator of Sect. 3.5, the relationship for the pulse of force can be expressed as

$$F = t < Tp ? Fp : 0 \tag{6.8}$$

The pulse duration parameter Tp should be very short and the strength Fp such that their product is equal to one. The duration of the pulse we take as 0.0001 s, and the pulse strength equal to 1000 N.

Let the response to such a pulse be denoted as $h(t)$. Then the frequency response of the system is given by the Fourier transform

$$H(j\omega) = \int\limits_{-\infty}^{\infty} h(t)e^{-j\omega t}dt \tag{6.9}$$

We evaluate the Fourier transform using the Fast Fourier Transform (FFT) method [3] and find the amplitudes and phase as functions of frequencies in Hz.

The FFT treats the function as periodic. To approximate an aperiodic function, we can extend the function values by adding zeroes [2]. As more zeroes are added, better the resolution of the frequency response is obtained. However as shown in [3], it is also important that the function is zero-valued over the interval where the zeroes are added. What we will really do is to integrate the system well beyond time when the transients practically died out. Because we expect that this happens well before 0.5 s (see Fig. 6.22) we will select the simulation interval of 1 s, and take the time window also equal to 1 s. This gives a frequency resolution of 1 Hz.

To cover a large enough frequency range we chose the output interval of 10^{-5} s, which is one tenth of the pulse width. This gives a Nyquist frequency of $10^5/2 = 5 \times 10^4$ Hz. As is well known the Nyquist frequency must be higher than the highest frequency of interest in the problem. However, to achieve good accuracies of the plots, in particularly the phase plot, we choose a relatively short output interval. The impulse response of the body position is given in Fig. 6.28.

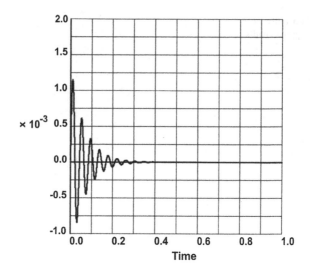

Fig. 6.28 The impulse response of the body position

To create the corresponding frequency response, we click using the right mouse button on the plot and select from the drop-down menu the *Continuous Fourier Transform* command. A new magnitude-frequency plot window opens together with dialogue which is used to define what part of time response plot would be selected (Fig. 6.29). We can accept the offered parameters which cover the

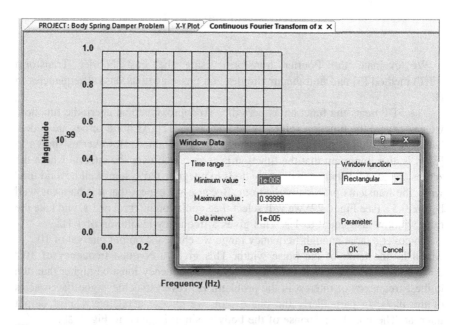

Fig. 6.29 Time window dialog

complete time range and simple rectangular window function. Thus the all generated time data are passed to the FFT routine to generate a continuous Fourier
transform in the form of a collection of amplitude- and phase-frequency data. The
plot that program generates is magnitude-frequency plot. Because it uses the linear
scales we will convert the magnitudes into the ratios of the magnitudes and a
convenient constant, which is expressed in decibels (dB). Right clicking the mouse
over the plots we select the *Amplitude Ratio in dB* from the drop-down menu. The
three subcommands appear:

- To DC Value
- To Maximum Value
- To Optional Value

We will choose the last subcommand and as the optional value we simply use
the value of 1.0. The new plot is generated with magnitude ratios in dB plotted
along the ordinate axis. We change the frequency scale to logarithmic clicking by
the right mouse key the plot and selecting from the drop-down menu the command
Logarithmic Frequency. We will select only a part of this diagram from frequency 0
to 100 Hz. We can achieve this by right clicking the mouse and selecting the
command *Expand*. Next we will position the mouse cursor slightly to the left of the
magnitude axis and press the left mouse button, drag the cursor to the vertical line
slightly left of 100 Hz, and then release the mouse button. Resulting plot is shown
in Fig. 6.30.

In a similar way we may obtain the phase response. We will return to the first
frequency plot, we right click the button and select *Show Phase*. On the phase

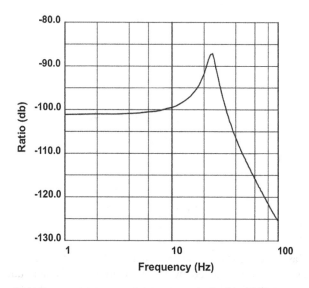

Fig. 6.30 Magnitude-frequency response of the system obtained by FFT

Fig. 6.31 Phase-frequency
response of the system
obtained by FFT

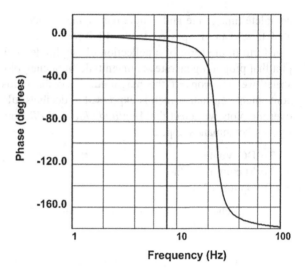

frequency plot we right click the mouse and select *Unwrap Phase*, and finally
expand part of the plot from 0 to 100 Hz. The resulting phase frequency response is
shown in Fig. 6.31.

We again compare simulation values with the exact solution. Recall from (6.3)
that the amplitude of the frequency response to the unit impulse force is given by
(see e.g. [1])

$$|H(j\omega)| = \frac{1/m}{\sqrt{\left(\omega_n^2 - \omega^2\right)^2 + 4\zeta^2\omega_n^2\omega^2}}$$

After normalization of the frequencies, we get

$$|H(j\omega)| = \frac{1/k}{\sqrt{\left(1 - \left(\frac{\omega}{\omega_n}\right)^2\right)^2 + 4\zeta^2\left(\frac{\omega}{\omega_n}\right)^2}} \tag{6.10}$$

Similarly for the phase we find

$$\tan\varphi = -\frac{2\zeta\left(\frac{\omega}{\omega_n}\right)}{1 - \left(\frac{\omega}{\omega_n}\right)^2} \tag{6.11}$$

Using the last expressions, we can calculate the amplitude and phase at the same
frequencies as that were obtained by the simulation. The magnitudes and phase
angles obtained by FFT and by (6.10) and (6.11), rounded to eight figures, are given
in Table 6.3 and show good agreement.

Table 6.3 Comparison of magnitudes and phase angles calculated analytically and by FFT

Frequency [Hz]	Magnitude [dB] by (6.10)	Magnitude [dB] by FFT	Phase angle [°] by (6.11)	Phase angle [°] by FFT
1.0002	−101.00810	−101.00810	−0.4809289	−0.49164163
10.0002	−99.391821	−99.391836	−5.8016807	−5.9096765
20.0004	−91.703357	−91.703415	−29.336570	−29.552562
24.00048	−87.102075	−87.102158	−93.042766	−93.301957
30.0006	−97.028864	−97.028993	−156.54251	−156.86650
40.0008	−106.31123	−106.31146	−169.49645	−169.92843
60.0012	−115.57470	−115.57521	−174.59915	−175.24713
80.0016	−121.23905	−121.23996	−176.25157	−177.11553
99.00198	−125.22338	−125.22476	−177.06871	−178.13787

Note that the Fourier transform of the pulse input is not equal to one, as was assumed in (6.10) and (6.11). The magnitudes are given by well-known *sinc* function [3]

$$|F(j\omega)| = \left| \frac{\sin(\omega T_p/2)}{\omega T_p/2} \right| \tag{6.12}$$

and the phase angle by

$$\angle F(j\omega) = -\omega T_p/2, |\omega| < 2\pi/T_p \tag{6.13}$$

and is restricted to $(-\pi, \pi)$ by the common convention. Thus, we find that in the frequency range $0 \leq f < 100$ Hz the magnitudes are between 1.0 and

$$\left| \frac{\sin(2\pi \cdot 100 \cdot 0.0001/2)}{2\pi \cdot 100 \cdot 0.0001/2} \right| = 0.9998$$

and the phases between zero and

$$0.01 \cdot \pi \text{ rad} = 1.8°$$

These figures explain why we obtain such a good approximation of the Fourier transformation using the pulse input and FFT. If the pulse width is much larger or much higher frequencies are considered the expected approximation would be worse, in particularly the phase.

We can also find the Fourier transform of the input pulse. We will repeat the simulation using e.g. a simulation interval of 0.05 s and the output interval of 10^{-6}. Applying the *Continuous Fourier Transform* on the plot of the input force and expanding the first part of the magnitude plot up to 5×10^4 Hz we obtain the diagram shown in Fig. 6.32. Clearly the absolute value of *sinc* function is clearly seen. The value of the magnitude at zero frequency is 0.988197; theoretically it is 1.0.

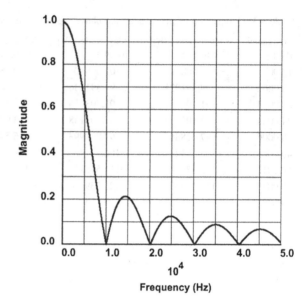

Fig. 6.32 Magnitude of Fourier transform of the input force pulse

6.3 Effect of Dry Friction

We continue with the study of vibration systems by analysing the influence of *dry friction*. This is usually treated by studying the motion in the positive and negative directions separately, and the conditions under which the motion ceases [1]. We will develop an integral model of dry friction that can be used for prediction of such a motion. The model can be used for the analysis of more general systems in which this type of friction is important.

6.3.1 The Model of Dry Friction

So-called *dry friction* occurs when one solid slides over another. The laws governing such friction date back to Leonardo da Vinci (1452–1519), but are better known from the work of Coulomb in 1785 (Fig. 6.33). A modern exposition of the theory is given in [4]. In the bond graph literature there were also attempts to model dry friction. Thus in [5], the friction around zero velocity is modelled as a force dependent on motion and, outside of this region as a function of velocity only. In [6] discontinuous laws of friction were modelled by sinks of fixed causalities. The modelling of friction was analysed also in [7], motivated by the physical theory of [4]. The model proposed consists of capacitors and resistors interconnected by transformers, whose ratios change smoothly from one to zero, depending on the state of the body motion.

Fig. 6.33 The dry friction
law

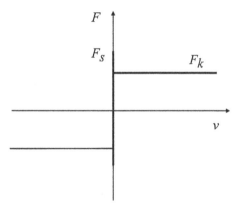

Fig. 6.34 Solid joining of the
bodies at the contact points

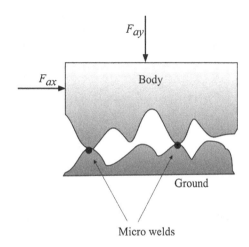

We also use the theory of [4], but friction will be modelled by a single element that imposes restrictions on the body motion. A suitable element for this is the switch element (Sect. 2.5.8), as it is capable of imposing the zero flow (sliding velocity) condition before motion commences, and also when the motion ceases. It also accounts for constant—or possibly variable—effort (friction force) during the motion.

The contact area between bodies generally is not smooth, but is actually rough Fig. 6.34. Thus, when a body is pressed onto another, the real contact starts at the tips of the highest asperities. These deform until the area of contact is large enough to support the load without yielding. Owing to high pressures and temperatures at the contacts, solid junctions are created at the contact places known as *micro welds*.

To slide one body over another body (ground), it is necessary to apply a force in the direction of the sliding that is large enough to break the micro welds. Until the

Fig. 6.35 The forces acting
on the body

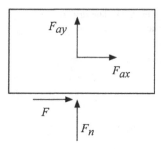

motion commences, the body is in static equilibrium under the action of the applied
forces, including the other body's reaction force (Fig. 6.35).

F_{ax} and F_{ay} are components of the applied force. The component of the reaction
force F_n that is normal to the sliding direction is the usual *normal reaction*, and the
component in the direction of sliding is the *friction force*. It is denoted simply as
F. Thus, the conditions of equilibrium can be stated as

$$\left. \begin{aligned} F_{ax} + F &= 0 \\ F_{ay} + F_n &= 0 \end{aligned} \right\} \tag{6.14}$$

and

$$\left. \begin{aligned} v_x &= 0 \\ v_y &= 0 \end{aligned} \right\} \tag{6.15}$$

where v_x and v_y are the velocity components of the body relative to the other body
(ground).

The limiting force value at which the welds break and hence the sliding com-
mence is termed the *static friction F_s*. Thus (6.14) is satisfied if the friction force
satisfies the condition

$$|F| \leq F_s \tag{6.16}$$

This limiting value, corresponding to the strength of the micro welds, depends
also on the time that the bodies are in the contact. When the bodies are in contact
there is migration of particles over the junction by diffusion, and this needs some
time. If there is enough time, intimate junctions are established and the two bodies
behave as single body. In this way, the static force increases to a maximum value
that the micro junctions can sustain. When we speak of static friction, we mean this
full strength value. The Coulomb law relates this limiting force to the normal
reaction by

$$F_s = \mu_s F_n \tag{6.17}$$

The μ_s is the well-known static friction coefficient.

When motion starts, there is deformation of the asperities that are micro welded until they break. Perhaps before the complete breakdown the other asperities come into contact and the new micro welds form. They, in turn, break and yet other micro welds form and the process continues as the body slides. During the motion there is not enough time for migration across junctions, hence the force of asperities breaking during the body motion is somewhat less than the static friction value. According to experimental evidence, this force is more or less independent of the sliding velocity and has the sense opposite to the velocity. During sliding, friction satisfies the relationship

$$\left.\begin{array}{ll} F + \mu_k F_n = 0, & v > 0 \\ F - \mu_k F_n = 0, & v < 0 \end{array}\right\} \tag{6.18}$$

Here the μ_k is the *kinetic coefficient of friction*, which is somewhat less than the static coefficient, typically about 20–25 %.

To complete the model of friction at this stage, we need to define the state at the very moment when the motion commences, which ideally occurs at zero velocity (Fig. 6.33). This is not covered by (6.18) and hence we add this condition to (6.16). The friction characteristics valid from no sliding until the sliding commences read now as

$$\left.\begin{array}{l} v = 0, |F| \le \mu_s F \\ F - \mu_k F_n = 0, F > \mu_s F_n \\ F + \mu_k F_n = 0, F < \mu_s F_n \end{array}\right\} (v = 0) \tag{6.19}$$

This equation is compatible with (6.18) that is valid for the sliding. The (6.18) and (6.19) are constrains on the body motion imposed by the mechanism of dry friction. We have not tried to find an explicit relationship for the friction force, as this is difficult to find in the general case. We don't know if the force at the contact is known or not, and similarly for the velocity. Hence, the effect of friction is represented by a simple linear implicit equation, different for the body sliding and not sliding.

We dwell a little more on (6.19). To find out whether it is applicable for a given state of the body motion, we need to test the sliding velocity against the zero value. In computer arithmetic it is, in general impossible to find out the exact moment when some variable attains a definite value. Thus, some tolerance on the velocity around zero is necessary. In parallel to the fact that physically there is no such thing as "rest". The theory of [4] proves that the static friction behaviour holds also when the sliding is very slow. Thus, we reformulate the above relations to be valid if |

$v| < tol$. The *tol* represents a tolerance, which we take to be much smaller than the error tolerance used for simulating motion and close to the machine *epsilon*.[4] In the BondSim there is a predefined parameter BG_EPSILON, defined as four times the machine epsilon (Table 4.1). This can be used if such a low tolerance is needed.

The relationships given by (6.18) and (6.19) can be represented in the form that is used when describing the element constitutive relations (Sect. 3.5). The relations can be compactly expressed using the question ('?:') operator. The relationship reads

$$abs(v) < tol?(abs(F) < = mus * F_n?v : (F > = 0?F - muk * F_n : \\ F + muk * F_n)) : (v > = tol?F + muk * F_n : F - muk * F_n) = 0 \qquad (6.20)$$

In the equation instead of the Greek symbol μ, *mu* is used, and also the indices are written on the same line. The program supports only Latin characters, without any formatting. The statement can be understood as:

> if (the absolute value of v is less than the tolerance) then
>> if (the absolute value of F is less than or equal to the static friction) then
>>> $v = 0$
>> else if (F is greater than or equal to zero) then
>>> F- *muk*F*n* = 0
>> else
>>> F+*muk*F*n* = 0
> else if (v is greater than or equal to the tolerance) then
>> F+*muk*F*n* = 0
> else
>> F-*muk*F*n* = 0

The left side of (6.20) looks like a C language statement. It states in a compact form the conditions imposed by dry friction on the rest of the system. The mechanism of friction is not a trivial one. There are five possible states through which the system goes during its motion. Unfortunately, this is not the end of the story. We have to define also the conditions for motion stopping. Otherwise, such a model is not of much use in simulations.

In books on vibrations such as [1, 8], the motion stops when the amplitude of the spring force acting on the body is less than the static friction. This is not a precise enough statement. According to [4, 7], when the body stops, friction does not achieve the static value again, but at most the kinetic value. To find a more precise statement of the stopping condition, we need to express it in terms of the variables at the interfaces of the two bodies, i.e. the force and the velocity at contact.

---

[4]By definition the machine epsilon is the smallest positive number ε such that $1 + \varepsilon \neq 1$ in machine arithmetic and for the given floating-point number representation. It is equal about 2.2×10^{-16} for double precision numbers.

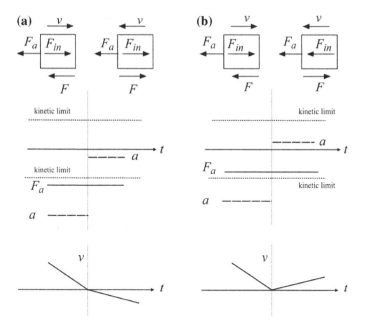

Fig. 6.36 Effects of velocity sign change when applied force is **a** outside or **b** inside the friction limits

We look more carefully at what happens when the body velocity changes its sign. During the motion there is dynamic equilibrium of the forces acting on the body, including the body inertia force. Suppose that the active force is acting in a manner that reverses the body motion. When the velocity changes its sign the sense of the friction force changes too. Hence, the body acceleration changes abruptly. If the active force on the body is outside the kinetic friction limit, the acceleration will not change its sign and hence the body will continue to move in the opposite direction (Fig. 6.36a).

If, on the other hand, the active force is within the kinetic friction limit, the body acceleration will change its sign and the body will try to move back (Fig. 6.36b). This, in effect, means that the friction force tries to push the body back, which is impossible because dry friction is not capable of delivering positive power to the body. The dynamic equation of motion and the characteristics of dry friction as given by (6.19) thus are incompatible. Looking from the viewpoint of the asperities, there is no abrupt change in the friction force. However, owing to the asperities stiffness and internal damping, the body first slows down and stops. It continues to move in the opposite direction only if the absolute value of the total active force on the body is greater than the asperities breaking force, i.e. the kinetic friction.

Thus, to solve this problem we apply additional requirements: if the power delivered by dry friction is *positive* then the sliding velocity should be equal to zero. This is stated as follows:

Fig. 6.37 The ground
component with dry friction
(a) and its model (b)

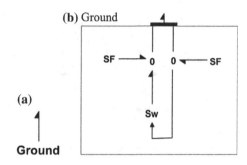

(b) Ground

SF → 0   0 ← SF

Sw

(a)

Ground

$$F \cdot v > 0 \Rightarrow v = 0 \tag{6.21}$$

Adding this requirement, the final form of the dry friction constitutive relation reads (see 6.20)

$$abs(v) < tol?(abs(F) < = mus * Fn?v : (F > = 0?F - muk * Fn : F + muk * Fn)) :$$
$$(F * v < = 0?(v > = 0?F + muk * Fn : F - muk * Fn) : v) \tag{6.22}$$

The model of a ground body that acts on another body by the mechanism of dry friction can be represented by Ground component as shown in Fig. 6.37a. Its model is shown in Fig. 6.37b. To model dry friction we use a Switch element with a constitutive relation given by (6.22). There are two branches in the model.

The left branch corresponds to the interactions in the sliding direction, and the other describes the interactions in the normal direction (Fig. 3.35). The 0-junction on the left simply states that the body velocity (the upper port) is equal to the sum of the ground body velocity—defined by the flow source SF on the left—and the body sliding velocity. The flow junction variable is the friction force. Positive power flow at the port corresponds to the force and velocity components taken in the direction of the coordinate axis. This way, power transfer from the ground by the mechanism of dry friction is positive if the friction is positive. Clearly the (6.18) shows that it is negative, and hence the real power transfer is negative. The other flow junction asserts that the velocities of the bodies in the normal direction are equal. The junction variable is the normal reaction, which is fed back to the control port of the switch element supplying it with the necessary information.

Based on this model we can define a component representing the dry friction interactions between the sliding bodies (Fig. 6.38a). The component implements the model represented by the central part of Fig. 6.37 (without the flow sources), as shown in Fig. 6.38b.

We will now test the model on various problems in which the friction is important. We start with vibrations of the single degree of freedom system of Sec. 6.2, replacing the linear friction by dry friction. We then analyse some problems in which *stick-slip* appears.

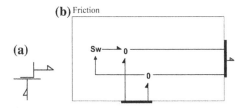

Fig. 6.38 The component representing dry friction (**a**) and its model (**b**)

6.3.2   Free Vibration of a Body with Dry Friction

We return to the problem of Sect. 6.2.1 and analyse the free motion of the body under the influence of dry friction at the contact with the ground (Fig. 6.1). The corresponding bond graph model is given in Fig. 6.39.

The model is similar to that of Fig. 6.15. The main difference lies in replacing the linear mechanical damper with the model of the ground in Fig. 6.37. This includes the effect of dry friction, but the external force is removed. We are no longer interested in the wall force, and thus the wall is represented simply by a zero flow source. The model of the body is somewhat more complex than that in Fig. 6.11 because we need a model of the body motion in two dimensions, i.e. along the ground and normal to the ground (Fig. 6.40).

The left effort junction of Fig. 6.40 describes the translation of the body along the ground surface. Similarly the other junction represents the summation of the forces in the normal direction. The weight of the body is represented by the source effort SE. The I elements represent the inertia of the body. To monitor the motion of the body, the velocity (v) is taken from the junction and fed out to the corresponding display port (Fig. 6.39). The position x is obtained by integrating the v (velocity) and is fed out and connected to the other display port.

To compare the behaviour of the two models, the basic parameters used are the same. Thus, the parameters of the model are taken as

- The body mass $m = 5$ kg
- The spring stiffness $k = 112.5$ kN/m
- The coefficients friction: static $\mu_s = 0.5$, kinetic $\mu_k = 0.4$
- Gravitational acceleration $g = 9.80665$ m/s^2 (Table 4.1)

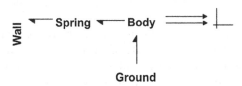

Fig. 6.39 Free motion of a body under dry friction

Fig. 6.40 Model of the body motion in two dimensions

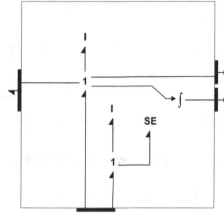

Fig. 6.41 The transient of the body position under the action of dry friction

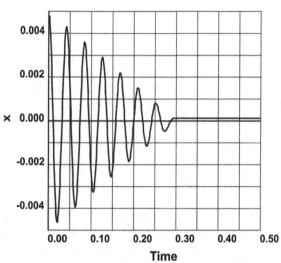

The initial displacement of the body is 5.0×10^{-3} m and the initial velocity is 0. The complete model can be found in the project library under the name *Body Motion with Dry Friction*.

The model is more complex than in Sect. 6.2; it consists of seventeen equations of index 2 type (Chap. 5). During the simulation the structure of the model frequently changes, which could pose a problem for the solver. The method described in Chap. 5, however, copes with this rather well. The simulation was run for 0.5 s with an output interval of 0.005 s.

Results (Fig. 6.41) show that amplitudes of the body position decrease linearly until the body settles down after about 0.3 s. The period of vibrations is equal to $2\pi/\omega_n$, where ω_n is the natural frequency of vibration. The theoretical value of the decrease in amplitude per cycle is equal to 4Δ, where $\Delta = \mu mg/k$, is based on the same value for both friction coefficients [1].

Table 6.4 Simulation statistics for body with dry friction

Name	Value
The order of the method on the last step	1
The number of the steps taken	965
The number of the function evaluations	2340
The number of the partial derivative matrix evaluations	1477
The relative error tolerance	10^{-6}
The absolute error tolerance	10^{-6}
The elapsed time s	0.04

Using the parameters given above the value of the amplitude drop obtained by this formula is 0.0006976 m/cycle. By simulation the amplitude drop per cycle of 0.000699 m is obtained. The body settles down at a position of 0.000114846 m, which corresponds to a spring force of 12.92 N, or 66 % of the kinetic friction. The simulation statistics are given in Table 6.4 (for the output interval 0.005 s).

In comparison with Table 6.2 there are apparently many more integration steps, and the function and the matrix evaluations. However, the simulation takes nearly the same time. Thus, despite of a more complex mathematical model, the simulation was conducted successfully and efficiently.

6.3.3 Stick-Slip Motion

In the previous problem the static friction is important only at the beginning of the motion. Later, when oscillations begin and finally settle down, the kinetic friction is the main influence. This and similar behaviour in engineering equipment often leads to neglecting the difference between static and kinetic friction and thus simplifying the model of dry friction (see e.g. [8]). This is not the case, however, in processes in which intermittent (*stick-slip*) motion appears as a consequence of the difference between these two friction coefficients. This is analysed experimentally and theoretically in detail in the classical reference on friction [4].

The stick-slip motion occurs between the sliding surfaces of two bodies in contact when one body is driven with constant, a fairly low velocity, and the other has a certain degree of elasticity. This is the case with the bodies in Fig. 6.39 if the ground body is driven with a constant and very low velocity and the other body is initially at rest. At the beginning when the spring force is less than the static friction, the ground drags the body, thereby increasing the tension in the spring. This is the "stick" phase. It lasts until the tension in the spring reaches a value at which the spring force overcomes the static friction. At that moment, friction between the body and the ground drops to the kinetic friction value, which generally is lower. Hence, the body under the action of the spring slips quickly back over the ground. The "slip" phase lasts until the two velocities again are equal. A new "stick" phase then begins.

Intermittent motion can be a problem in applications in which smooth motion is important. As discussed in [4], if there is some damping in the system, intermittent motion may not occur at all, even if there is a finite difference between static and kinetic friction. Nevertheless, we will show the behaviour of the system of Fig. 6.39 under very slow ground motion conditions.

In the problem of Fig. 6.39, we assume the ground is driven at a constant velocity $V_0 = 0.001$ m/s and the body is initially at rest. Thus in the left flow source of the Ground component (see Fig. 6.37) we change the value of the flow source velocity to 0.001. Also, in the capacitive element of the Spring (see Fig. 6.12), we set the initial displacement to zero. In the inertial elements of the Body model (Fig. 6.40), we retain zero initial conditions, but change the initial position in the integrator to zero. We build the simulation model and run the simulation for 0.5 s with an output interval of 0.001 s, accepting the default values for the other simulation parameters. The results are given in Figs. 6.42 and 6.43.

Figure 6.42 shows the position of the body and Fig. 6.43 its velocity with time. The body sticks to the ground body until the spring tension is greater than the static friction. This occurs 0.218 s after the start of the motion. The body then detaches from the ground and, at first, continues to move in the same direction. Shortly after, at about 0.219 s, its velocity drops to zero and it then moves in the opposite direction. Reattachment occurs at 0.241 s, when its velocity catches up with the velocity of the ground.

We can easily find the displacement of the body during the slip. Let y_1 be the position of the body at the start, and y_2 at end, of slip. During slip the friction force is equal to kinetic friction F_k. Applying the law of kinetic energy change during the slip, we easily find the relationship

Fig. 6.42 The simulation of the stick-slip motion of the body

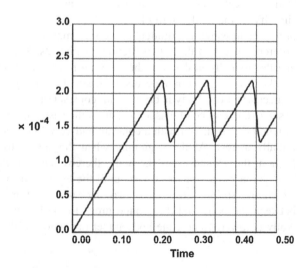

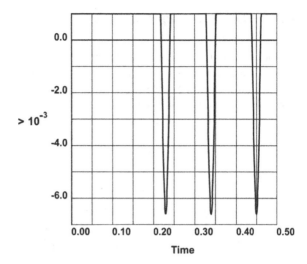

Fig. 6.43 Change of body velocity during stick-slip motion

$$\frac{1}{2}ky_1^2 - \frac{1}{2}ky_2^2 = F_k(y_1 - y_2)$$

or

$$y_1 + y_2 = 2F_k/k \tag{6.23}$$

On the other hand, at the start of the slip, we have $y_1 = F_s/k$. Thus, we obtain

$$y_2 = 2F_k/k - y_1 = (2F_k - F_s)/k$$

and the displacement during the slip is given by

$$s = y_1 - y_2 = 2(F_s - F_k)/k \tag{6.24}$$

Taking into account that $F_s = \mu_s \cdot mg$ and $F_k = \mu_k \cdot mg$ we get

$$s = 2(\mu_s - \mu_k)mg/k \tag{6.25}$$

This shows that slip is possible if there is a difference between the static and kinetic coefficients of friction. Using the values of the parameters given earlier (Sect. 6.3.2) we find $s = 0.872 \times 10^{-4}$ m. From the simulation we find that slip in Fig. 6.42 is 0.875984×10^{-4} m. Thus, the value obtained is somewhat greater, which can be explained by the fact that detachment and reattachment of the body to the ground occurs shortly before and after the zero velocity positions (Fig. 6.43).

6.3.4   The Stick-Slip Oscillator

As the next example, we simulate the behaviour of the stick-slip oscillator described in [9]. The system that we analyse consists of a body which has one elastic degree of freedom and which can slide over the ground under the action of an external force (Fig. 6.44). On the body there is placed another body, which can also slide over the first under the action of a force.

The forces change with time harmonically, i.e.

$$F_i = F_{i0} \cos(\omega_i t + \varphi_i), (i = 1, 2) \tag{6.26}$$

We assume that dry friction exists at both slipping surfaces.

A detailed analysis of the possible states of sticking or slipping, along with phase portraits obtained by numerical integration, is described in [9]. We apply here the more direct approach of bond graph modelling and simulation, as in the previous problems.

The bond graph model of the system in Fig. 6.44 can be developed easily using the model of dry friction developed in Sect. 6.3.1. We develop the model using the mechanical schematics. Dry friction is modelled using the corresponding component from the library (Fig. 6.38). The complete model is shown in Fig. 6.45. The main components are similar to the model of Fig. 6.39. Components F1 and F2 are source efforts described by (6.26). The component Body1 differs from Body2 by having four ports, two for connecting to the spring and the external force source and two for the interactions with the ground and the upper Body2. There are also several signals used for monitoring. Thus, from the first body the inertial force, velocity and position in the sliding direction are taken out and connected to the X − Y displays shown on the right. Similarly, the velocity and position signals of the upper body motions are taken from the Body2 component and connected to the third display. The details of the model can be found in the library under the project name *Stick-Slip Oscillator*.

The parameters of the model are as in [9]:

- The body masses are $m_1 = m_2 = 1$ kg
- The spring stiffness $k = 150$ N/m
- The coefficients of friction $\mu_s = \mu_k = 1$ (the same for both surfaces)

Fig. 6.44 The stick-slip oscillator

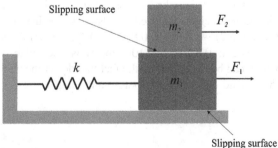

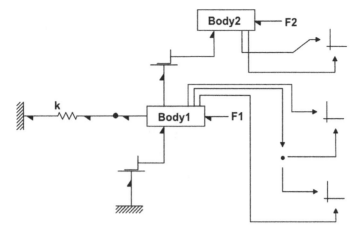

Fig. 6.45 The bond graph model of the stick-slip oscillator

- The driver amplitudes $F_{01} = 60$ N and $F_{02} = 60$ N
- Driver frequencies $\omega_1 = \omega_2 = 2\pi$ rad/s
- The drive phase angles $\varphi_1 = 0$ and $\varphi_2 = \pi$ rad
- The acceleration due to gravity $g = 10$ m/s^2 (different than in Table 4.1)

Note that gravity is different from the value defined at the level of the program. Thus, to overrule this value a new value given above is defined at the system level

The simulation was run for 5 s, which corresponds to five cycles of vibration. To more accurately simulate discontinuities in inertial forces (accelerations here), a fairly small output interval of 0.001 s was chosen as well as a tighter error tolerance of 10^{-8}. Even with these values, the simulation time is not too long (about 0.9 s on an i7 laptop). The results are shown in Figs. 6.46, 6.47 and 6.48.

Fig. 6.46 The phase portrait of Body1

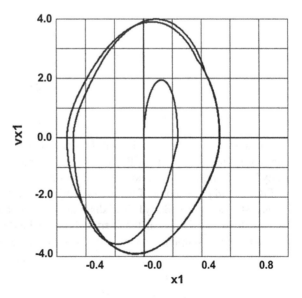

Fig. 6.47 The phase portrait
of Body2

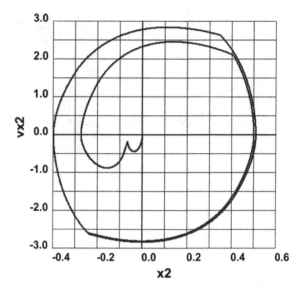

Fig. 6.48 The acceleration–
velocity diagram of Body1

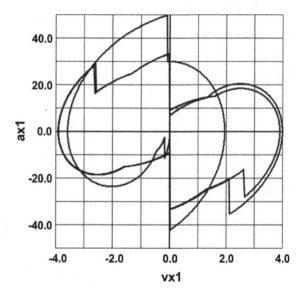

The system establishes stable limit cycles after two or three cycles (Figs. 6.46 and 6.47). The oscillations of Body1 are fairly symmetrical about the origin and of Body2 are displaced a little to the right (about 0.04 m).

Figure 6.48 gives a plot of the inertial force—or acceleration, as the mass is 1 kg —against velocity for Body1. The diagram is not smooth as in the previous case, because there are jumps in the inertial forces when either body changes motion from sticking to slipping, or vice versa. The bodies are coupled by friction; hence, these changes are reflected in both bodies. By looking carefully, we can see the limit cycle (the inner loop).

Discontinuities in accelerations occur at zero velocity and at a velocity of ± 2.626 m/s. The limit cycles closely agree with the phase diagrams of [9]. In addition, Figs. 6.46, 6.47 and 6.48 also show how the cycles gradually develop.

6.4 Bouncing Ball Problems

The next type of problem that we analyse is one of the simplest problems dealing with impact. We analyse the impact of a ball dropping on a massive table (ground). We analyse first the case where the table is at the rest. But more interesting is the problem when the table vibrates with constant frequency in which case chaotic vibrations can occur [10]. We first develop a simple model of impact, which is then used in the bouncing ball problems. This model can be used as a basis for solving much more complicated systems involving multibody motion with impact. It also can be combined with the model of dry friction developed in the previous section to model complex interactions at the contacts.

6.4.1 Simple Model of Impact

The dynamics of multibodies with contacts are studied in detail in [9]. This uses an approach based on the classical Newton theory and the Poisson laws of impact. It is applied to the solution of a range of problems. We will not follow such an approach here, but use the relatively simple model of Fig. 6.49 in which the contact between two bodies in the direction normal to the contacting surface is represented by a spring-dashpot component. Such a model is also presented in [9], but we analyse it here in more detail. The model is convenient because it enables us to treat processes at the impact of two bodies as a component that imposes certain conditions on the rest of the system. It offers the possibility of a more rigorous treatment of the properties of the materials at the contact. It is usually argued that such models lead to very stiff systems, which are not easy to solve. But other approaches, such as in [9], also are not simple because the interactions at impact are rather complicated. The BDF integration method used in the *BondSim* is capable of solving such stiff models efficiently.

We formulate the equation of motion of the ball and the spring-dashpot component as in [9]. First we define the kinematic relationships

$$v = \frac{dy}{dt} \tag{6.27}$$

Fig. 6.49 A simple model of impact

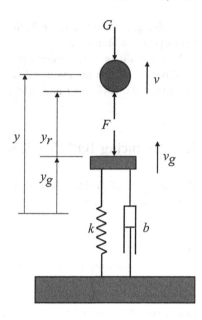

and

$$v_g = \frac{dy_g}{dt} \tag{6.28}$$

The relative velocity of the ball with respect to another body (ground) is

$$v_r = v - v_g \tag{6.29}$$

Hence, the relative distance of the ball to the point of contact is given by

$$y_r = y_{r0} + \int_0^t (v - v_g)\, dt \tag{6.30}$$

If $y_r > 0$ there is a free fall of the ball, hence we have the equations

$$\left. \begin{array}{l} m\frac{dv}{dt} = -mg + F \\ b\frac{dy_g}{dt} + ky_g = -F \\ F = 0 \end{array} \right\} \quad (y_r > 0) \tag{6.31}$$

The ball hits the other body when $y_r = 0$. From that moment on, the ball and the spring-dashpot move as single system. Thus we have

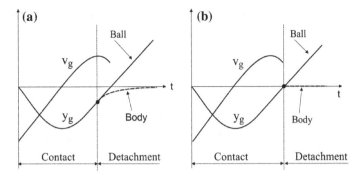

Fig. 6.50 Detachment of the ball at **a** zero force, **b** zero compression

$$
\left.
\begin{aligned}
m\frac{dv}{dt} &= -mg + F \\
b\frac{dy_g}{dt} + ky_g &= -F \\
v_r &= 0
\end{aligned}
\right\} \quad (y_r \le 0) \tag{6.32}
$$

At the moment of impact the ball has a velocity v_0. The acceleration of the ball changes abruptly at that moment to satisfy (6.32). Throughout the contact phase, the motion the ball-spring-damper is governed by the equation

$$
m\frac{d^2y_g}{dt^2} + b\frac{dy_g}{dt} + ky_g = -mg \tag{6.33}
$$

In the first phase of the motion there is *compression* of the spring until the ball velocity drops to zero (Fig. 6.50a). This is followed by an expansion phase during which the ball starts rebounding. The question is: At what moment does the ball detach from the other body? According to [9] this occurs when the force F drops to zero. At that moment the velocity of the ball and of the spring and damper ends are equal; they continue to move in the same direction gradually separating one from the other (Fig. 6.50a).

Such detachment is difficult to detect numerically. What we need is an *abrupt* change of the ball motion and of the other body at the point of contact. This occurs at zero compression where the velocity is $v_1 > 0$. At that moment by (6.31) the velocity of the spring-damper end drops abruptly to zero and its motion ceases (Fig. 6.50b). Simultaneously, the force on the ball abruptly changes from the value $-b \cdot v_1$ by (6.32) to the value 0 by (6.31). Thus, there is a positive impulse, which changes the ball acceleration from value $-g - b \cdot v_1/m$ to $-g$. In effect the body rejects the ball.

We thus assume that the contact with the other body is established when the ball's relative displacement is less or equal to zero, and that the detachment occurs when it is positive again. It is possible that the spring extension never drops to zero in a finite time. This is the case when there is high damping in the system, which corresponds to an inelastic impact.

To complete the description of this simple model of impact, we calculate the ratio of the ball velocity after and before impact. Solving (6.33) we get

$$\frac{v_1}{v_0} \approx -e^{-\frac{\zeta\pi}{\sqrt{1-\zeta^2}}} \tag{6.34}$$

where ζ is damping ratio of the ball-spring-damper system. In the classical theory of impact, this ratio is known as the coefficient of restitution α. Hence,

$$\alpha \approx -e^{-\frac{\zeta\pi}{\sqrt{1-\zeta^2}}} \tag{6.35}$$

This is just the *decrement* known from theory of vibrations. For near-elastic impact, i.e. $\zeta \ll 1$, we have

$$\alpha \approx 1 - \zeta\pi \tag{6.36}$$

Table 6.5 gives values of the coefficient of restitution for various values of damping ratio.

Now we can formulate a bond graph component that describes the interactions between the two bodies during impact. This is basically a two-port component (Fig. 6.51a). The ports are used to connect the bodies experiencing impact. The model of impact is shown in Fig. 6.51b. There are two effort junctions 1, which represent the velocities of bodies in the normal direction (see 6.27 and 6.28). Summator s evaluates their difference according (6.29); and the integrator evaluates the difference between the positions of the bodies, as in (6.30).

The first equations in (6.31) and (6.32) describe the motion of the body (the ball); thus, these are not included in the model of contact. This is represented by the capacitive element C and the resistive R elements, which in conjunction with the third effort junction, describes the second equations in (6.31) and (6.32). The switch element Sw changes the condition represented by the third equations depending

Table 6.5 Restitution coefficient as a function of the damping coefficient	Damping ratio ζ	Coefficient of restitution α
	0	1
	0.02	0.9391
	0.04	0.8818
	0.06	0.8279
	0.08	0.7771
	0.1	0.7292
	0.2	0.5266

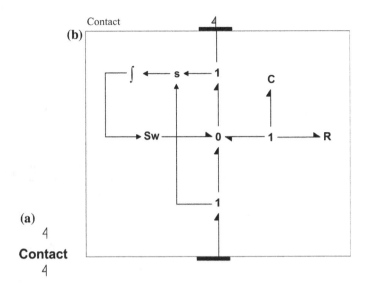

Fig. 6.51 The component representing impact between two bodies

upon the body relative displacement evaluated by the integrator. The constitutive relation of the switch is simply

$$yr > 0?F : vr = 0 \tag{6.37}$$

6.4.2 A Ball Bouncing on a Table

We now analyse the motion of a ball, which is dropped from a height h onto a table that is at rest (Fig. 6.52). The ball drops under the action of gravity, hits the table and bounces back. It continues to bounce until it eventually reaches rest.

We treat the ball as a particle moving in a vertical direction under the action of gravity. The model of the system consists of three components Ball, Ground and Contact (Fig. 6.53). The Ball is represented in the usual way by an inertial element representing the ball inertia in the vertical direction and a source effort describing its weight connected to a common effort junction. The Ground is described by a zero flow source. The last component models the impact of the ball with the table, as described in the last section.

We extract several signals such as the position and velocity of the ball and the position of the table (ground). In this problem this last signal is, of course, constant (zero) since the table is at rest. Details of the model can be found in the library under the project name *Bouncing Ball Problem*.

Fig. 6.52 A ball bouncing on
a table at rest

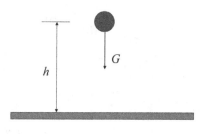

Fig. 6.53 Bond graph model
of the ball bouncing on the
table

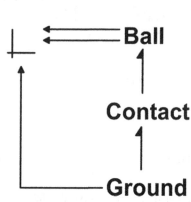

We assume that the table is relatively rigid and that the restitution coefficient is
about 0.9. Thus the parameters used for the simulation are:

- The ball mass $m = 1$ kg
- Spring stiffness $k = 1 \times 10^6$ N/m
- Damper velocity coefficient $b = 60$ N s/m
- Initial ball height above the table $h = 1$ m

These values correspond to a natural frequency of 1000 rad/s and a damping
ratio of 0.03. From (6.35) we get a coefficient of restitution value $\alpha = 0.9100$. The
interval of simulation was chosen as 5 s and the output interval 0.01 s. The complete simulation lasts 0.11 s of processor time. The results are shown in Figs. 6.54
and 6.55.

In Fig. 6.54 we see the characteristic, partially elastic, pattern in which the
bouncing height gradually diminishes until the ball comes to rest (not shown).
Figure 6.55 shows sudden changes of the ball velocity when it hits the table.

We can estimate the coefficient of restitution by comparing the height of the ball
above the table following every rebound. If v is the rebound velocity, the corresponding height of the ball after rebound is $v^2/(2\ g)$. Thus, the ratio of two successive rebound heights diminishes as α^2. From the simulation result it is found that
the ratio of heights is about 0.828, thus the coefficient of restitution is 0.910 as
expected.

Fig. 6.54 Change of the ball height during the bouncing

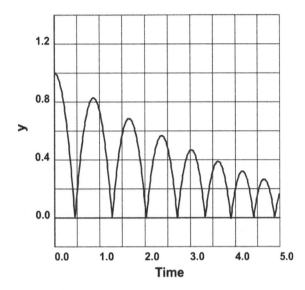

Fig. 6.55 Change of velocity of the ball

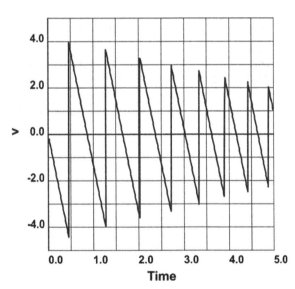

6.4.3 A Ball Bouncing on a Vibrating Table

We continue with analysing the motion of a ball bouncing on a table, which is not at rest, but vibrates with a constant frequency (Fig. 6.56). For low velocities of vibrations, a stable ball motion appears similar to the case of the fixed table analysed in Sect. 6.4.2. On the other hand, if the velocity increases an irregular, chaotic motion occurs. The dynamics of such motions are analysed in [10]. We will

Fig. 6.56 A ball bouncing on
a vibrating table

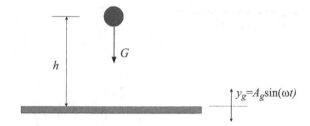

not repeat this here, but show by simulation some of characteristic motions that can
appear.

We use the model developed in the last section (Fig. 6.53). We also use the same
parameters, but change the velocity generated by the source flow of the component
Ground. We fix the amplitude of the table displacement to the value $A_g = 0.1$ m and
change its frequency of vibrations. The velocity of the table, as generated by the
source, is defined as

$$v_g = -A_g \omega \cos(\omega t) \tag{6.38}$$

That is, it starts moving down. We show two characteristic bouncing ball motion
patterns.

The simulations are run using a simulation interval of 10 s and an output interval
of 0.01 s. Figure 6.57 shows the simulation of the ball motion when the table is
vibrating slowly at frequency of 3 rad/s. It can be seen that the ball bouncing
follows the table and finally comes to rest with respect to the table.

Fig. 6.57 The ball bouncing
of a table vibrating with
3 rad/s

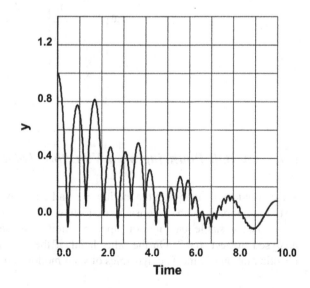

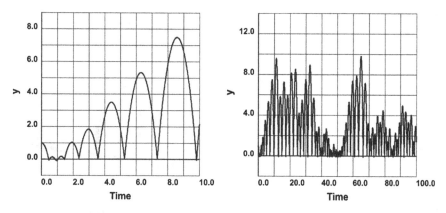

Fig. 6.58 The ball bouncing of a table vibrating with 15 rad/s

A quite different ball motion appears at higher frequencies. Figure 6.58 left shows the ball motion during the first 10 s when the table vibrates at 15 rad/s. To see if the ball comes to rest with respect to the table, Fig. 6.58 right illustrates the simulation of the same ball motion during the first 100 s. It apparently doesn't stop bouncing!

From the figures it can be seen that the interval between two bounces steadily increases until chaotic motions develop. In [10] it was shown that the existence of stable and unstable orbits of various periods can occur. Under certain conditions the ball bouncing from the table at the uppermost position can double its period of motion. As pointed out in [11], doubling the period leads to chaos. This is clearly seen in Fig. 6.58. The reader interested in this and other chaotic systems behaviour can consult [10, 11] for a detailed exposition.

6.5   The Pendulum Problem

As the last example of simple mechanical systems we return to the see-saw problem of Sect. 2.7.3 (Fig. 6.59). This problem was treated as a simple example of a multibody system, which we will discuss in more detail in Chap. 9. It was shown that it is possible to develop a bond graph model by systematically decomposing the system into its components.

Following the approach described in Sect. 2.7.3 and using *BondSim*, we can develop a corresponding model. In Fig. 6.60 the basic level of the model is shown in which the main components and the interactions between them are depicted. The model consists of several levels and can be reviewed in the project library under the title *See-saw Problem*.

The model developed is based on the direct application of the laws of Dynamics. The corresponding   mathematical   model   consists   of   a   system   of

Fig. 6.59 See-saw problem
of Sect. 2.7.3

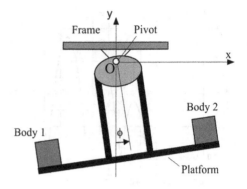

Fig. 6.60 Basic level of the
see-saw model

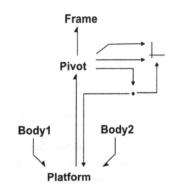

differential-algebraic equations (DAEs). It corresponds to multibody models using
Lagrange multipliers with constraints on velocities [12, 13]. It is well known that
such models are of *index* 2 type [12, 14]. This formulation lies between index 3
formulation when using constraints on the positions of connected bodies and index
1 formulation where all such constraints are eliminated e.g. by differentiation. It
should be pointed out that the index 2 formulation appears here quite naturally as a
result of the application of the bond graph modelling approach.

The see-saw can be looked at on as a pendulum, as has already been pointed out
in Sect. 2.7.3. In [12, 14] various formulations of pendulum problem are examined
in the context of higher index DAEs. Our formulation is a little different from those
in the references citied. Thus, we examine it in some detail.

We use two approaches. The first is simulating the system behaviour by the
BondSim program using the default method (Chap. 5), i.e. the BDF variable
coefficient method and the analytical Jacobian matrix. Also we excluded all algebraic variables from the error tests, as is the usual practice when solving higher
index systems by BDF methods. We also scale the algebraic equations, as suggested in [14].

Another approach, which we will use for comparison, consists of reducing the
multibody model to a pendulum. The pendulum will be represented in classical
state-space form and solved using the implicit Runge-Kutta code RADAU5 of [12].

Fig. 6.61 Oscillation of see-saw about its equilibrium position

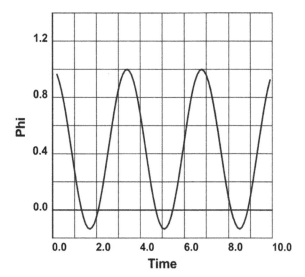

To simulate the see-saw problem using BondSim, we retrieve the problem from program library (double clicking the project and selecting the command *Copy to the Workspace*), open it and then build the model. We use the following model parameters:

- Body1 mass $m_1 = 80$ kg
- Body2 mass $m_2 = 20$ kg
- Platform mass $m_3 = 40$ kg
- Platform mass moment of inertia (centroidal) $I_3 = 90$ kg m^2
- Geometric parameters (Fig. 2.20) $a = 1.5$ m, $b = 1.125$ m, $c = 0.375$ m
- Initial angle of the platform = 1 rad

The simulation interval was taken equal to 10 s and the output interval to 0.5 s, the error tolerances to default values. Some simulation results are shown in Fig. 6.61 and 6.62. From Fig. 6.61 it can be seen that the see-saw oscillates about the equilibrium position that is at $\varphi = 0.4324$ rad (24.77°). The period of oscillation is 3.4 s. The amplitude of the force on the frame in the horizontal direction is 349.2 N (Fig. 6.62). The force in the vertical direction (not shown) doesn't change too much, i.e. it oscillates between -1152 N and -1613 N about a value corresponding to the weight of the platform and the bodies on it. Some numerical values also are given in Table 6.6 (the second and fourth columns) and the simulation statistics in Table 6.7 (the second column).

Next we develop a simplified model of the see-saw as a pendulum (Fig. 6.63). First we note that the see-saw's centre of mass in not on the body axis. Using the parameters of Fig. 2.20 we can calculate its coordinates in the body fixed frame as

Fig. 6.62 Change of the
horizontal force of the
see-saw on the frame

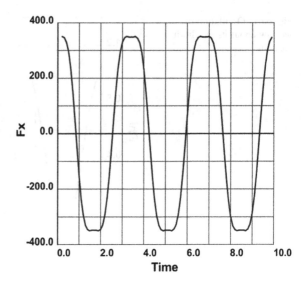

Table 6.6 Comparison of results obtained by the BondSim and the RADAU5 routine

Time (s)	φ (rad)		F_x (N)	
	BondSim	RADAU5	BondSim	RADAU5
0.1	0.99043873	0.99043865	348.15083	348.15018
0.5	0.77540086	0.77540064	293.32720	293.32710
1.0	0.27476494	0.27476482	−153.46435	−153.46446
1.5	−0.098058291	−0.098058331	−349.23567	−349.23563
2.0	−0.049639565	−0.049639515	−345.01367	−345.01384
2.5	0.38312370	0.38312303	−49.423719	−49.424294
3.0	0.85591563	0.85591521	330.02007	330.01993
3.4	0.99998092	0.99998105	347.16784	347.16789

Table 6.7 Simulation
statistics for the BondSim and
RADAU5 solutions

	BondSim	RADAU5
Equations	49	5
Jacobian matrix elements	123	25
Steps	459	975
Function evaluations	967	8144
Jacobian evaluations	502	879
Decompositions	502	975
Error tolerances	1e−6	1e−10
Elapsed time s	0.04	−

Fig. 6.63 The see-saw as a pendulum

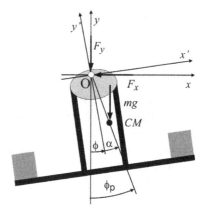

$$\left.\begin{array}{l} x'_C = -(m_1 - m_2)a/m \\ y'_C = -[(m_1 + m_2)(b+c) + m_3 b]/m \end{array}\right\} \tag{6.39}$$

Here m is the total mass of the see-saw and the bodies on it, i.e.

$$m = m_1 + m_2 + m_3 \tag{6.40}$$

The angle made by a line drawn from the origin through the mass centre CM to the y'-axis is given by

$$\alpha = \operatorname{atan}\left(-x'_C/y'_C\right) \tag{6.41}$$

This line is taken as the pendulum axis. Denoting by φ_p its angle with the vertical axis, we find the angle of the see-saw by

$$\phi = \phi_p - \alpha \tag{6.42}$$

Finally, the length from the origin to the centre of mass is given by

$$l_C = \sqrt{x'^2_C + y'^2_C} \tag{6.43}$$

Now we can formulate the equation of rotation of the see-saw about the origin (the pin axis) as

$$\left.\begin{array}{l} \dot{\phi}_p = \omega \\ \dot{\omega} = -\dfrac{mgl_C}{I_O}\sin\phi_p \end{array}\right\} \tag{6.44}$$

Here the I_O is its mass moment of inertia about the axis of rotation and is given by Fig. 2.25

$$I_O = (m_1 + m_2)\left(a^2 + (b+c)^2\right) + I_3 + m_3 b^2 \tag{6.45}$$

We also need expressions for the components of the force of the see-saw that acts *on* the frame. We calculate such a force because, in the bond graph model of Fig. 6.60, the efforts associated with the bond connecting the platform and the pin is the force acting *on* the pin, not its reaction.

Hence by applying the law of the mass centre motion we have

$$\left.\begin{aligned} m\frac{d^2 x_C}{dt^2} &= -F_x \\ m\frac{d^2 y_C}{dt^2} &= -F_y - mg \end{aligned}\right\} \tag{6.46}$$

Taking also into account that

$$\left.\begin{aligned} x_C &= l_C \sin\phi_p \\ y_C &= -l_C \cos\phi_p \end{aligned}\right\} \tag{6.47}$$

we can calculate the force components by

$$\left.\begin{aligned} F_x &= -m l_C \left(\dot{\omega}\cos\phi_p - \omega^2 \sin\phi_p\right) \\ F_y &= -m\left(l_C\dot{\omega}\sin\phi_p + l_C\omega^2 \cos\phi_p + g\right) \end{aligned}\right\} \tag{6.48}$$

The system (6.44) is in the state-space form, and (6.42) and (6.48) are the output equations. These equations can be solved by many methods. We use the RADAU5 method of [12] implemented as C++ library of reference [15] for the comparison. This method is considered by many as one of the powerful general-purpose methods for solving higher index DAEs.

We use the same parameters as given at the beginning of this section. The other necessary parameters are calculated by (6.39)–(6.41), (6.43) and (6.45). We also use the same simulation parameters as before (the simulation period 10 s, output interval 0.1 s, but error tolerances are 1×10^{-10}). The results for the angle of rotation and x-component of the force are given in Table 6.6 (third and fifth column).

Comparing the results obtained by the BondSim simulation and by the RADAU5 solution we see that there are minor differences that are of the order of the error tolerances used. It is interesting to note, in particular, that the accuracy of the force component obtained by the BondSim is in the same range of accuracy as that obtained by RADAU5. This is important because it is an index-2 variable.

The simulation statistics are given in Table 6.7. The solution of the reduced set of equations is expected to be more efficient. However, the performances of BondSim, which is based on BDF integration method, are excellent. We should take into account that the semi-explicit index-2 equations are solved, and that the system of equations is relatively large in comparison with the simplified model. Number of matrix evaluations and decompositions are comparable in both methods. On the other hand RADAU5 needs much larger number of function evaluations, which is to be expected from a Runge-Kutta type solver.

References

1. Ginsberg JH (2001) Mechanical and structural vibrations, theory and applications. Wiley, New York
2. Smith SW (1999) The Scientific and engineer's guide to signal processing, 2nd ed. California Technical Publishing. Electronic form http://dspguide.com
3. Oran Brigham E (1988) The fast fourier transform and its applications. Prentice-Hall, Upper Saddle River, NJ
4. Bowden FP, Tabor D (1986) The friction and lubrication of solids. Clarendon Press, Oxford
5. Karnopp D (1985) Computer simulation of stick-slip friction in mechanical dynamic systems. Trans ASME 107:100–103
6. Borutzky W (1995) Represeting discontinuities by sinks of fixed causality. In: Cellier FE, Granda JJ (eds) 1995 international conference on bond graph modeling and simulation, Las Vegas, Nevada, pp 65–72
7. Mera JM, Vera C (1999) Dry Friction modelling by means of the bond graph technique. In: Granda JJ, Cellier FE (eds) 1999 international conference on bond graph modeling and simulation, San Francisco, California, pp 30–35
8. Rao SS (1995) Mechanical vibrations, 3rd edn. Addison-Wesley, Reading
9. Pfeiffer F, Glocker C (1996) Multibody dynamics with unilateral contacts. Wiley, New York
10. Guckenheimer J, Holmes P (1997) Nonlinear oscillations, dynamical systems, and bifurcations of vector fields. Springer, New York
11. Acheson D (1997) From calculus to chaos, an introduction to dynamics. Oxford University Press, Oxford
12. Hairer E, Wanner G (1996) Solving ordinary differential equations II, stiff and differential-algebraic problems, 2nd Revised edn. Springer Berlin Heidelberg
13. Karnopp D (1997) Understanding multibody dynamics using bond graph representations. J Franklin Inst Eng Appl Math 334B:631–642
14. Brenan KE, Cambell SL, Petzold LR (1996) Numerical Solution of initial-value problems in differential-algebraic equations, classics in applied mathematics. SIAM, Philadelphia
15. Ashby B (2002) ItegratorT.zip. bmashby@stanford.ed. 15 Nov 2002

Chapter 7
Electrical Systems

7.1 Introduction

This section shows that the component model approach developed in Part 1 can be readily used to model electrical and electromechanical components and systems. This is important, as both the mechanical and electrical part of mechatronic and recently evolved micro-mechanical systems [1] can be modelled and analysed on the same basis, i.e. from the bond graph point of view.

It is well known that the electrical systems can be modelled in terms of bond graphs [2, 3]. But this approach may appear strange to engineers used to the electrical schematics. The models in terms of bonds are usually simplified, e.g. by removing the ground nodes, resistor, capacitor, and other ports. Such models are quite abstract, and often it is not easy to correlate them with the devices of which they are models. In addition, the causality relations discussed in Sect. 2.10 restrict models that could be used for modelling general electrical, as well as of other engineering systems.

The approach developed here is based on the component models, the constitutive relations of which can be freely defined. This enables a systematic approach to development of the electrical components and system models. It is not necessary to represent the electrical components in terms of word models only but, as already discussed in Sect. 2.7.2, these can be depicted using graphical electrical symbols. In this way, the bond graph point of view is retained, but on the surface they are represented as the electrical schemes. At a deeper level, bond graph elements are employed to model the processes in the components.

The models of electronic components are usually described in the form of SPICE models [4]. These are developed for use in SPICE type programs specifically developed for electrical and electronic systems (Sect. 1.7).[1] This type of model can be used in bond graphs, as well. But here it is not necessary to use the SPICE

---

[1]There are various versions of SPICE such as SPICE 2 and 3, HSPICE, PSPICE etc.

© Springer-Verlag Berlin Heidelberg 2015
V. Damić and J. Montgomery, *Mechatronics by Bond Graphs*,
DOI 10.1007/978-3-662-49004-4_7

language, and there are no pre-built models of resistors, diodes, transistors, etc. The component models are constructed from the ground up from simpler ones. The same equivalent circuits, constitutive relations, parameter names, and values can be used as in SPICE. The component models can deal easily with the macro models that play an important role in the development of complex SPICE models. Once developed, the component models can be moved into a library and used as needed.

An added advantage of our approach is that component models are transparent, i.e. they can be opened for overview, and modified if required, or maybe a different model of the same component can be developed and tested. The BondSim program is not designed to be a replacement for specially designed programs, such as SPICE, or others, but more as an open environment for developing the models and simulating the behaviour of the mechatronic and micro-mechanics systems (MMS). The analysis currently supported is essentially in the time-domain.

This section starts with a relatively detailed description of a typical approach to developing the bond graph models of the electrical systems and their simulation using the electrical schematics. This is explained in an example of the simple RLC circuit already described in Sect. 2.7.2. The approach is very similar to that used in Sect. 6.2 for the mechanical systems. We then show how SPICE models of electrical circuit elements, such as resistors and capacitors, are developed. We continue with modelling some basic semiconductor components. The chapter ends with the analysis of an electro-magnetic system.

7.2 Electrical Circuits

7.2.1 The Problem

The electrical circuits can be modelled in a similar way to the mechanical systems in Sect. 6.2. A typical approach to modelling the electrical circuits is illustrated by an example of the RLC circuit of Sect. 2.7.2, repeated here in Fig. 7.1.

The circuit consists of the series connection of a resistor, inductor and capacitor driven by a voltage source. We use the electrical component template tools to create

Fig. 7.1 The scheme of the RCL circuit

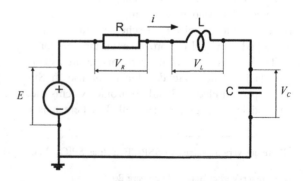

the bond graph models, which model these components represented in the form of electrical symbols. Using these tools and standard bond graph techniques we can systematically develop corresponding bond graph models.

7.2.2 The Bond Graph Model

7.2.2.1 System Level Model

To begin with, we launch BondSim as described in Sect. 6.2.2. Using the command *New* on the *Project* menu, or *New Project* toolbar button, we define a new project, *RCL Circuit* in the dialogue that appears and then click the *OK* button. The dialogue closes and a new empty project document appears. Because we will define the RLC model using the electrical symbols we expand the *Template Box* at the right of the new document window, and then the *Electrical Components* branch as shown in Fig. 7.2. The model structure at the left shows only the *RLC circuit* project entry.

We will create the word model components represented by the electric components symbols corresponding to the components shown in the circuit in Fig. 7.1. This can be achieved by dragging the tools from the electrical components list in the Template Box and dropping them in the RLC circuit document window at suitable positions. We start by dragging the *Voltage Source* tool and dropping it near the left document border, but leaving some space (Fig. 7.3 left). The component created appears in the form of the circuit symbol for the voltage source. Near the symbol a component name VS (Voltage Source) appears; the caret shows that we are in the

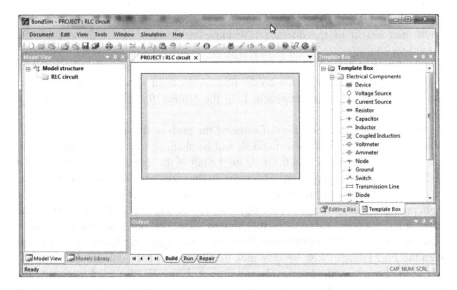

Fig. 7.2 RLC circuit empty document window and selected electrical components toolbox

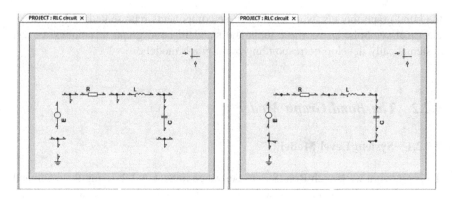

Fig. 7.3 The circuit components corresponding to Fig. 7.1

text-editing mode. We can edit the name if we wish, e.g. to *E*. To end the editing, we click outside of the component, as is usual with word component models. After the component has been created, its two power-in and power-out ports are created as well.

We create the other circuit components in a similar way. By default, the resistors, inductors and capacitors are drawn horizontally. To draw the capacitor vertically, we need to click the *Vertical* in the *Editing Box, Text Orientation* branch. After dropping the component we select again the *Horizontal* text orientation to return to the default drawing mode. To complete the circuit, we create the *Ground* component. We also create a *Node* between each of the components. All of these are not always necessary, but we add them anyway to ease the connection of the measuring instruments. The left bottom node is used for the connection of the circuit to the ground. During the addition of the component we can also adjust the width and height of the document window by dragging the right and bottom document edges.

In addition to the basic components such as nodes, resistors, etc., we create also the X-T display at the right top corner of the document window. (Remember that the tool for creating this component is in the *Editing Box* tab, under *Continuous Signal Components*.)

We can see that the positions of some of the ports in the nodes, as well as their power directions, are not correct and should be changed (see Fig. 7.3 right). Thus, we remove the bottom port of the node on the left of the resistor. Hence, we select it and then press the *Delete* key. The same is the case with the node between the resistor and the inductor. In the node at the right of the inductor we will remove the right port. Similarly, we remove the right and the bottom port of the node below the capacitor. However, to connect the bottom capacitor port, we need to add a power-in port at the top of the node. To do this, we expand the *Editing Box* tab, and then the *Port* branch. Now we may pick the *Power-out* port tool and drag it to the top of the node bounding rectangle and drop it there. We also need to change the power flow sense of the left port by selecting and pressing the *Change Port* toolbar button. A similar operation we apply to the node below the Voltage Source component.

Now that all the components are created and their ports have the proper power senses, we interconnect the components by drawing bond lines between corresponding ports. To do this we simply pick a port and by holding the left mouse button pressed drag the cursor to the port with which we wish to connect. As we move the cursor, a line is drawn. We release the mouse button when we are above the port with which we wish to connect. The both ports then disappear and the bond line appears with a half-arrow pointing to the corresponding node. The bond having been drawn we can reshape at will by dragging it. Thus we obtain the bond graph which defines the basic model of the circuit as shown in Fig. 7.4 left.

It is necessary also to define the variables that will be observed during the simulation. Otherwise the simulation would be useless. This is similar to what is encountered when dealing with the real circuit. To measure or record the processes in the circuits we use instruments. In this example we monitor the voltages across all components and the currents throughout the circuit. This requires the model equivalents of "voltmeters" and "ammeters", respectively.

To create a voltmeter word model component to measure the voltage drop across the resistor, we drag *Voltmeter* from the *Electrical components* branch of the *Template Box* (Fig. 7.2) and drop it just above the resistor. The voltmeter appears with two control input ports and a single output port (Fig. 7.4 right). We drag the output port around the *Voltmeter* and drop at the top to point upwards. To connect its ports to nodes at the front and rear of the resistor we need to add the corresponding *Control-out* ports to these nodes (using the *Editing Box, Ports* branch). The voltmeter, like other components, is created as an empty component that has to be defined. In a similar way we create voltmeters for measuring voltages across the other components.

Next we create the ammeter. Note that the ammeters are commonly connected serially. We may insert the ammeter in the bottom line. Thus, we select the bottom bond and click the *Delete* key to delete it. Next, we drag the *Ammeter* and drop it

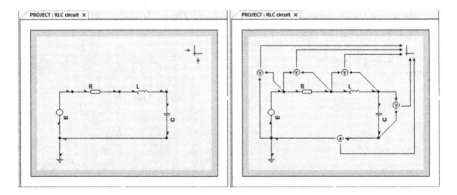

Fig. 7.4 The system level bond graph model of the RCL circuit

somewhere between the bottom nodes. Note that the ammeter component that appears has one power-in and one power-out ports and a single control out port. We will drag the power input port to the right side of the ammeter, and the power-out one to the left side. The output port is already placed at the bottom of the component. Now, we may connect the right port of the ammeter to the corresponding port of the node on the right, and the left power port to the node on the left (Fig. 7.4 right).

We now connect the output ports of all voltmeters and ammeters to the X–Y display component created earlier. Because by default it has only two input ports, we first create three additional *control-in* ports at the component periphery at convenient places. At their creation the ports have default names, which are used to label signals displayed along the *x or y* axes. We can change the names of the signals after the display ports are connected. All the signals are, by default, plotted along the *y*-axis, the *x*-axis serving as the *time* axis. This can be changed easily by double-clicking the plotter ports (Fig. 7.5). The dialogue that opens can be used to define the signal names and choose the display axis.

We may select only one variable to be used as the *x*-axis variable. All others must be *y*-axis variables. We can also choose the *None* option, in which case the variable is not displayed. We define *E* for the source voltage, *eR* for the voltage across the resistor, *eL* for the voltage across the inductor, and *eC* for the voltage across the capacitor, and the *I* for the current throughout the circuit. Because the expected range of the currents (*mA*) is quite different from the range of the voltages (V), we check *None* for the current axis, so that only the voltages will be displayed. (Later, we can display the time history of the current in the circuit.)

7.2.2.2 The Component Models

To complete the RCL circuit model, we next define the model of every one of the circuit components including the voltmeters and the ammeter. We start with the voltage source *E*. To open it we double-click it. We can open it also by right clicking the component icon *E* in *Model Structure* under the project *RLC circuit*. In the menu that drops-down select the command *Open*. Because the model of the

Fig. 7.5 The output variable dialogue

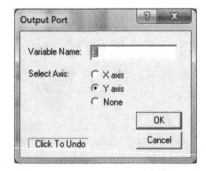

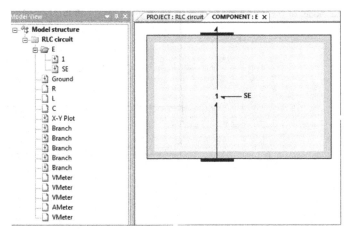

Fig. 7.6 The model of the `Voltage Source`

voltage source has not already been defined, a dialogue appears asking if we would like to create the model of the component. If we answer by clicking the OK a new document opens in the form of a new document page labelled *Component: E*. The document has the already created document ports corresponding to the component ports. To create the model we will expand the *Bond Graph Components* branch in the *Editing Box*. Next we drag 1 junction tool and drop it into the new document window, and then do the same with SE (Source Effort) tool. We will rearrange the inserted components and their ports and connect them as shown in Fig. 7.6. The 1–junction defines the voltage difference across the Source Voltage ports, and SE generates the electromotive force of the source.

   The constitutive relation describing the voltage generated by the SE is stored in its port. To edit it, we double-click the port. A dialogue appears that is used to assign the port variable name, define the model parameters, and edit the constitutive relation that describes the *emf* at the port (Fig. 7.7). We retain *e* as the name of the

Fig. 7.7 The dialogue for editing the voltage source constitutive relation

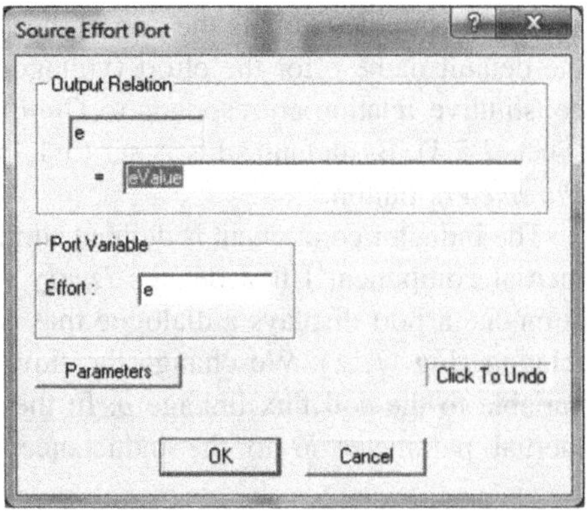

Fig. 7.8 The model of the
Resistor

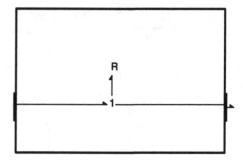

Fig. 7.9 Dialogue for
defining the resistor
constitutive relation

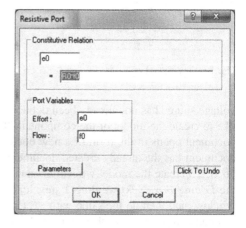

effort variable (voltage). The *eValue* is the parameter defining the constant voltage
generated at the port. At the port creation it is set to zero; we now change it to 10
(V) by use of the *Parameters* button (see Sect. 4.8.2).

Next, we define the simple low frequency models of the resistor, inductor and
capacitor, ignoring the parasitic effects. The model of the resistor (Fig. 7.8) consists
of an effort 1-junction, which determines the voltage drop across the resistor and a
resistive component R. We edit the constitutive relation for the voltage across the
resistor by double-clicking the port of the resistor component R (Fig. 7.9). We retain
the default name *e* for the effort (voltage) and change the flow variable to *i*. The
constitutive relation corresponds to *Ohm's Law*, with the resistance parameter *R0*
(Sect. 2.5.4). By default, this is set to 1, but we change it to 200 (ohm) using the
Parameters button.

The Inductor component is defined similarly (Fig. 7.10). This component uses an
inertial component I that defines *Henry's Law* (Sect. 2.5.2). Double-clicking the
component port displays a dialogue that is used to define the inductor constitutive
relation (Fig. 7.11). We change the flow variable to the current *i*, and the state
variable to the coil flux linkage *p*. In the constitutive relation box we change the
inertial parameter *I0* to the inductance parameter *L*. Likewise, we delete the

Fig. 7.10 The inductor component model

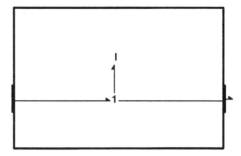

Fig. 7.11 The dialogue for defining the inductor constitutive relation

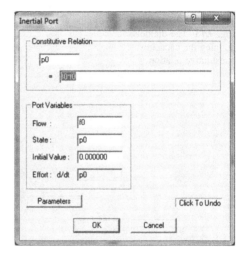

parameter *I0* from the parameter list and insert the parameter *L* with a value of 0. 010 (H).

Finally, we define a simple model of the Capacitor (Fig. 7.12). Again, the effort junction describes the voltage across the capacitor. We use the capacitive component C to describe the processes in the capacitor (Sect. 2.5.3).

The default constitutive relation can be changed by double-clicking the component C port (Fig. 7.13). We change the effort variable (the voltage across the capacitor) to *e*, and the state variable to the capacitor charge *q*. The parameter *C0*, the capacitance, is set to 2.0×10^{-6} (F).

The last group of components that we have to create are the Voltmeters and Ammeter (Fig. 7.4). We implement the Voltmeter simply as an instrument, which gives as its output the difference between the signals fed to its inputs. Thus, to create the model of the voltmeter, we open the component, e.g. by double clicking the voltmeter object, and in the document that appears we insert a summator (+) by dragging and dropping the corresponding tool from the *Editing Box*, *Continuous signal Components*. By default at the summator creation its ports have assigned the *plus* sign. To change the sign we double-click the port and from the drop-down

Fig. 7.12 The model of the
capacitor

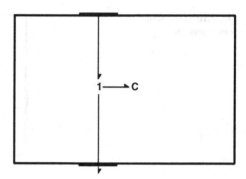

Fig. 7.13 The dialogue for
defining the capacitor
constitutive relation

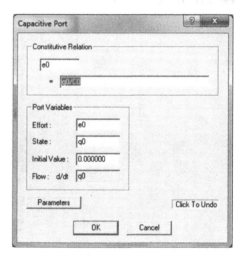

menu select the appropriate sign, e.g. the *minus* sign. If we wish to scale the output
we may insert a function (gain) component (Fig. 7.14).

To define the scaling we open the output port (by double-clicking) and in the
dialogue that opens change the gain parameter $k0$ (Fig. 7.15). By default its value is
pre-set to 1.

The ammeter we may define in an analogue way. This time we will insert a 1-
junction and connect it to the ammeter power-in and power-out ports (Fig. 7.16).
We delete the third power port and instead of it insert a control-out port, which
gives the access to the common junction flow, i.e. the current in this case. Finally,
as in the case of the voltmeter, we may insert a function component between the
junction output and the ammeter document output port. This component can serve
to scale the ammeter output up or down, e.g. to mA.

Thus, we have defined the models of all the circuit components. The models of
the nodes are already defined, as they are just flow junctions. The ground is a source
effort component and, by default, the voltage at its port is set to 0 (V). It could be
changed, if required, as is the case with the other source efforts, by double-clicking
its port and using the corresponding dialogue.

Fig. 7.14 The model of the voltmeter

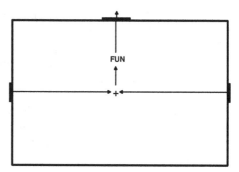

Fig. 7.15 The dialog for defining the function output relation

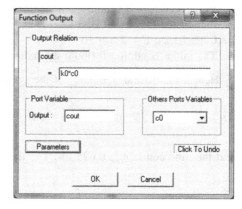

Fig. 7.16 The model of the ammeter

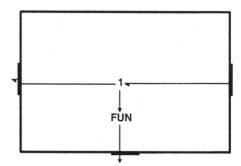

7.2.3 Analysis of the System Behaviour by Simulation

Before we can start the simulation of the circuit we have first to build its mathematical model by clicking the *Build* button or selecting the *Build Model* command on the *Simulation* menu. After the model is built, the mathematical model of the circuit can be reviewed.

During the *Build* all the component document windows are closed and saved. Only the main document window remains opened and active. In addition for each of the X-Y display components a separate document tab is opened. These documents are used to display the plots generated during the simulation. Thus, we can activate the display page by clicking the corresponding tab to see how the simulation advances.

We start the simulation by clicking the *Run* button on the toolbar, or by choosing the *Run Simulation* command in the *Simulation* menu. This opens a dialogue in which we must enter the simulation interval (Fig. 6.21). The values of the circuit parameters set during model development were:

1. Source voltage E = 10.0 V
2. Resistance R = 200 ohm
3. Inductance L = 0.010 H
4. Capacitance C = 2.0 × 10^{-6} F
5. The initial condition for both the inductor and the capacitor was set to zero

Hence, the natural frequency of the system is

$$\omega_n = \frac{1}{\sqrt{L \cdot C}} = 10,000 \ 1/s = 1592 \ \text{Hz}$$

and the time constant is 0.628 ms. The damping ratio is

$$\zeta = \frac{1}{2} R \sqrt{C/L} = 1.414$$

which implies that the transients in the circuit are highly damped.

We start the simulation by setting the interval to 0.002 s, the output interval to 10 μs, and accept the defaults for all other parameters. The results of the simulations are shown in Fig. 7.17. The diagram shows the transients across the resistor, inductor, and capacitor to a 10 V step in the Source voltage.

The legend in Fig. 7.17 is added directly to the plot by right-clicking the mouse and choosing *Insert description*. The cursor changes to a cross, and we can select the position where we wish to type the text of a legend. When we are finished with text editing we then click somewhere outside it. We can move a text across the plot, or delete it. We can also add arrows using the right mouse button, selecting the *Insert Arrow* command from the drop-down menu, and then drawing the arrow. The arrow can be moved or stretched in the usual way. We can also add or remove the markers used to distinguish between plotted curves. This is done by clicking the right mouse button and checking the *Set Curve Marks*.

We can also display the plot of the current in the circuit. We double-click the input ports of the display component and set all the voltages to *None*, and the current to the *y-axis*. By activating the X-Y display tab the resulting diagram is shown (Fig. 7.18).

Fig. 7.17 Transient voltages across the resistor, inductor, and capacitor

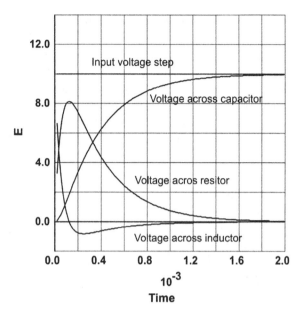

Fig. 7.18 The transient of the current in the circuit

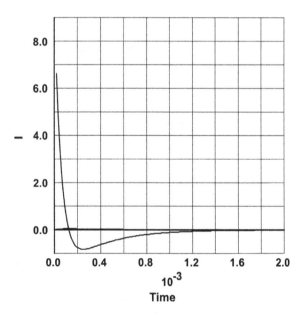

From these plots we can see that at the start of simulation the complete source voltage appears across the inductor coil; then, as the current starts flowing through the circuit and the capacitor accumulates the charge, it drops steeply. After the transients die out, there is no current in the circuit and the capacitor is fully charged to a value corresponding to the source voltage.

7.3 Models of Circuit Elements

In this section the models of basic circuit components, such as Resistors, Capacitors, and Inductors are developed. These can be stored in the library. Formulation of these models follows a SPICE-type description [4–7]. We use the same basic mathematical forms, including similar parameter names, but there is no "SPICE scripts". We treat the electrical components as objects that can be connected to other components by bonds, just as the real components are wired.

7.3.1 Resistors

The resistor is a two-terminal component usually represented schematically as in Fig. 7.19a. It can be represented by a two-port component in Fig. 7.19b. Its model is very simple and consists of an effort junction and a resistive elementary component (Fig. 7.20). Thus, it is described by the following constitutive relations:

$$\left.\begin{array}{c} e_1 - e_2 - e_R = 0 \\ e_R = R \cdot i_R \end{array}\right\} \tag{7.1}$$

Here, e_1 and e_2 are the efforts (voltages) at the ports, e_R the voltage across the resistor, i_R the current through the resistor, and R its resistance.

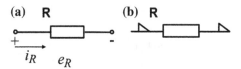

Fig. 7.19 The resistor: **a** the circuit symbol, **b** the bond graph representation

Fig. 7.20 The model of the resistor

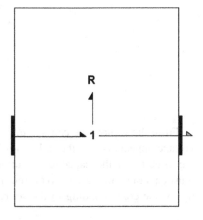

The resistance is typically a constant expressed in *ohm*. The SPICE software has a built-in model of a resistor that can exhibit the temperature dependence. The bond graph models used here are much more flexible in this respect.

It is a simple matter to describe the dependence of the resistance on temperature as used in SPICE. We can define two temperatures, *TNOM* and *TEMP*, as the parameters at the resistor document level using the *Parameters* button, or the *Model Parameters* command on the *Edit* menu. The nominal temperature *TNOM* is usually set to 300 K (27 °C), and the value used for *TEMP* corresponds to the resistor's working temperature. We define (similar to SPICE) two temperature coefficients, *TC1* and *TC2*. SPICE-like temperature dependence can be formulated as

$$R = R_0 * (1 + TC1 * (TEMP - TNOM) + TC2 * (TEMP - TNOM)^{\wedge}2) \quad (7.2)$$

Such an expression can be defined in the resistive component R, or at the resistor document level. We are not restricted to such a linear or quadratic relationship, but other forms can be used as well, e.g. an exponential dependence. We could go a step further and introduce a separate thermal port at the resistor boundary (Sect. 7.4.1). In this way, dependence on temperature can be used instead of the constant temperature parameter in the constitutive relation of (7.2).

At high frequencies the impedance of resistors drops due to the parasitic capacitance effects [6]. This can be included in the resistor model by adding a parallel capacitive element using a flow 0-junction (Fig. 7.21).

7.3.2   Capacitor

The capacitor is a two-terminal component usually represented schematically as in Fig. 7.22a. We can represent it by the two-port component shown in Fig. 7.22b. Its model consists of an effort junction and a capacitive element (Fig. 7.23).

Fig. 7.21 The model of a resistor with a parasitic capacitance

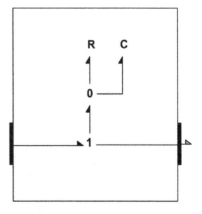

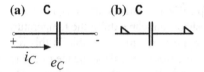

Fig. 7.22 The capacitor: **a** the circuit symbol, **b** the bond graph representation

The constitutive relations for the capacitor can be found easily (Sect. 2.5.3)

$$\left.\begin{aligned}
e_1 - e_2 - e_C &= 0 \\
e_C &= q/C \\
i_C &= dq/dt
\end{aligned}\right\} \tag{7.3}$$

Here, e_1 and e_2 are the efforts (voltages) at the ports, e_C the voltage across the capacitor, q the charge, i_R the current through the capacitor, and C its capacitance.

The dependence of the capacitance on temperature can be described essentially in the same way as we did for the resistor. A non-linear capacitor can be described using a second equation in Eq. (7.3) that expresses voltage across the capacitor as a function of the charge. In this way, polynomial, exponential, or other non-linear dependence can thus be specified readily.

The capacitor model given above is a reasonably good approximation for frequencies up to about 1 kHz [6]. At higher frequencies a series resistive and inductive element gives a better approximation to real capacitors. The model of Fig. 7.23 can be modified to include such effects (Fig. 7.24). The parallel resistive element models the capacitor leakage, which can be usually neglected.

The capacitor model of Eq. (7.3) is based on the charges; it differs from the SPICE model, in which the charge is eliminated. This model simplifies the description of the accumulation of the charges in the circuits and their conservation. There is also some evidence that this type of model behaves better when simulating the complex electrical circuits [8].

Fig. 7.23 The model of the capacitor

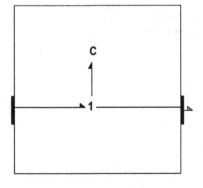

Fig. 7.24 The capacitor
model with parasitic effects

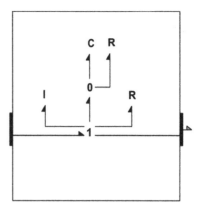

7.3.3 Inductors

7.3.3.1 Simple Inductor

The inductor is a two-terminal component usually represented schematically as in
Fig. 7.25a. We represent it by the two-port component shown in Fig. 7.25b. Its
model consists of an effort junction and an inertial component (Fig. 7.26).

The constitutive relations for the inductor read (Sect. 2.5.2):

$$\left.\begin{array}{l} e_1 - e_2 - e_L = 0 \\ p = L \cdot i_L \\ e_L = dp/dt \end{array}\right\} \tag{7.4}$$

where e_1 and e_2 are the efforts (voltages) at the ports, e_L the voltage across the
inductor, p the flux linkage of the coil, i_L the inductor current, and L the inductance
parameter.

The inductor's temperature dependence can be described in the same way as we
did for the resistors. The model above corresponds to the case in which the
inductance is constant. This can be modified to include non-linear dependence on
the current. We return to this problem in Sect. 7.5 when analysing the electro-
magnetic systems.

The model of the inductor given above is a reasonably good approximation at
medium frequencies [6]. Its behaviour at low frequencies is determined by the

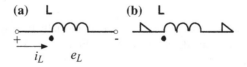

Fig. 7.25 The inductor: **a** the circuit symbol, **b** the bond graph representation

Fig. 7.26 The model of the inductor

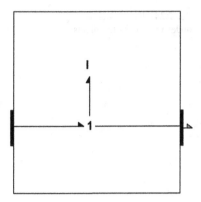

Fig. 7.27 The model of the inductor with parasitic effects

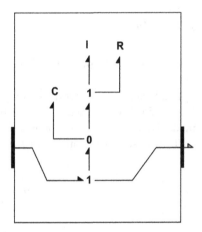

resistance of the inductor coils. At very high frequencies, the winding capacitance comes into effect. The model of Fig. 7.26 can be modified easily to include such effects (Fig. 7.27).

The model of (7.4) is flux based and differs from the SPICE model in which the flux linkage is eliminated from the constitutive relations. It is used here similarly as in the case of the capacitor because it describes the inductor as a dynamic component with explicitly defined internal state variable. There also is evidence that such models are better suited to simulating the electrical circuits [8].

7.3.3.2 Coupled Inductors

The coupled inductor is a four-terminal component usually used for modelling ideal transformers. It is represented schematically in Fig. 7.28a. We represent it by the four-port component of Fig. 7.28b.

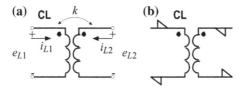

Fig. 7.28 Coupled inductors: **a** the circuit symbol, **b** the bond graph representation

The model of the coupled inductor, an extension of the simple inductor of Fig. 7.26, is given in Fig. 7.29. This model consists of two effort junctions at every pair of external (document) ports. These are connected internally to a two-port inertial element.

To describe the mathematical model of the component, we use the following variables:

- e_1 and e_2 are the efforts (voltages) at the left port, and e_3 and e_4 are those at the right
- e_{L1} is the voltage difference at the left port and e_{L2} is that of the right port
- i_{L1} and i_{L2} are the flows (currents) through the left and the right ports
- p_1 and p_2 are the flux linkages of the left and the right inductor coils

The equations now read:

$$\left.\begin{array}{l}
e_1 - e_2 - e_{L1} = 0 \\
p_1 = L_1 \cdot i_{L1} + M \cdot i_{L2} \\
e_{L1} = \frac{dp_1}{dt} \\
e_3 - e_4 - e_{L2} = 0 \\
p_2 = L_2 \cdot i_{L2} + M \cdot i_{L1} \\
e_{L2} = \frac{dp_{21}}{dt}
\end{array}\right\} \tag{7.5}$$

Here, L_1 and L_2 are the coil inductances defined as parameters. Likewise, M is the mutual inductance of the coils. It is usually expressed as

Fig. 7.29 The model of the coupled inductor component

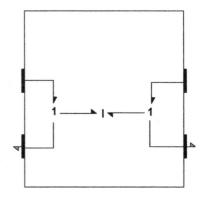

$$M = k \cdot \sqrt{L_1 L_2} \tag{7.6}$$

where k is the coupling coefficient. The relation (7.6) can be defined as a parameter expression at the document level of the component, or inside the inertial component.

The model presented above is linear with constant values of the impedances. It could be modified in a way similar to that of the simple inductive component.

7.3.4 Sources

7.3.4.1 Voltage and Current Sources

The independent voltage source is a two-terminal component that generates a voltage across its port (electromotive force) independently of the current drawn from the source. The circuit symbol used for such a component is shown in Fig. 7.30a. The bond graph component representing the independent voltage source is given in Fig. 7.30b.

It is a two-port component with a half-arrow showing the sense of the power delivery. The model used for independent voltage source components is similar to other bond graph models of circuit components and consists of an effort junction and a source effort component (Fig. 7.31).

Likewise, an independent current source is a two-terminal component that generates a current that is independent of the voltage across its terminals. The electrical circuit symbol used for such a component is shown in Fig. 7.32a. The corresponding bond graph representation is given in Fig. 7.32b. This is a two-port component with the half-arrows pointing in the sense of power delivery.

The model of the current source is similar to that of the voltage source and consists of an effort junction and a *source flow* component (Fig. 7.33).

These simple components can be modified to produce better models of the real sources, e.g. by adding a resistive element in series or in parallel.

(a)   **(b)**

Fig. 7.30 The independent voltage source: **a** the circuit symbol, **b** the bond graph representation

Fig. 7.31 The model of the independent voltage source component

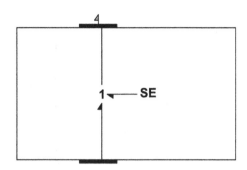

Fig. 7.32 The independent current source: **a** the circuit symbol, **b** the bond graph representation

Fig. 7.33 The model of the independent current source

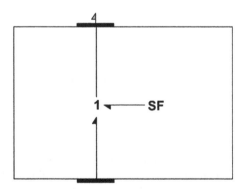

7.3.4.2   Constitutive Relations

There are no built-in functions for voltages or currents generated by the sources. Voltages or currents, instead, are defined by the constitutive relations of the corresponding SE or SF components (Sect. 2.5.5). We now show how some important relationships can be described for voltage sources, but the same is valid for current sources, as well.

The sinusoidal emf can be described simply as (Fig. 3.34)

Fig. 3.34 The sine source
function

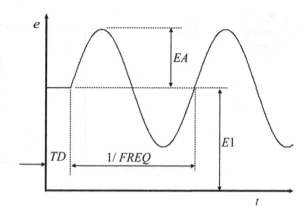

Fig. 7.35 The exponential
source function

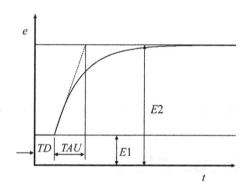

$$e = t < TD ~?~ E1 ~:~ E1 + EA * \sin(2 * PI * FREQ * (t - TD)) \qquad (7.7)$$

This function assumes that the voltage has a constant offset $E1$ for times less than TD, and then oscillates with a frequency $FREQ$ (Hz). The constants $E1$, EA, TD, and $FREQ$ are defined as the default parameters at the source document level of the component. The specific values that override them are input in the SE or SF components by using the corresponding dialogues.

An exponential function can be expressed as (Fig. 7.35)

$$e = t < TD?E1 : E1 + (E2 - E1) * (1 - \exp(-(t - TD)/TAU)) \qquad (7.8)$$

Here, TD is a time delay and $E1$ is an offset, as above. $E2$ is the voltage strength and TAU the time rise constant which are defined by the default values, e.g. $E2 = 0$, $TAU = 1$. The specific values that override these defaults can be defined in the respective source effort or source flow components.

The law as given by (7.8) can be extended to include a dropping exponential

$$e = t < TD? \; E1 :$$
$$(t < TD2?(E1 + (E2 - E1) * (1 - \exp(-(t - TD)/TAU))) : \quad (7.9)$$
$$(E3 + (E1 - E3) * (1 - \exp(-(t - TD2)/TAU1)))$$

The time parameter $TD2 > TD$ and $E3$ corresponds to the voltage at the beginning of the fall, i.e.

$$E3 = E1 + (E2 - E1) * (1 - \exp(-(TD2 - TD)/TAU)) \quad (7.10)$$

The last parameter in (7.10) is the time constant, defined by a default value, such as $TAU = 1$. The time constant must not be zero, nor should it be too small; either situation will cause an arithmetic fault and the program may crash.

It is not much more difficult to define a pulse train of arbitrary waveform, such as in Fig. 7.36. To define the pulse, we need the time τ measured from the start of the pulse. This can be found as the *mod* (%) of $t - TD$ with respect to the pulse period *PER*, i.e. the remainder after dividing by the pulse period,

$$\tau = (t - TD)\%PER \quad (7.11)$$

This expression can be used as the argument of a function defining the pulse shape

$$e = t < TD?E1 : ((t - TD)\%PER < TR? \; E1 + KR * ((t - TD)\%PER):$$
$$((t - TD)\%PER < TR + PW? \; E2:$$
$$((t - TD)\%PER < TR + PW + TF? \quad (7.12)$$
$$E2 - KF * ((t - TD)\% \; PER - TR - PW):E1)))$$

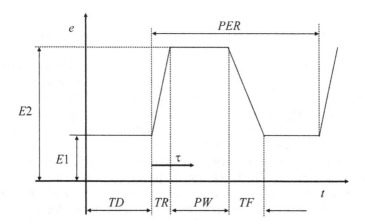

Fig. 7.36 The pulse function

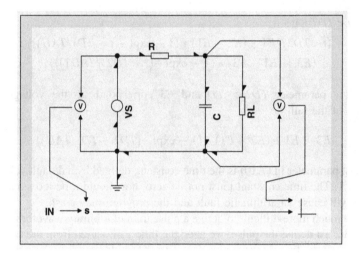

Fig. 7.37 Simulation of noise filtering

$E1$ and $E2$ are the pulse offset and pulsed value, respectively, and

$$KR = TR > 0?(E2 - E1)/TR : 0$$
$$KF = TF > 0?(E2 - E1)/TF : 0 \qquad (7.13)$$

The effect of such pulses on the circuit of Fig. 7.37 is described next. This circuit model can be found in the library project *Filtering of Noise Pulses*. The resistor R and capacitor C are used as a high frequency filter of noise transmitted to the load resistor RL [6].

Signal noise is simulated by a voltage source that generates a rectangular pulse train with the following parameters: $TD = 5$ ns, $TF = TR = 0$, $TW = 2$ ns, $PER = 10$ ns, $E1 = 5$ V and $E2 = 6$ V. The pulse is defined by

$$e = t < TD?E1 : ((t - TD)\%PER < TF + PW + TR)?E2 : E1) \qquad (7.14)$$

The resistors are as in Fig. 7.20, with the resistance of the filter resistor $R = 75$ ohm and that of the load $RL = 10$ kohm. The capacitor includes the parasitic inductance and resistance, and neglects the leakage (Fig. 7.24). The capacitance C is 1.5 mF. It initially is charged to 7.44417 mF, which corresponds to 4.96278 V. The voltages generated by the source and across the load resistor are measured by the voltmeters and fed to the plotter. Because they are in the same range, the source voltage is displaced by 2.5 V using a summator s and a signal generator IN.

Two simulations were run. In the first, the parasitic resistance and inductance of the capacitor were set to zero. The simulation interval was set to 50 ns and the output interval was a relatively short 10 ps, the better to display the transients

Fig. 7.38 Filtering by an ideal capacitor

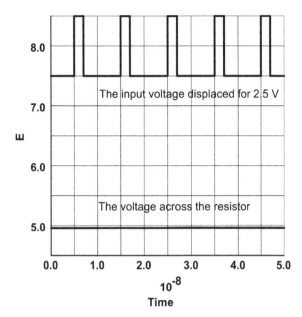

during the short pulse (2 ns). The results (Fig. 7.38) show that the ideal capacitor efficiently removes the noise.

In another simulation the parasitic parameters of the capacitor were set to $L_C = 10$ pH and $R_C = 0.001$ ohm. Simulation parameters were as in the previous run.

The results in Fig. 7.39 show the sharp peaks at the rear and the front edges of the pulses. This is to be expected, as the Fourier transform of a pulse contains all frequencies (Fig. 6.32). Hence, they surely will excite the higher frequency modes of the capacitor.

7.3.4.3 Dependent Sources

Dependent sources are often used to describe the dependence of the voltages or currents supplied by the sources on the other voltages or currents in an electrical circuit. Electric circuit design tools like SPICE define four types of such sources: voltage- and current-controlled voltage sources, and voltage- and current-controlled current sources. We don't need such special devices, because we can use the independent sources of the previous section, to which we add the control ports (Fig. 7.40). These are a direct extension of the controlled components discussed in Sect. 2.5.8.

These models are similar to those of ordinary sources. We need only to add a control port to the corresponding source effort or source flow bond graph

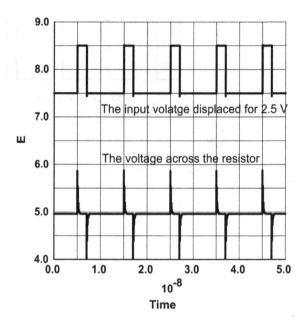

Fig. 7.39 Filtering by a real capacitor

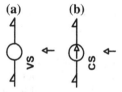

Fig. 7.40 The bond graph representation of controlled sources: **a** the voltage source **b** the current source

component and connect it externally. Figure 7.41 shows a model of a controlled voltage source (see Fig. 7.41).

Sources can be controlled in different ways, depending upon the type of control variable used. If this variable is a current—taken, for example, from an effort junction—it is a current-controlled source; if the signal is a voltage, perhaps taken from an electrical node, it is a voltage-controlled source. It could, of course, be some other variable type, such as the position of a wheel of a manually controlled source.

The voltage generated by the source is defined in the SE or SF component. This can be a linear or non-linear function of the control variable (Sect. 2.5.8), e.g.

Fig. 7.41 The model of the
controlled voltage source

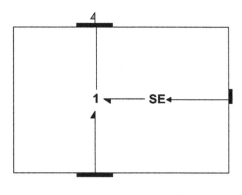

$$e = k * c \tag{7.15}$$

The control c can be the difference of two voltages, or a current flowing somewhere in the circuit.

Using the controlled voltage or current source components has the advantage that the signal serving as the control of the source is clear on the first glance.

7.3.5   Switches

Switches are components often found in electrical circuits. A typical circuit symbol used for switches is shown in Fig. 7.42a. We represent electrical switches by the component given in Fig. 7.42b.

A switch toggles between an open position, in which there is no electrical connection between its terminals; and a closed position, in which the terminals are short-circuited. In SPICE, the switch is modelled as a controlled resistor that toggles between a high resistance value *ROFF* and a low resistance *RON*, depending on the value of the control input. Thus, a controlled resistor can be used to model a switch (Fig. 7.43).

The logic of the switch can be described as

$$c < EON?e = ROFF * i : RON * i \tag{7.16}$$

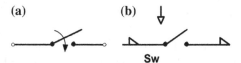

(a)   **(b)**

Sw

Fig. 7.42 The switch: **a** the circuit symbol, **b** the bond graph representation

Fig. 7.43 The model of the
switch as a variable resistor

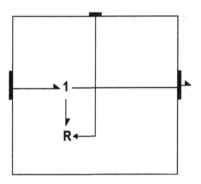

c is the control signal, e is the voltage across the switch, and i is the current
flowing through it. *EON* is the value of the voltage when switching occurs.

The constitutive relation can also be defined in such a way that there is a
continuous transition from the low to the high resistance, e.g.

$$c < EON - EW ? e = ROFF * i :$$
$$(c < EON ? (RON - (ROFF - RON)/EW * (c - EON)) * i : RON * i)$$
$$(7.17)$$

The parameter *EW* is the width of the zone where the transition between the open
and closed switching occurs.

The (7.16) describes a switch that is initially open, i.e. when there is no signal. It
is easy to define a switch that is initially closed, e.g.

$$c < = EON?e = RON * i : ROFF * i \qquad (7.18)$$

Switches are sometimes used to model logical gates. Figure 7.44 shows a circuit
for simulating an OR logical gate. The circuit model can be found in the library
projects under the name *OR Gate Test Circuit*.

As BondSim currently has no predefined symbol for the logical gates, a device
symbol can be used. Inputs to the gate are generated by the two voltage sources
connected to the ground through 1 kohm resistors. The sources generate voltage
pulses according to (7.12) (with *TD* = 0). The parameters of the pulse are given in
Table 7.1. The input signals to the gate are taken from the nodes between the
voltage sources and the resistors. These are fed to the input ports of the gate. The
gate is supplied from a separate 5 V source. The IN input components displace the
inputs of the gate for the better displaying.

Figure 7.45 shows a model of the OR gate implemented using the ideal switches.
The switches are connected to the ground port by a 1 kohm resistor. The switching
logic of the *OR* gate switches are defined by (7.16), with *EON* = 2.5 V,
RON = 1 ohm and *ROFF* = 1 Mohm.

Fig. 7.44 The circuit with an OR gate

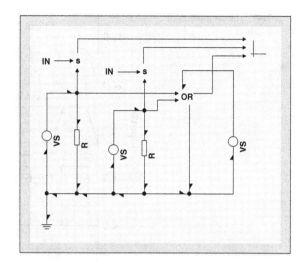

Table 7.1 Parameters of the input pulses of Fig. 7.44

Parameters	Input 1	Input 2
PER (ns)	50	100
TF (ns)	10	10
TR (ns)	10	10
PW (ns)	10	20
E1 (V)	0.0	0.0
E2 (V)	3.0	3.0

Fig. 7.45 The model of an OR gate using switches

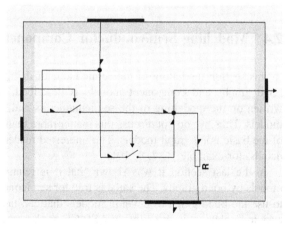

When the model is built, a warning appears informing the user that the model is purely algebraic, i.e. without derivatives. It is, however, acceptable to the numerical solver, so we proceed with the simulation. The results of a simulation run for 200 ns with an output interval of 100 ps are shown in Fig. 7.46.

Fig. 7.46 The simulation of an OR gate

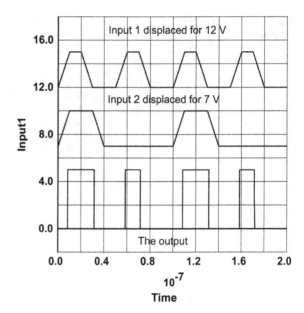

The models of logical gates based on the switches offer an idealized picture of the real devices. They are sometimes useful for representing gates in a simple way; but they often create problems for the simulators. The logical gates are usually built with transistors. Because these introduce some delays, the switching behaviour of the real gates is often slightly different from that predicted by the idealised models.

7.4   Modelling Semiconductor Components

In this section modelling of the semiconductor devices from the point of view of bond graphs and component models is described. Numerous books have been written on the modelling of the semiconductors and, in particular, on their SPICE models. Thus, we do not discuss this matter more than necessary, but describe some of the basic bond graph models. The interested reader can consult the literature for details, for example [1, 7, 9, 10].

In the last section it was shown that it is relatively easy to describe SPICE models by bond graphs. The same is true for semiconductors. But it is also possible to use the bond graph to develop models that can take care of effects that are not easy to implement in programs like SPICE, such as thermal effects. The component model approach of this book, supported by the language used for the description of the underlying physical relations, offers a good basis for developing such models.

We start by developing suitable models of diodes with their bond graph representation. Then, three of the main types of transistors are considered: the Bipolar Junction Transistor (BJT), the Junction Field Effect Transistor (JFET), and the

Metal Oxide Field Effect Transistor (MOSFET). The models developed correspond to SPICE large signal level 1 models. More complex models can be developed in a similar way. The component model approach is shown to be a powerful method for modelling the semiconductors.

7.4.1 Diodes

7.4.1.1 Static Model

We start with diodes, which are fundamental to the functioning of practically all semiconductors. The electrical circuit symbol used for the diode is shown in Fig. 7.47a. The bond graph component model of a diode is quite similar in appearance and is shown in Fig. 7.47b.

The diodes are the resistive components described by a non-linear constitutive relation

$$i_d = Is \cdot (e^{e_d/V_T} - 1) \tag{7.19}$$

Is is the diode saturation current and V_T the thermal voltage, defined by

$$V_T = k \cdot T/q \tag{7.20}$$

where $k = 1.3806 \times 10^{-23}$ JK^{-1} is Boltzmann's constant, T the temperature in degrees Kelvin (K), and $q = 1.6022 \times 10^{-19}$ C the electron charge. Under the nominal working temperature of 300 K (27 °C), the thermal voltage is 0.0258 V. A diode can be represented by a non-linear resistor, as shown in Fig. 7.48a.

Models of diodes are sometimes simplified by representing them internally as modulated switches, instead of non-linear resistive elements (Fig. 7.48b). The constitutive relation of such a switch ensures the correct switching between the *on* (the forward biased) and *off* (the reverse biased) states, e.g.

$$e_d < = 0 \ \& \ i_d < = 0 \ ? \ i_d : e_d = 0 \tag{7.21}$$

When the voltage across the diode is less than or equal to zero *and* the same is true for the current through the diode, the ports are disconnected and there is no current through the diode. Otherwise, its ports are short-circuited and the voltage across the diode is zero.

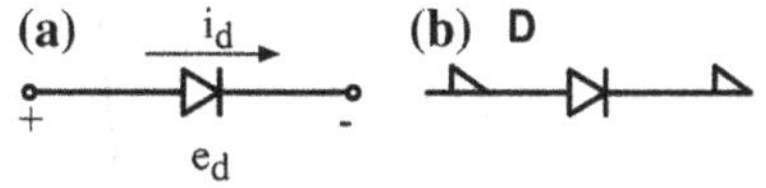

Fig. 7.47 The diode: **a** the circuit symbol, **b** the bond graph representation

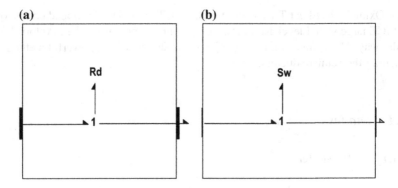

Fig. 7.48 The diode as: **a** a non-linear resistor, **b** a switch

The simplified models of the diodes as switches are of a limited value, as they do not show some of their important features, such as the voltage drop. This could be included in the model, of course, but at the cost of increased complexity. From the numerical point of view, they are less convenient than resistor models. This is because they model the diode by discontinuous algebraic constraints on the voltages across diodes or currents through them.

Experience shows that there is a departure of the behaviour of the real diodes from the ideal law given by (7.19). This occurs over the greater part of the forward and reverse regions, as illustrated in Fig. 7.49 (see e.g. [7] for a detailed explanation). To account for such departures, SPICE models diodes as the current sources with non-linear diode characteristics and a series resistor. The non-linear characteristic is similar to (7.20), but introduces the *emission constant n* in the exponent

Fig. 7.49 The ideal and real diode characteristics

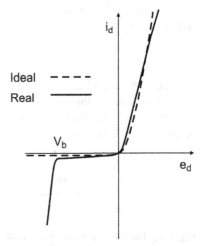

Fig. 7.50 The static model of a real diode

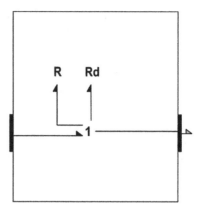

$$i_d = Is \cdot (e^{\hat{e}_d/(nV_T)} - 1) \tag{7.22}$$

The default value of this parameter is 1, which corresponds to the ideal diode, but can be in the range 0.7–3 for the real diodes. This coefficient describes the departure of the diode at a low forward bias. The series resistor, on the other hand, describes the departure at a high forward bias. The voltage $\hat{e}_d$ appearing in (7.22) is the *effective* voltage, equal to the voltage across the diode terminals less the voltage drop across the series resistor, i.e.

$$\hat{e}_d = e_d - R_s i_d \tag{7.23}$$

We follow the same approach but, instead of the current source, a non-linear resistive element Rd is used (Fig. 7.50). The series resistor effect is modelled by a resistive element R added to the resistor effort junction. In this way, the non-linear resistive element effort variable is the effective diode voltage, as given by (7.23).

The constitutive relation of the non-linear resistive element is given by (7.22)

$$id = IS \cdot (\exp(ed/(N * VT)) - 1) \tag{7.24}$$

The saturation current *IS*, the thermal voltage *VT* and the emission constants N are parameters of the model. These parameters can be defined at the diode document level by the default values similar to that of SPICE. These values can be overridden inside the resistive element.

In the reverse bias region the diode current is practically equal to the saturation current. Some diodes show a stronger dependence on the reverse voltage. This can be taken into account by modelling the diode characteristics separately for the forward voltages, as per (7.24); and using a similar expression, but with a different emission parameter, for the reverse region.

One point that remains to be addressed is the reverse diode breakdown. This occurs when the reverse voltage reaches some specific value V_b (Fig. 7.49). The reverse current then suddenly increases to very high values. The process is not necessarily destructive but, owing to the high power dissipated in the diode, it is often critical to its useful life. We describe diode breakdown using a similar approach as in SPICE [4, 6, 7]. The diode characteristics in the breakdown region can be described by

$$i_d = -IS \cdot e^{-(BV0 + \hat{e}_d)/V_T} \tag{7.25}$$

When the voltage is less than $-BV0$ the current starts rising without bound. This critical value is found from the specification of the breakage voltage value BV and the corresponding current $-IBV$, i.e.

$$-IBV = -IS \cdot e^{-(BV0 - BV)/V_T} \tag{7.26}$$

Solving, we get

$$BV0 = BV - V_T \cdot \ln(\frac{IBV}{IS}) \tag{7.27}$$

$BV0$ is found once the breakdown point, defined by BV and IBV, is known. The shape of the function in (7.25) ensures a smooth transition from the characteristics given by (7.22) at the breakdown.

Taken together, the constitutive relation of the resistive element of Fig. 7.50 can be described as

$$
\begin{aligned}
id = ed > {}&- BV0 ? IS \cdot (\exp(ed/(N * VT)) - 1) : \\
&- IS \cdot \exp(-(BV0 + ed)/VT)
\end{aligned} \tag{7.28}
$$

When applying such a relationship for the simulation of diode behaviour, a few points of caution should be noted. One of these relates to the behaviour of the exponential term at high voltages. If the series resistance is zero, then the entire voltage drop occurs across the diode resistive element. Owing to the exponential character of the diode behaviour, the current through the diode could be very high. Such a diode would surely melt down. Thus, the series resistor is an essential guard against this possibility.

Because of the finite precision of floating-point arithmetic, there may be overflow during the numerical calculation. The order of the maximum number that can be represented in double-precision mode typically is 10^{308}. Thus, the maximum value of the exponential term before overflow is about 709, or a voltage across the diode of about 18 V. To prevent this from occurring, it is possible to approximate the exponential function for high values of exponents by a linear function, e.g.

$$
\begin{aligned}
ed/(N * VT) > {}&\text{MEXP?} \\
&IS * (\exp(\text{MEXP}) * (1 + ed/(N * VT) - \text{MEXP}) - 1) : \\
&IS \cdot (\exp(ed/(N * VT)) - 1)
\end{aligned} \tag{7.29}
$$

Fig. 7.51 A simple rectifier circuit

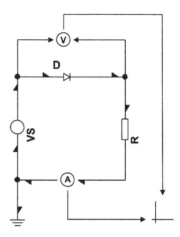

The MEXP parameter is set at 50, but can be changed. It is possible, of course, to limit high currents by using a series resistance [4].

Another important issue is the low conductance at reverse biases, as the exponential term in (7.22) very quickly becomes almost zero. This may cause problems during simulation because the partial derivative matrix of the system equations can become badly conditioned. To remedy this, the approach employed in SPICE is used; that is, to add to (7.29) a constant conduction term of the form $GMIN * ed$, where is $GMIN$ is very low, e.g. 10^{-12}. This term will not change the diode behaviour appreciably, but can help to solve this problem. The constitutive relation of (7.28) now reads

$$id = ed > \ - BV0 \ ? \ IS \cdot (\exp(ed/(N * VT)) - 1) \ + GMIN * ed :$$
$$- IS \cdot \exp(-(BV0 + ed)/VT) + GMIN * ed \tag{7.30}$$

To illustrate the behaviour of a diode based on the model developed, we analyse the *Rectifier Circuit* project of Fig. 7.51. It consists of a voltage source generating a sine voltage of amplitude 50 V at a frequency of 50 Hz, a diode and a 100 ohm resistor. The parameters of the diode are given in Table 7.2 and correspond to the *Motorola 1N4002* general purpose rectifying diode at 25 °C [6]. The breakdown voltage is taken at the default BG_MAX value,[2] i.e. there is no the diode breakdown.

In the first simulation the interval is set to 10 ms, i.e. to the half period of the voltage generated by the source when the diode is forward biased. The goal is to simulate the diode I-V characteristics. The output interval is chosen as 10 μs. The voltage across the diode was plotted along the *x*-axis and the current along the *y*-axis. The characteristic obtained is shown in Fig. 7.52. It can be seen that the current starts rising at a voltage of about 0.6 V.

---

[2]It is set to *DBL_MAX*.

Table 7.2 Model parameters of the diode in Fig. 7.51

Parameter	Value
IS	46.5×10^{-12} A
RS	0.123 ohm
N	1.35
VT	0.0255 V
BV0	BG_MAX

Fig. 7.52 The diode forward voltage versus current

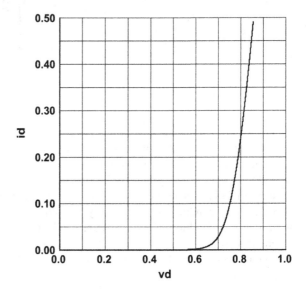

In the second experiment the simulation interval is set to 0.100 s, i.e. five periods of the voltage. Figure 7.53 shows the current in the circuit. During the first half-of the period (0.010 s) the current through the resistor reaches its maximum 0.491 A. When the voltage changes its polarity, the diode becomes reverse biased and there is an extremely low current through the circuit.

To analyse the behaviour of the circuit when the diode breakdown occurs, the VB0 was set to 40 V and the simulation repeated. Figure 7.54 shows a plot of the diode current. During the breakdown there is an appreciable reverse current through the diode (−0.0944 A).

It should be noted that the program failed when the diode breakdown characteristics were described according (7.30). When the breakdown voltage was reached, there was an error message indicating that the program failed to evaluate the partial derivative matrix. To cure the problem, the exponential term was changed as given by (7.29).

Fig. 7.53 The half-rectified
current in the circuit

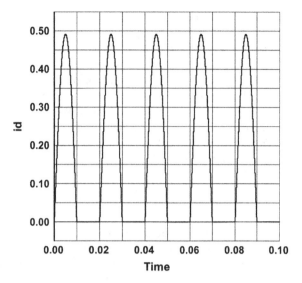

Fig. 7.54 The current in the
circuit when there is diode
breakdown

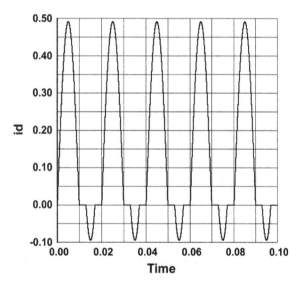

7.4.1.2 Dynamical Model

The model developed so far is quasi-static. Such a model, from the perspective of
the dynamical applications, would be infinitely fast. This is not the case in real
diodes, as the charge-storage effects limit the velocity of the response. To account
for the dynamical effects, we take a similar approach as in the circuit simulators
such as SPICE [4, 7]. Our approach is charge-based, however.

There are two mechanisms of charge storage in diodes: the charges stored in the depletion layer and the charges injected across the layer into neutral regions. The depletion region is an area between n-type and p-type regions. Under no bias voltage, the depletion region consists of the fixed dopant ions, and acts as an isolator. This layer behaves as a plate capacitor of capacitance C_{j0}. Under negative bias (and under positive bias that is less than the built-in potential), the width of this layer changes, as does the capacitance. We use the charge formulation of such capacitor effects instead of the more common capacitance formulation.

The charge of the depletion region can be expressed as a function of the applied voltage using the formula

$$Q_j = \frac{CJ0 \cdot VJ}{1 - M} \left[1 - (1 - (e_d/VJ))^{1-M} \right] \tag{7.31}$$

where CJ0, VJ and M are the zero-bias capacitance, built-in potential and grading coefficient, respectively. The last coefficient can have a value between 0.5 for an abrupt, and 0.33 for the linearly graded, junction. The corresponding capacitance is given by

$$C_j = \partial Q_j / \partial e_d \tag{7.32}$$

From (7.31) we get

$$C_j = \frac{CJ0}{(1 - (e_d/VJ))^M} \tag{7.33}$$

The junction charge and capacitance functions, as given by (7.31) and (7.33), are only approximate and agree with the more exact values at voltages less than about $VJ/2$ [7]. At a voltage equal to the built-in potential, the capacitance is infinite. At such high bias the charges generated by the injection dominate. Thus, it is usually assumed that some error in the charge and the capacitance near the built-in voltage is acceptable. If better accuracy is required, the formula of [11] can be used. We note that, in spite of the fact that we are only interested in the charges, the capacitances naturally appear when the program calculates the partial derivatives of (7.31) by symbolic differentiation. We thus must take care of the behaviour of the capacitances, as well.

We follow the approach in [7] that approximates junction charges by (7.31) up to $FC \cdot VJ$, and employs quadratic interpolation beyond. The corresponding formula reads

$$Q_j = \frac{CJ0 \cdot VJ}{1 - M} \left[1 - (1 - (e_d/VJ))^{1-M} \right], \ e_d < FC \cdot VJ \tag{7.34}$$

Fig. 7.55 The dynamical model of the diode

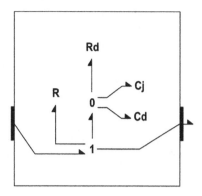

and

$$Q_j = CJ0(\frac{VJ}{1-M}[1-(1-FC)^{1-M}] + \frac{1}{(1-FC)^M} \cdot (e_d - FC \cdot VJ)$$
$$+ \frac{M}{2VJ(1-FC)^{M+1}}(e_d - FC \cdot VJ)^2), \ (e_d \geq FC \cdot VJ)$$

(7.35)

The effect of these charges can be represented by a capacitive element Cj in parallel with the diode's non-linear resistive element (Fig. 7.55). The constitutive relation of the element is defined by (7.34) and (7.35), i.e.

$$q = e_d/VJ < FC \ ? \ CJ0 * VJ * (1 - (1 - e_d/VJ)^\wedge(1-M))/(1-M) :$$
$$CJ0 * VJ * (F1 + F2 * (e_d/VJ - FC) + F3 * (e_d/VJ - FC)^\wedge 2)$$

(7.36)

where *VJ* is the built-in potential, *CJO* the initial junction capacitance value, and

$$F1 = (1 - (1 - FC)^\wedge(1 - M))/(1 - M)$$
$$F2 = (1 - FC)^\wedge(-M)$$
$$F3 = 0.5 * M * (1 - FC)^\wedge(-1 - M)$$

(7.37)

The *FC* typically has a value of 0.5.

The minority-carrier charges injected into the neutral sections also influence the diode dynamics. These are termed the diffusion charges [4, 7] and are given by

$$Q_d = TT \cdot i_d$$

(7.38)

where *TT* is the average transit time parameter of the diode. This large minority charge current occurs for only a short period of time, until access carriers in n- and p-type sides are exhausted. The accumulation of the diffusion charges can be represented by a capacitive element Cd in parallel with the junction charges capacitor Cj, as shown in Fig. 7.55.

Table 7.3 The dynamical
parameters of the diode

Parameter	Name	Value
Zero-biased capacitance	CJ0	51.5×10^{-12} C
Grading coefficient	M	0.333
Junction potential	VJ	0.381
Transit time	TT	5.77×10^{-6} s

Fig. 7.56 Response of the
diode current to a pulse in the
supply voltage

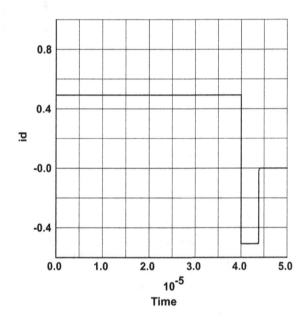

The constitutive relation for the charge is given by (7.24) and (7.38)

$$q = TT * IS \cdot (\exp(ed/(N * VT)) - 1) \tag{7.39}$$

To find the time response of a diode we simulate transients in the circuit of Fig. 7.51. The voltage is pulsed from 0 to 50 V, and then to −50 V. The diode dynamical parameters are given in Table 7.3. The parameters of the generated pulse are (Fig. 7.36): $TD = 0$ s, $TR = 10$ ns, $PW = 40$ μs, and $TF = 20$ ns. $BV0$ is set to the maximum value (BG_MAX). The corresponding model can be found in the library under the project name *Diode Recovery*. The simulation was run for 50 μs with an output interval of 2 ns. The results are shown in Fig. 7.56.

When the pulse switches to a negative voltage, the injected charges cause a large reverse current of practically the same intensity as the forward current (Fig. 7.56). Following the analysis given in [9], the *storage delay time* is approximately equal to 3.91 μs. Only when these charges are removed does the current return to the saturated (or breakdown) value. The necessary transition time depends on the depletion capacitance and circuit resistance. The corresponding time constant can be approximated by Cj0 · R ≈ 0.005 μs. Thus, the total diode recovery time is dominated by the storage delay time. The value found by the simulation is 3.96 μs, which agrees well with these figures. It can be checked (by opening the diode voltage plot) that the diode voltage is ≈0.8 V until the charges are removed.

7.4.1.3   Diode Self-heating

Diodes and other semiconductor devices are known to be notoriously dependent on temperature. It is thus important to model the thermal processes as well as the electrical processes.

The temperature is explicitly contained in the exponent of the diode I-V characteristics through the thermal voltage kT/q. But this is not the only temperature effect because the other parameters, such as the saturation current, built-in voltage, and others are strongly influenced by the temperature. It is thus of interest to take into account this temperature dependence when modelling the static and dynamic behaviour of the diodes.

In SPICE, all circuit parameters are defined at the nominal working temperature of 300 K (27 °C). This temperature can be changed to some other *specified* value and all parameters re-evaluated, but the simulations are run at a fixed temperature. Other languages, such as VHDL-AMS [12], are better equipped to deal with the thermal side of the device modelling. Bond graphs can also be used to model the thermal effects of the devices [2, 13]. The approach taken in this book is particularly appropriate for that purpose.

During the diode's operation a part of the electrical power is transformed to heat. This heat flows out of the diode to the surroundings; but part also accumulates in the diode, thereby changing its temperature. Depending upon the net heat balance, the diode can heat up or cool down. A change of temperature, on the other hand, influences the electrical characteristics of the diode, both static and dynamic.

To account for such interactions, we add an additional port that serves for the heat transfer with the environment (Fig. 7.57). We label such a diode as *DTh* (*Diode Thermal model*).

We develop a thermal model of the diode based on the static model of Fig. 7.50. The proposed model is shown in Fig. 7.58. In addition to the components that model electrical processes, the model also contains a component termed

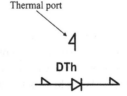

Fig. 7.57 The representation of a diode with self-heating

Fig. 7.58 The diode model
with self-heating

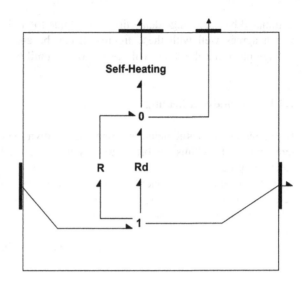

`Self-Heating`, which models the thermal processes. These components interact
through their thermal ports.

Heat generation occurs in elements `R` and `Rd`, where electrical power is transformed into heat that flows out. To represent these heat flows, a thermal port is
added to each resistive element. The thermal processes are modelled using *pseudo-
bond graphs* (Sect. 1.3). In accordance with Table 1.1, the effort-flow pair of
variables at these ports is the temperature *temp* and heat flow *fQ* at the port.

These pseudo bond graphs are convenient for modelling thermal processes. It is
also possible to use true bond graphs that are based on *temperature* and *entropy
flow* pairs. True bond graphs are more general and also much more complicated
than pseudo-bond graphs. They are not really needed here.

The constitutive relations at thermal ports are of the form

$$fQ = e \cdot i \qquad\qquad (7.40)$$

where e and i are the voltage and current, respectively, at the electrical ports. The
heat flow is equal to the input power.

The temperature at the port is defined relative to the nominal temperature
TNOM. Thus, the absolute (thermodynamic) temperature is

$$T = temp + TNOM \qquad\qquad (7.41)$$

The program is designed to start from the state in which the system is not yet
activated; hence, all variables, with the possible exception of certain state variables,
are zero. Regarding temperature, we assume that the system is not at absolute zero,
but has some predefined value *TNOM*, which is taken equal to 300 K.

The temperature at a thermal port affects the processes at electrical ports. In the diode resistive element this is through the thermal voltage defined by (7.20). Thus, we change the constitutive relation (7.24), replacing the thermal voltage parameter *VT* by the expression *BOLTZMANN* · (*temp* + *TNOM*)/*ECHARGE*, where the electron charge and Boltzmann's constants are globally defined (Sect. 4.8.2):

$$ECHARGE = 1.6022 \times 10^{-19}(\text{C})$$

$$BOLTZMANN = 1.3806 \times 10^{-23}(\text{J K}^{-1})$$

Practically all the diode parameters depend on temperature [7]. Temperature dependence of the saturation current is given by

$$\text{is}(T) \ = is(T_0) \cdot \left(\frac{T}{T_0}\right)^{\text{XTI}/\text{n}} \cdot e^{-\frac{q \cdot Eg(T_0)}{nkT}(1 - T/T_0)} \tag{7.42}$$

where T_0 is the nominal temperature and *XTI* is a parameter that has a value 3 (for *pn* diodes); $Eg(T_0)$, the energy gap at nominal temperature, is defined (for *Si*) by the parameter *EGNOM* = 1.115 V.

We describe the right side of the above equation as

$$\begin{aligned} \text{IS} * (1 + \text{temp/TNOM})^{\wedge}(\text{XTI/N}) * \exp(\text{ECHARGE} * \text{EGNOM} \\ * (\text{temp/TNOM})/(\text{N} * \text{BOLTZMANN} * (\text{temp} + \text{TNOM}))) \end{aligned} \tag{7.43}$$

and simply substitute it for the saturation current *IS* in (7.24).The resulting expression is quite complicated, but this is not a problem for the program. We need only to write it correctly.

We can do it in another way, too. The total heat generated is equal to the sum of the heat that flows out of thermal ports, which is represented by a flow junction in Fig. 7.58. The junction variable is the temperature *temp* of the diode junction (relative to *TNOM*). This can be used to define the saturation current using a function component and to return its value through a control input port to the resistive element.

The thermal model of the diode is defined in the component Self-Heating. In general, thermal models of the diodes, as well as those of the other semiconductor devices, are represented using thermal circuits [14]. That is, the continuous thermal processes are discretised and represented using electrical analogies. We do not need such analogies, for we can deal with them directly by bond graphs or, in particular, the pseudo bond graphs [2, 13].

Two basic elements are used to build the thermal models. The first is the *thermal resistor*, represented by a resistive bond graph element (Fig. 7.59a). The constitutive relation of such a resistor is

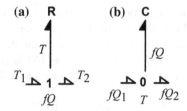

Fig. 7.59 The thermal elements. **a** Resistor. **b** Storage

$$T_1 - T_2 - T = 0$$
$$fQ = fQ(T) \tag{7.44}$$

i.e. the heat flow is a function of temperature. In the linear case, the constitutive relation of the element, is simply

$$fQ = T/R \tag{7.45}$$

where R is the thermal resistance expressed in K/W (or °C/W).

The other element represents the heat storage (Fig. 7.59b) and is defined by

$$fQ_1 - fQ_2 - fQ = 0$$
$$fQ = \frac{dE}{dt} \tag{7.46}$$

where E is the thermal energy. The accumulated energy is

$$E = C \cdot T \tag{7.47}$$

The parameter C is the thermal capacitance and has units of J/K. Strictly speaking, it corresponds to specific heat at constant volume. But in solids and liquids, including semiconductor materials, the work done by the expansion of the material is very small compared to the net heat inflow fQ; hence it is usually referred to simply as the material's specific heat. (7.47) is the constitutive relation of the thermal capacitive element in Fig. 7.59b.

The Self heating component of the diode can be represented by the RC bond graph in Fig. 7.60. The resistive element represents heat that flows due to the temperature gradient between the junction and the diode's outside surface. A part of this heat is accumulated in the diode. This is represented by the capacitive element.

The diode thermal port is normally connected to the components that model the diode environment. This is usually the *heat sink* that removes heat from the diode, ensuring that its temperature is held within the acceptable margins (Fig. 7.61). The other side of the sink is usually at ambient temperature, here represented by a SE component. Thermal models of such sinks can be represented in a similar way using a RC bond graph.

Fig. 7.60 The thermal model
of the diode

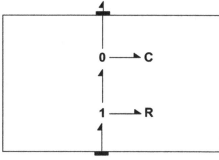

Fig. 7.61 The diode
thermally connected to a heat
sink

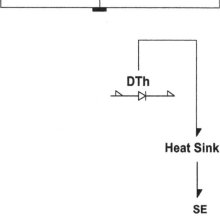

Heat sinks are usually made of extruded aluminium and for which the manufacturers provide the thermal data, such as the thermal resistance, volume, and material. The complete thermal models also should include the resistance of the heat path to the sink, which depends on the design of the semiconductor components. A component is usually enclosed in a case, and between the case and the heat sink there is some layer of isolation. The thermal model of the diode and heat sink, as shown in Fig. 7.61, is a simplified one. For proper thermal modelling, details of the design should be taken into account. Detailed models usually consist of several RC segments connected in series [14].

To illustrate the application of models of the diode, including the self-heating, we return to the rectifier circuit of Fig. 7.51. We replace the diode model with its thermal model (Fig. 7.62). Information on the diode temperature is taken from the heat flow junction (Fig. 7.58) and sent to the display component. The corresponding project can be accessed from the BondSim project library under the title *Rectifier circuit with self-heating*.

We simulate the temperature rise in the diode following a sudden voltage change from 0 to 50 V. Parameters of the diode thermal model and of the heat sink are given in Table 7.4. The other side of the sink is held at nominal temperature. These parameter values have been chosen to illustrate the behaviour of the diode under

Fig. 7.62 The rectifier circuit
with the thermal diode model

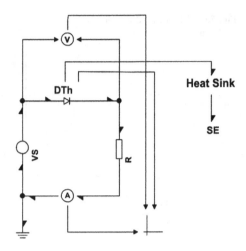

Table 7.4 Thermal
parameter of the circuit

Parameter	Value
Diode thermal resistance	0.1 K/W
Diode thermal capacitance	0.0001 J/K
Heat-sink thermal resistance	10 K/W
Heat-sink thermal capacitance	0.01 J/K

changing temperature conditions; real values could be derived from, for example,
the cooling tests of the real devices. Simulation results are presented in Fig. 7.63.

7.4.2   Transistors

The bipolar junction transistor was invented by Bardeen, Brattain, and Shockley in
1947 while working at Bell Laboratories. Rightly considered to be one of the
greatest inventions of our time, it earned them the Nobel Prize in physics in 1956.
For a long time after its invention the BJT remained one of most important
three-terminal devices. It is found in amplifiers and drivers and, even today, serves
as one of the most important devices in a wide array of applications.

The field effect transistor (FET) emerged shortly after appearance of the BJT.[3] In
fact, many different types of FETs were developed, their difference depending upon
how isolation between the gate and channel was implemented. These include the

---

[3]FET was patented in 1925 by Julius Edgar Lilienfeld in Canada. However, Lilienfeld did not
publish any research articles about the device. The production of high-quality semiconductor
material was decades away. MOSFET was invented by Dawon Kahng and John Atalla at Bell Labs
in 1959.

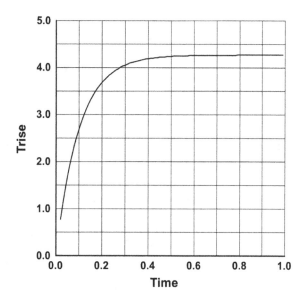

Fig. 7.63 Transient of diode temperature with change in the source voltage output

junction FET or JFET, metal semiconductor FET or MESFET, metal-oxide FET or MOSFET, as well as others.

We will not go into the details of the design and functioning of BJTs or FETs, for this is outside the scope of this book. The interested reader can consult the specialised books on the semiconductors, such as [7, 9, 10]. We will, however, show how models of some basic types of transistors can be developed as components that can be used to study complex mechatronic or micro-mechanic systems. We follow the modelling approach of SPICE. The models we present are fundamental ones that correspond roughly to *Level 1* models. More advanced component models can be developed in the same way.

We wish to stress that the aim here is *not* to develop the special bond graph models that are, perhaps, energetically correct, but often not so easy to comprehend by those who are not experts in bond graphs. Our models will be visually close to the electrical schemas used to describe SPICE models, but they are the complete component models without anything hidden.

7.4.2.1 Bipolar Junction Transistor

The Bipolar Junction Transistor (BJT) is a three-terminal device. The terminals commonly are denoted as the *emitter* E, *base* B, and *collector* C. There are two main types of BJTs: the *npn* and the *pnp*. The electrical circuit symbol used for npn transistors is shown in Fig. 7.64a.

The emitter arrow shows the direction of the current under normal operations. The positive directions of currents at the other terminals are shown. These are slightly different from those in SPICE [4, 7]. The corresponding bond graph

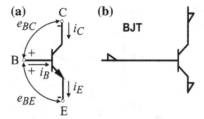

Fig. 7.64 npn BJT: **a** the circuit symbol, **b** the bond graph representation

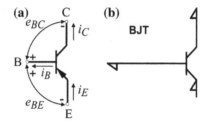

Fig. 7.65 pnp BJT: **a** the circuit symbol, **b** the bond graph representation

component representation is shown in Fig. 7.64b. It is assumed that power flows into the component at the base and the collector ports, and flows out at the emitter port. A component represented by such a symbol can be created in BondSim by choosing the *Bipolar Junction Transistor* tool from the *Electrical Component* branch in the *Template Box* (Fig. 7.2). The text *BJT* is simply a label used for reference and can be changed at the component construction time or later.

In the *pnp* type of *BJT*, the positive direction of the currents is just the opposite (Fig. 7.65a). Hence, the *pnp* bond graph assumes positive senses of power-flow at all ports to be the opposite to that of the npn type (Fig. 7.65b). Thus, to create an empty *pnp* BJT component, the *npn* component is created first, then the power direction of each of the ports is changed. This is accomplished by selecting the component and applying the *Change Ports All* command on the *Edit* menu.

We develop a large signal model of a BJT based on the Ebers-Moll model, specifically the transport version [7]. The *npn* transistor model as given (Fig. 7.66) consists of two diodes that model two back-to-back pn junctions, and a current source that models the transport of current from the collector to the emitter. The models used for the diodes are slightly modified models of Sect. 7.4.1. They consist of the non-linear resistive and capacitive elements, but without the series resistive element (Fig. 7.67). In addition a 1-junction is inserted between nonlinear diode resistor and 0-junction to extract information on the diode current. The transistor ohmic resistances are accounted for by the parasitic base Rb, collector Rc, and emitter Re resistors (Fig. 7.66). The constitutive relations in the non-linear resistive elements of the base-collector diode and base-emitter diode are:

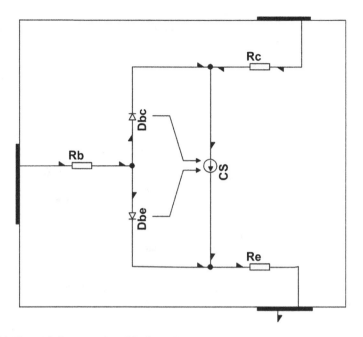

Fig. 7.66 Ebers-Moll transport model of npn BJT

$$i_{dBC} = (Is/\beta_R)(e^{e_{BC}/(n_R VT)} - 1) \tag{7.48}$$

$$i_{dBE} = (Is/\beta_F)(e^{e_{BE}/(n_F VT)} - 1) \tag{7.49}$$

where I_s, β_R and β_F are the transistor saturation current, reverse and forward current gains, respectively. The current source represents the transport of charges given by

$$I_{CT} = \beta_F I_{dBE} - \beta_R I_{dBC} = Is(e^{e_{BE}/(n_F VT)} - e^{e_{BC}/(n_R VT)}) \tag{7.50}$$

To generate a current given by (7.50), it is necessary to supply the current source in Fig. 7.66 with the information on two diode currents given by (7.48) and (7.49). It is now an easy matter to define the current source that implements the constitutive relation (7.50). This is shown in Fig. 7.68.

The function component calculates the right side of (7.50). The generated output controls the SF component, which generates the transport of charges according to (7.50).

The charge storage is modelled by the capacitive elements that are a part of the collector and emitter junction diode models (Fig. 7.67). Constitutive relations of these capacitors are as described in Sect. 7.4.1. The model developed doesn't include secondary effects, which could be included by following a similar approach as in the SPICE [7].

Fig. 7.67 Model of the
base-collector and
base-emitter diodes

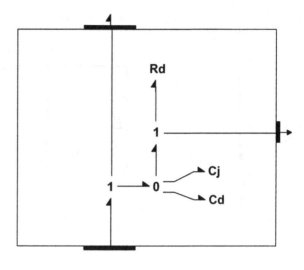

Fig. 7.68 Model of the
current source

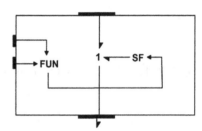

To simulate the transistor characteristics, we create the project *npn Transistor Characteristic*. The system level model shown in Fig. 7.69 represents the model of a set-up for the measurement of the *npn* bipolar junction transistor characteristics in the common emitter configuration.

The transistor base terminal is fed by a current source and the collector terminal is supplied by a voltage source Vcc, which is a sinusoidal with an amplitude 10 V and a frequency of 10 Hz. The voltage of the collector terminal with respect to the emitter terminal is measured using voltmeter V, and the current flowing through the collector terminal is measured by ammeter A. Their outputs are connected to the plotter. We wish to determine a plot of collector current *Ic* versus the collector to emitter voltages *VCE* for the base current in the range of 0–50 μA. The parameters of the transistor are given in Tables 7.5 and 7.6.

The default values of parameters of Table 7.5 are defined at the transistor document level; those of the diodes in Table 7.6 are defined at the document level of the respective diodes. The values that override these defaults must be defined on a higher level (closer to the components). Thus, for example, the parasitic resistances are defined in the respective resistors. What is a default and what is a specified value is not predefined, as in SPICE; this decision is left to the modeller.

Fig. 7.69 Measurement
set-up for a npn BJT

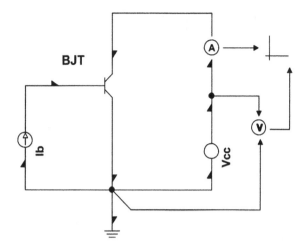

Table 7.5 Parameters of the transistor of Fig. 7.69

Parameter	Symbol	Name	Value
Saturation current	I_s	IS	10^{-16} A
Thermal voltage	V_T	VT	0.0258 V
Reverse gain	β_R	BR	1
Forward gain	β_F	BF	80
Base resistance	R_b	RB	5 ohm
Collector resistance	R_c	RC	1 ohm
Emitter resistance	R_e	RE	0.01 ohm

Table 7.6 Forward and reverse diode parameters of the transistor

Parameter	Name	Value
Zero-bias junction capacitance	CJ0	0.0 F
Junction potential	VJ	0.75 V
Grading coefficient	m	0.333
Coefficient for forward-biased depletion capacitance	FC	0.5
Average transient time	TT	0.0 s

In the display object (Fig. 7.69) we assign the collector current to the y-axis and label it as *Ic*, and the voltage to the *x*-axis and label it as *VCE*. During the simulation the outputs are normally stored in, and plotted by, the plotter object. To have a family of curves plotted on the same plot, we must perform several simulation runs and store all the resulting values in the same object, then plot them. This can be done relatively simply.

We start, for example, with a base current of 10 μA, build the model, and run the simulation for 0.1 s. We select the output and maximum interval to be 0.0001 s. Next, we change the current value to 0 μA. We need not rebuild the model, as it is

Fig. 7.70 Simulation Options
dialogue with Restart and
Reset controls

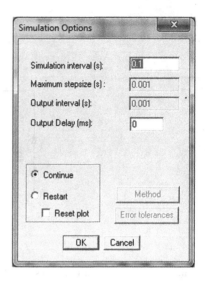

updated automatically when we click the *OK* button after changing the current. We
then rerun the simulation, but in the *Simulation Option* dialogue (Fig. 7.70) we
select the *Restart* button and uncheck the *Reset plot* check box. The first option
forces the simulation to restart from zero; the second prevents release of the data
from the previous run. We generate transistor characteristics for base currents up to
50 µA.

The resulting graph is shown in Fig, 7.71. In the graph the curves are labelled by
the value of the base current in µA, which are added using the right mouse button as
explained in Sect. 7.2. The diagram shows the familiar emitter characteristics for
different values of the base current. Because of different current gains for the
forward and reverse diodes (80–1), they are asymmetrical. When the base current is
50 µA the corresponding collector current is 4 mA. This is in accordance with the
forward current gain of 80.

The pnp BJT is similar to the npn type with holes replacing the roles of the
electrons. Thus, the directions of currents and polarities of the voltages across the
emitter, base, and collector are opposite to that of the *npn* BJT. The corresponding
component representation already has been discussed (Fig. 7.65); its model is
developed in a similar way.

We can also create a model of the *pnp* BJT directly by using a copy of the
corresponding *npn* model. Starting with the *npn* BJT (Fig. 7.65b), we select the
component and change the sense of all ports by the *Change Ports All* toolbar button
(or by the corresponding command in the *Edit* menu). This changes the external
ports of the component and also the ports of all components inside its document that
are connected to them, i.e. ports of resistors Rb, Rc, and Re (Fig. 7.66).

We also need to change the power-flow direction of the ports on the other side of
the resistors. Next, we select the diodes and the current source and change power
directions of all their ports. A model of a *pnp* BJT can be found in BondSim's

Fig. 7.71 Characteristics of
npn BJT

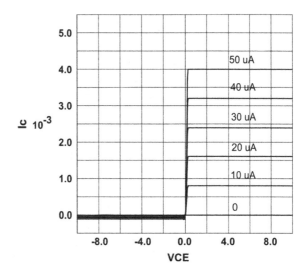

Component Library. The *pnp Transistor Characteristics* project parallels the *npn* transistor project discussed above and is also found in the *BondSim* projects library.

7.4.2.2   Junction Field Effects Transistor

The Junction Field Effect Transistor (JFET) is a three-terminal voltage-controlled device. Its operation involves an electric field that controls the flow of charge through it. JFET terminals commonly denoted as the *source* S, *drain* D and the *gate* G and are analogous to the emitter, collector, and base terminals of the BJT. The source and drain are ohmic contacts. There is a conductive channel between the drain and the source, through which current flows. The third terminal, the gate, forms a reverse-biased junction with the channel. The conductivity of the channel is modulated by a potential applied to the gate. As the conduction process involves predominantly one type of carrier, JFETs also are called unipolar transistors.

Depending on the type of material involved, a JFET may be referred to as an *n-channel* or a *p-channel* JFET. The electrical circuit symbol used for n-channel JFETs is shown in Fig. 7.72a. The corresponding bond graph component representation is shown in Fig. 7.72b.

It is assumed that power flows into the component at the gate and the drain ports, and flows out at the source port. A component represented by such a symbol can be created by the *n-channel JFET* tool of the *Electrical Component* branch in the *Template Box* (Fig. 7.2). The text JFET is simply a label used for reference and can be changed when the component is constructed or later.

The polarities are just the opposite in a *p-channel JFET* (Fig. 7.73). Thus, the p-channel JFET component can be created from an n-channel component by

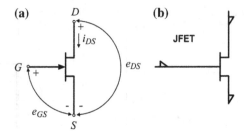

Fig. 7.72 n-channel JFET: **a** the circuit symbol, **b** the bond graph representation

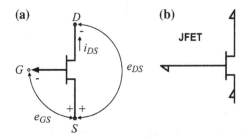

Fig. 7.73 p-channel JFET: **a** the circuit symbol, **b** the bond graph representation

reversing the power flow direction of all ports. This can be done by selecting the component and using the command *Change Ports All*.

It is also possible to change only the gate port. In this case, the drain and the source of the p-channel JFET change places with respect to the n-channel JFET. The n-channel and p-channel JFETs differ in the sense of the gate port power flow. The drain of the n-channel component is a port where the power flows *in*; for the p-channel, the power flows *out*. For the sources, the opposite is true.

The direction of power flow through JFETs is taken to correspond to the normal mode operation in which the drain of the n-channel device is at a higher potential with respect to the source. Thus, electrical power flows in at the drain and flows out at the source. The power flow direction in the p-channel JFETs is just the opposite. The source is normally at a higher potential, so the current flows from the source to the drain.

We develop the model of the n-channel JFET (Fig. 7.74) that consists of the two diodes, DS and DD, which model the reverse-biased junctions between the gate and the channel, and the voltage-controlled resistor. The resistors RD and RS represent the ohmic resistances of the drain and source. This model corresponds to the large-signal SPICE model with the controlled resistor R replacing the dependent current source [4]. The constitutive relation of the resistor is given by [4]

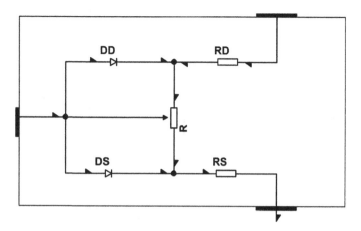

Fig. 7.74 Model of n-channel JFET component

$$
i_{DS} = \begin{cases} 0,\; e_{GS} - VT0 \le 0 \\ \text{BETA} \cdot e_{DS} \cdot (2(e_{GS} - \text{VTO}) - e_{DS})(1 + \text{LAMBDA} \cdot e_{DS}), 0 < e_{DS} < e_{GS} - VT0 \\ \text{BETA} \cdot (e_{GS} - \text{VTO})^2(1 + \text{LAMBDA} \cdot e_{DS}), e_{DS} \ge e_{GS} - VT0 > 0 \end{cases}
$$

$$(7.51)$$

The *VT0* parameter is the threshold (pinch-off) voltage. This determines the gate bias at which the channel is completely pinched-off and there is effectively no current through the device [4, 9]. If *VT0* < 0, then at $e_{GS} = 0$ the device is in the *on* condition and, under a positive drain to source voltage, current will flow from drain to source. To cut off the device it is necessary to apply a negative gate to the source voltage. Such device operation is known as the *depletion mode*. In the *enhanced mode VT0* > 0, the device is pinched-off initially and it is necessary to apply a positive voltage larger than the threshold to enable the device to conduct the current. The parameter BETA is a trans-conductance parameter, and LAMBDA is the output conductance at saturation.

Equation (7.51) shows that at a relatively small drain-to-source voltage e_{DS} the current increases with the voltage until the saturation voltage (equal to $e_{GS} - VT0$) is reached. This region is called the linear region of the operation. When the drain-to-source voltage increases above the saturation voltage, the drain-source current is practically independent of the voltage and the device is saturated.

There is an additional current component due to the pn junction current, characterized by saturation current *IS*. The gate pn junctions are reverse biased, and thus the pn junction current is negligible.

The diode models consist of a non-linear resistor and a capacitor (Fig. 7.75). Because in JFETs the diodes normally are reverse-biased, there is no diffusion charge. Hence, the charges consist of fixed ions in the depletion region and are represented by a junction capacitor. Following (7.31) the charge of the depletion region can be expressed as a function of the applied voltage using the formula

Fig. 7.75 Model of the junction diodes

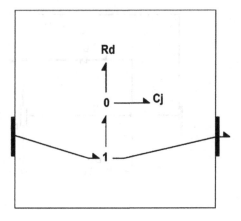

Fig. 7.76 Model of controlled resistor

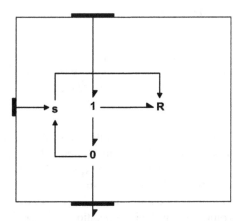

$$Q_{jDS} = \frac{CGS \cdot PB}{1 - M} \left[1 - (1 - (e_{GS}/PB))^{1-M} \right]$$
$$Q_{jDD} = \frac{CGD \cdot PB}{1 - M} \left[1 - (1 - (e_{GD}/PB))^{1-M} \right] \tag{7.52}$$

Where *CGS*, *CGD*, *PB* are the zero-bias gate-source capacitance, zero-bias gate-drain capacitance, and built-in junction potential, respectively. The grading coefficient *M* is set to default value of 0.5 [4].

The controlled resistor defined by (7.51) can be described as shown in Fig. 7.76. Its model is similar to other resistors, i.e. it consists of a resistive element R connected to an effort junction, the ports of which also are connected to the document (external) ports of the resistor. In this way, the effort and flow variables of the resistive element are the drain-to-source voltage e_{DS} and the channel current i_{DS}.

The element has also a control port to collect information on the gate bias. Note that the input signal at the document left port is the gate potential (Fig. 7.74). Thus,

Fig. 7.77 Measurement
set-up for n-channel JFET

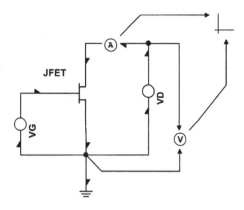

to create the gate-to-source voltage e_{GS} a 0-junction is inserted, the junction variable of which is the source potential. A summator is used to evaluate the difference of these two potentials e_{GS}, and its output is connected to the resistive element control-input port.

The constitutive relation (7.51) is valid for drain-to-source voltage $e_{DS} \geq 0$. This corresponds to the normal mode of operation. In the inverted mode ($e_{DS} < 0$), the drain and source ports switch roles and the current flows in the opposite direction. The corresponding constitutive relation is the symmetric [4, 7]. Thus, instead of the e_{GS} voltage, the e_{GD} voltage is used in (7.51). Similarly, e_{DS} is changed to $-e_{DS}$ and the sign of the current is changed.

As was the case with the BJT, we can find the i_{DS}-e_{DS} characteristics of a JFET by simulation. For this purpose we create a project *n-Channel JFET Characteristics*, the system level model of which is given in Fig. 7.77.

The drain of the JFET is supplied from the voltage source VD generating voltage ramp 0–5 V. The gate port is connected to a separate voltage source. The source of the JFET is grounded. We measure the voltage across the drain and source ports by the voltmeter V and the current through the component by ammeter A. The instruments outputs are fed to the plotter for display. The JFET parameters correspond to the 2N4416 JFET of reference [6] and are given in Table 7.7.

To generate the i_{DS}-e_{DS} characteristics of the JFET, several simulations were run with constant gate voltage ranging from 0 to −3.32 V. The simulation interval was set to 1 s, which corresponds to increasing the drain-to-source voltage from 0 to 5 V. The output interval was 0.001 s. The outputs of each simulation run are stored in the display component. The results are given in Fig. 7.78.

The transistor is in the depletion mode. As the gate bias becomes negative, conduction of the transistor drops and, at voltage of −3.32 V, is completely pinched-off. The curves also show a linear region in which the source voltage increases with the drain-to-source voltage; and the saturation region in which the current is independent of the drain-to-source voltage.

The discussion so far has been for n-channel devices, but a similar discussion is valid for p-channel devices. In p-channel devices all voltage and current polarities,

Table 7.7 Basic 2N4416
JFET parameters [6]

Parameter	Name	Value
Threshold (pinch-off) voltage	VT0	−3.32 V
Trans conductance parameter	BETA	0.05 A/V^2
Chanel length modulation parameter	LAMBDA	0.00928 1/V
Source ohmic resistance	RS	0.575 ohm
Drain ohmic resistance	RD	0.575 ohm
Gate junction saturation current	IS	5×10^{-12} A
Gate junction potential	PB	0.76 V
Zero-bias GS junction capacitance	CGS	3.37 pF
Zero-bias GD junction capacitance	CGD	3.37 pF

Fig. 7.78 Simulation of the characteristics of 2N4416 JFET

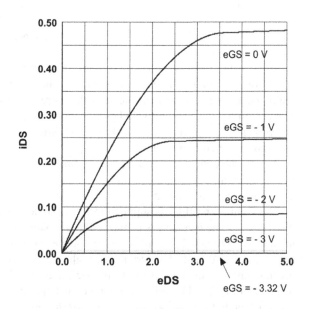

including the threshold voltage and the directions of two gate junctions are reversed [7]. Likewise, greater-than and less-than relations must be reversed. We retain, however, the same constitutive relation as that used for the n-channel device: Instead of terminal voltages e_{GS} and e_{DS}, we use reverse voltages e_{SG} and e_{SD}. Similarly, VT0 is now the negative of the threshold voltage.

Thus, a positive value of VT0 corresponds to a device that is off—in the *enhanced mode*—for both n-channel and p-channel devices. Similarly, the negative value corresponds to a *depletion* mode device, which is initially on, for both n-channel and p-channel devices. The change in the polarity of the voltages across the terminals and of the direction of the current is taken care of by a change in the power flow direction through the p-channel device (Figs. 7.74 and 7.75).

The model of the p-channel JFET is similar to the n-channel model of Figs. 7.74, 7.75 and 7.76, but with direction of the power flow between the external ports reversed. The change of the power flow direction should be applied only to the bonds through which power is transferred between the external ports. The others are not affected, e.g. the power flow direction of the non-linear resistive element and of the capacitive element of the diode model in Fig. 7.75, as well as that of the controlled resistive element of the resistor in Fig. 7.76. These are the same for both n-channel and p-channel devices. In addition, we need to change the signs of the summator inputs of Fig. 7.76, because these components are used to evaluate Vgs, the voltage used in (7.51).

This way, the p-channel JFET model can be created from the n-channel model by changing the port power flow directions. This is done by using the *Change Ports All* command for all component ports, or the *Change Port* command for a particular port, then changing the summator input signs.

7.4.2.3 Metal-Oxide Semiconductor Field Effect Transistors

The metal-oxide semiconductor field effect transistors (MOSFET) are so called because their gate is isolated from the channel by an oxide layer [9]. One of the major advantages of MOSFET technology is low cost and the possibility of dense packing. They also enable both the p- and n-channel devices to be made on the same substrate, leading to the so-called *Complementary* MOSFETs or CMOSs. Today MOSFETS are one of the most important semiconductor technologies.

Many different models of MOSFET have been developed and are used to describe the real devices—ranging from the digital to analog, and of different power capabilities and frequencies—more accurately. We limit our discussion to a basic MOSFET model corresponding to the SPICE model level 1. More complex models can be developed using a similar approach. Much of the earlier discussion on JFETs applies to MOSFETs as well and will not be repeated here.

MOSFETS, or MOS, for short, are basically *four*-terminal devices that, in addition to gate, drain, and source terminals, have a *bulk* terminal. Normally, this is connected to a terminal with the most negative potential for the n-channel MOSFETS, and to the most positive terminal in case of the p-channel MOSFETS. It can also be used for the additional control of the device [9].

The electrical circuit symbol used for the n-channel MOS (NMOS) is shown in Fig. 7.79a. The voltage polarities and the current direction correspond to *normal* operation. The corresponding bond graph component is shown in Fig. 7.79b. It is assumed that power flows into the component at the gate, bulk, and the drain ports; and flows out at the source port.

The polarities of the p-channel MOSFETs (PMOS) are just the opposite (Fig. 7.80a). Similarly, the port power flow senses of the corresponding bond graph components also are reversed (Fig. 7.80b).

The NMOS component can be created using the *n-channel MOSFET* tools of the *Electrical Component branch in the Template Box*e (Fig. 7.2). The text MOS is just

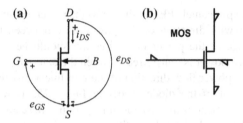

Fig. 7.79 n-channel MOSFET: **a** the circuit symbol, **b** the bond graph representation

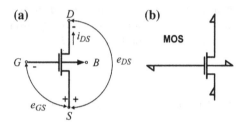

Fig. 7.80 P-channel MOSFET. **a** the circuit symbol, **b** the bond graph representation

a label used for the reference to the component and can be changed at this stage or later. The p-channel component can be created from an n-channel component by reversing the power flow direction of all ports. It is also possible to change only the base and bulk ports. In this case, the drain and source of the n-channel MOSFET effectively change places.

The n-channel MOSFET model (Fig. 7.81) is dynamic and corresponds to the Level 1 large-signal SPICE model. We use a controlled resistor to describe the static characteristics of NMOS, as we did in the JFET case. The constitutive relation of the resistor is defined by relation [4, 7]

$$i_{DS} = \begin{cases} 0, e_{GS} - VTH \leq 0 \\ 0.5KP \cdot e_{DS} \cdot (2(e_{GS} - VTH) - e_{DS})(1 + LAMBDA \cdot e_{DS}), 0 < e_{DS} < e_{GS} - VTH \\ 0.5KP \cdot (e_{GS} - VTH)^2 (1 + LAMBDA \cdot e_{DS}), e_{DS} \geq e_{GS} - VTH > 0 \end{cases}$$

$$(7.53)$$

where KP is the effective transconductance (scaled by the ratio of the device width and effective length). VTH is the threshold voltage in the presence of back-gate bias, $e_{BS} < 0$,

$$VTH = VT0 + GAMA \cdot (\sqrt{PHI - e_{BS}} - \sqrt{PHI}) \qquad (7.54)$$

VTO, GAMMA, PHI and *LAMBDA* are the threshold voltage, bulk threshold parameter, surface potential, and output conductance factor in saturation, respectively.

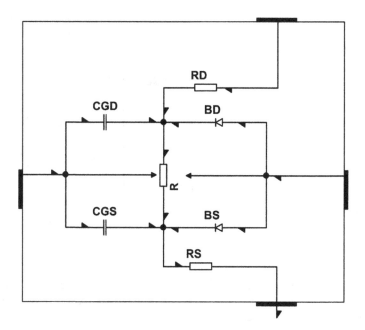

Fig. 7.81 Model of NMOS

In MOSFET there is no direct resistive current path between the gate and either the drain or source because of the oxide isolation layer. There is charge accumulation, however, that is modelled by capacitors CGD and CGS, between the gate and the drain and the gate and the source, respectively. In a first-order analysis these can be described by the constant capacitances. The capacitance between gate and body is neglected.

There also are junctions between the body, the source, and the drain that are represented by diodes (Fig. 7.81). These junctions are reverse-biased. We use the same model of the diodes as in JFET (Fig. 7.75, (7.52)). The capacitances Cj in that model describe the accumulation of the junction charges.

The discussion regarding direct and reverse operation of JFET applies here, as well. Thus, e_{GS} in (7.53) is replaced by e_{GD}, e_{BS} is replaced by e_{BD}, e_{DS} is changed to $-e_{DS}$, and the sign of the current is changed.

The same model applies to PMOSs. The direction of power transfer trough the component, however, is opposite to that of NMOS. Thus, the PMOS model can be created from the NMOS model in the same way as discussed for JFETs. The body port of the PMOS should, however, be connected to the port of higher potential in order that the body junction be the reverse-biased, to the drain for direct operation, and to the source for the reverse.

As a first application of the MOS model developed, we create a project named *NMOS Characteristics* that simulates the *iDS-eds* characteristics of an n-channel MOSFET. The corresponding system-level model is given in Fig. 7.82. The project

Fig. 7.82 Set-up for measurement of NMOS characteristics

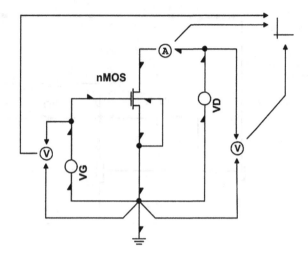

is similar to that of the evaluation of the JFET characteristics (Fig. 7.77). The drain junction of the NMOS is set at 0–5 V by the voltage source VD. Its gate port is connected to a separate voltage source VG.

Because ports of the same component cannot be directly interconnected, a separate branch node is inserted and is used to connect the NMOS source and body ports. The node is grounded. There are also instruments for measuring the gate and drain voltages and drain-source current. The outputs of these instruments are fed to a plotter for display. The NMOS parameters used for simulation are given in Table 7.8.

Several simulations were run with different values of the gate voltage 4–10 V to generate the i_{DS}-e_{DS} characteristics. The simulation interval was set to 1 s, with the output interval 0.001 s. Note that the transistor threshold voltage $VT0 = 3.5$ V (Table 7.8); hence, this is an enhanced mode NMOS and to conduct the current the gate bias must be greater than 3.5 V.

The results (Fig. 7.83) show a linear region in which the source voltage increases with drain-to-source voltage; and the saturation region in which the current is constant. According to (7.52) the linear range ends up at the value of the drain-to-source voltage $e_{DS} = e_{GS} - VTH = e_{GS} - 3.5$ V.

The program library also contains a project *PMOS Characteristics* that the reader is invited to analyse on his own.

We conclude this section with the analysis of the basic *CMOS inverter* (Fig. 7.84). It consists of a PMOS and a NMOS transistor, the drains of which are connected through a common node. The PMOS is supplied by a 5 V source; the source of the NMOS is grounded. The input voltage is applied to both transistors and the voltage is taken from the common node.

The parameters of the inverter are given in Table 7.9. Note that the threshold voltage of both transistors is 1 V; hence, they are in the enhanced mode.

Table 7.8 NMOS parameters

Parameter	Name	Value
Threshold (pinch-off) voltage	*VT0*	3.50 V
Transconductance parameter	*KP*	0.001 A/V^2
Chanel length modulation parameter	*LAMBDA*	0.01 1/V
Bulk threshold parameter	*GAMMA*	0.0 V
Surface potential	*PHI*	0.6 V
Source ohmic resistance	*RS*	10 ohm
Drain ohmic resistance	*RD*	10 ohm
Bulk junction saturation current	*IS*	1×10^{-14} A
Bulk junction potential	*PB*	1 V
Grading coefficient	*M*	0.5
Coefficient for depletion capacitance	*FC*	0.5
Emission coefficient	*N*	1
Zero-bias BS junction capacitance	*CBS*	5 pF
Zero-bias BD junction capacitance	*CBD*	5 pF
GD overlap capacitance	*CGD*	1 pF
GS overlap capacitance	CGS	1 pF

Fig. 7.83 Simulation of NMOS characteristics

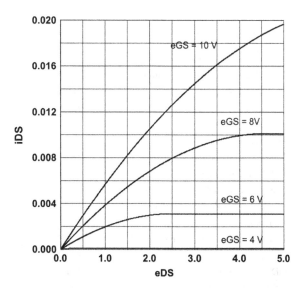

Fig. 7.84 Model of CMOS inverter

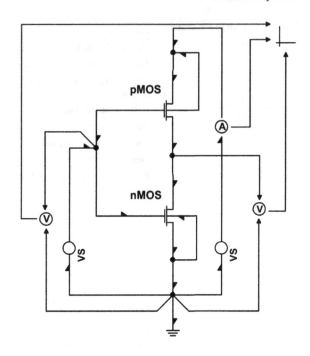

Table 7.9 CMOS parameters

Parameter	Name	Value (PMOS/NMOS)
Threshold (pinch-off) voltage	VT0	1 V
Transconductance parameter	KP	80 μA/V^2
Chanel length modulation parameter	LAMBDA	0.0 1/V
Bulk threshold parameter	GAMMA	0.0 V
Surface potential	PHI	0.6 V
Source ohmic resistance	RS	1 ohm
Drain ohmic resistance	RD	1 ohm
Bulk junction saturation current	IS	1×10^{-14} A
Bulk junction potential	PB	0.75 V
Grading coefficient	M	0.5
Coefficient for depletion capacitance	FC	0.5
Emission coefficient	N	1
Zero-bias BS junction capacitance	CBS	5 pF
Zero-bias BD junction capacitance	CBD	5 pF
GD overlap capacitance	CGD	8 pF/4 pF
GS overlap capacitance	CGS	8 pF/4 pF

Fig. 7.85 The CMOS
inverter simulation

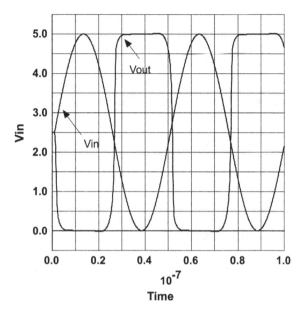

The inverter was driven by the voltage

$$V_G = V_0 + V_1 \cdot \sin(2\pi f) \tag{7.55}$$

where $V_0 = V_1 = 2.5$ V and $f = 20$ MHz. The simulation was run for 100 ns with an output interval of 100 ps for better resolution. The results are shown in Fig. 7.85.

When the input is low, e.g. equal to 0 V, the source-gate voltage of the PMOS is 5 V and is *on* (conducting). The NMOS, however, is *off* because its gate-source voltage is equal to 0 V, below the threshold. The current through both the transistors thus is zero. From the static characteristics of the PMOS it follows that the voltage between the source and drain must equal to zero. Thus, the inverter output is equal to the PMOS source port voltage, which is 5 V. On the other hand, when the gate voltage is high, e.g. equal to 5 V, the PMOS is off and current through the transistors is zero. But now the NMOS is on. This means that its drain-source voltage is 0 V. Because the source port of the NMOS is grounded, the inverter output is also 0 V. Thus, the inverter generates high voltage (5 V) when its input is low (0 V), and low voltage (0 V) when its input is high (5 V).

7.4.3 Operational Amplifiers

Operational amplifiers are important integrated circuit components widely used in analogue signal processing. We here consider how they can be represented as components in bond graphs.

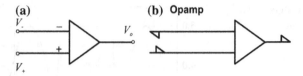

Fig. 7.86 Operational amplifier. **a** The circuit symbol. **b** The bond graph representation

Basically, the operational amplifiers—op-amps, for short—are represented in electrical circuits by the symbol shown in Fig. 7.86a. It has three terminals, two inputs and one output. The input terminal designated with a "−" is the *inverting* terminal; the other, denoted by a "+", is the *non-inverting* terminal. Bond graph components representing the op-amps are shown in Fig. 7.86b. This assumes that the power flows outward at the inverting port and inward at the non-inverting port. The op-amp generates power at its output, which is represented by the power-out port.

The simplest model of the operational amplifier represents it as an ideal voltage amplifier:

$$\left. \begin{array}{l} V_{diff} = V_+ - V_- \\ V_o = A_0 \cdot V_{diff} \\ i_+ = i_- = 0 \end{array} \right\} \tag{7.56}$$

Parameter A_0 is the so-called *DC open-loop gain*. It assumes that the currents drawn at the input ports are negligible and that there is no limitation on the voltage or current at the output port. Thus, the input resistance of the amplifier is infinite and its output resistance is zero. Such a simple model can be represented by the bond graph of Fig. 7.87. It consists basically of a source flow component SF that defines the zero input port current and a controlled source effort SE that generates an output voltage, as given by (7.56). The last component receives information on the voltage difference (efforts) at the non-inverting and inverting ports. This comes from the flow junction inserted between the effort junction and the source flow component.

Fig. 7.87 Simple model of op-amp

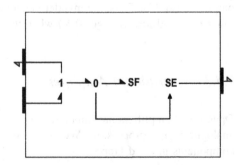

The power delivered at the operational amplifier output port is much larger than at the input. The excess power comes from voltage sources that supply the power to the op-amp through the supply terminals.

The voltage generated at the output port cannot rise above the voltages at these terminals. The supply terminals are not shown in the simple op-amp model of Figs. 7.86 and 7.87. However, we can take care of the limitation on the output voltage by defining the source effort in Fig. 7.87 as

$$eo = GAIN * ediff\, < \; = -VLIMIT? \; -VLIMIT :$$
$$(GAIN * ediff < VLIMIT?\; GAIN * ediff : VLIMIT) \tag{7.57}$$

where *GAIN* is the static gain of the op-amp and *VLIMIT* is the maximum value of the output voltage. This usually is 1–2 V less than the supply voltage.

We now simulate the characteristics of the operational amplifier defined by the models in Figs. 7.86 and 7.87 and (7.57). As usual, we create a project named *Ideal Op-amp Characteristics* and define the model of a set-up for simulating the static characteristics of the *Opamp*, as shown in Fig. 7.88.

The gain of the opamp is set to 1×10^5 and the output is limited to ±4.5 V. The op-amp is driven by the source voltage VS connected to the inverting port by a node. The source generates a sinusoidal voltage. The non-inverting port is grounded. The op-amp output port is connected to ground across a 1 kohm resistor. The output voltage reaches the limiting value at an input of $\pm 4.5 \times 10^{-5}$ V. To show the characteristics more clearly, we drive the op-amp with a low input voltage with amplitude of 5×10^{-4} V and frequency of 1 Hz. The frequency is not critical, as the op-amp model is static. To display the transition from the region of linear behaviour to saturation correctly, we must use a fairly small output interval, e.g. 0.001 s or less.

The results of the simulation are shown in Fig. 7.89, which shows that the input and output voltages are of the opposite sign. This follows from (7.56) because, for a grounded non-inverting port, the voltage difference $Vdiff = -V_-$ and, hence, the output voltage $V_o = -A_0 \cdot V_-$. The output voltage changes linearly over a narrow interval around the origin, and is constant elsewhere.

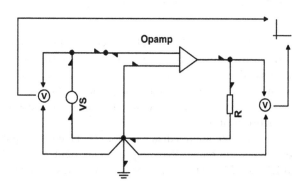

Fig. 7.88 Set-up for simulating opamp characteristics

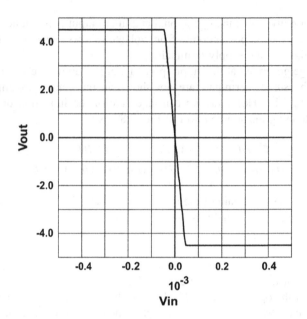

Fig. 7.89 Op-amp quasi-static characteristics

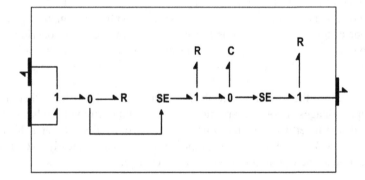

Fig. 7.90 Improved model of operational amplifier

We can improve the model by incorporating the finite input and nonzero output resistances, as well as the dynamics. The resulting model is given in Fig. 7.90. Compared to Fig. 7.87, the source flow in the input section is replaced by a resistive element that models the input resistance of the op-amp. Similarly, a resistive element is added in series between the source effort and output port. This models the output resistance.

Finally, an intermediate section is added that models the dynamics of the operational amplifier. Real amplifiers are built of many bipolar junction transistors, JFETs, or MOSFETs, so that the dynamics is quite complex. A capacitance is usually added to the op-amps high-gain node to control its dynamics over the useful range of the frequencies. In simplified models the amplifier dynamics is often approximated by a first-order dynamic circuit [1, 4, 5], as represented in Fig. 7.90 by the R and C elements.

Denoting by e_c the effort (voltage) on the capacitive element, the current is

$$i = \frac{d}{dt}(C_d \cdot e_C) = C_d \frac{de_C}{dt} \tag{7.58}$$

where C_d is the capacitance of the C element. The same current flows through the resistor. Thus, the dynamics of the operational amplifier of Fig. 7.90 is given by

$$R_d \cdot C_d \cdot \frac{de_C}{dt} + e_C = A_0 \cdot e_{diff} \tag{7.59}$$

where R_d is the resistance of the middle section R element. The natural frequency of this simple first-order system is

$$\omega_n = \frac{1}{R_d C_d} \tag{7.60}$$

and the dynamics of the op-amp can be described by

$$\frac{de_C}{dt} + \omega_n e_C = A_0 \cdot \omega_n \cdot e_{diff} \tag{7.61}$$

The resistance R_d and capacitance C_d are chosen to approximate the dominant low-order natural frequency of the opamp, which is taken as the bandwidth of the opamp, i.e. the frequency at which the output amplitude drops to $1/\sqrt{2}$, or -3 dB of the DC value. Thus, for an opamp with a bandwidth of 10 Hz = 62.83 rad/s, if we choose $R_d = 1$ kohm the capacitance is $C_d = 15.92 \times 10^{-6}$ F.

The product $A_0\omega_n$ on the right of (7.61), the *gain-bandwidth product*, is an important characteristic of the op-amp. It usually is expressed in Hz—as $A_0 f_n$, where $f_n = 2\pi\omega_n$—and typically is of order of MHz. In the example above, with DC gain 1×10^5 and bandwidth of 10 Hz, it is exactly 1 MHz.

In the model of Fig. 7.90 we limit the output voltage at the source effort in the output section by (7.57) using unity gain. The amplifier DC gain is defined in the source effort of the intermediate section. This is, in effect, similar to the diodes used in the output stage of the opamp models of [4].

Operational amplifiers are often used in circuits for analog operations on signals, such as amplifying, integrating, differentiating, and filtering. We analyse here only one such common circuit, the *Inverting amplifier* (Fig. 7.91).

The project *Inverting amplifier* is a slight modification of the operational amplifier circuit of Fig. 7.88; it has an added input resistor R and also a feedback resistor Rf. This connects the output port to the input node, which in turn is

Fig. 7.91 The inverting
amplifier set-up

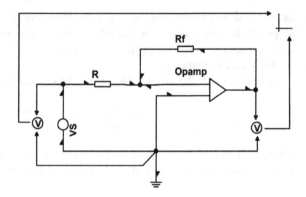

connected to the inverting input port of the op-amp. Loading the operational
amplifier output port is not needed here.

We may estimate the close loop gain of the amplifier by summing the currents at
the input node under the hypothesis of an ideal operational amplifier of very high
DC gain. Under this hypothesis the potential of the node is equal to the ground
potential, and the current drawn by the op-amp is zero. Hence we have

$$\frac{V_{in}}{R} + \frac{V_0}{R_f} = 0 \qquad (7.62)$$

or

$$V_0 = -\frac{R_f}{R} V_{in} \qquad (7.63)$$

The voltage gain of the inverting amplifier is, for sufficiently high open-loop
gain, given by the ratio of the feedback and input resistors. If we set R = 1 kohm
and Rf = 25 kohm, the gain is 25. This can be verified by simulation. We use a
sufficiently low input amplitude of 0.002 V, such that the amplifier is not saturated
(set VLIMIT = 5 V). The frequency chosen for the input voltage, 1 Hz, is much less
than the bandwidth frequency of 10 Hz. The parameters of the operational amplifier
are given in Table 7.10. The simulation was run for 5 s, corresponding to 5 periods
of the input voltage, and using an output interval of 0.01 s. The results (Fig. 7.92)

Table 7.10 Parameters of the
opamp model

Parameter	Name	Value
Input resistance	*Rin*	1 Mohm
Output resistance	*Rout*	100 ohm
Open-loop gain	*GAIN*	1×10^5
Output limit	*VLIMIT*	5 V
Dynamic resistance	*Rd*	1 kohm
Dynamic capacitance	*Cd*	15.92 μF

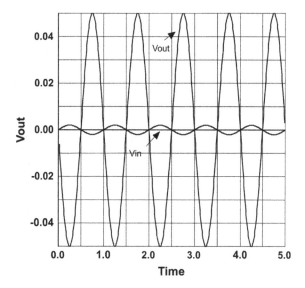

Fig. 7.92 The inverting amplifier response to 2 mV at 1 Hz

show that the output amplitude is 0.05 V, as expected (that is: 0.002 V·25 = 0.05 V). Also, the output phase is opposite to that of the input.

We expect that the output amplitude will roll off when the frequency reaches the amplifier bandwidth. Because of feedback around the operational amplifier, this frequency generally will be much higher than that of the open-loop operational amplifier, which is set to 10 Hz.

It can be shown [1] that the close loop bandwidth of the inverting amplifier can be approximated by

$$(f_n)_{close-loop} = \frac{A_0 f_n}{1 + \frac{R_f}{R}} \tag{7.64}$$

This relationship is valid if the open loop gain A_0 is much larger than $1 + R_f/R$, a criterion that normally is satisfied. In our example, the gain-bandwidth product $A_0 f_n$ equals 1 MHz. Hence, for the close-loop gain $R_f/R = 25$, we get, by (7.64), the close-loop bandwidth of about 38 kHz. The expected amplitude is 0.05/$\sqrt{2}$ = 0.03536 V.

We repeat the simulation with an input voltage of the same amplitude, but with a frequency of 38 kHz. The simulation interval was set to 0.0002 s and the output interval to 2×10^{-7} s. The results (Fig. 7.93) show that the output amplitude has dropped to 0.0354721 V. There also is a time lag of the output of about 16.5 μs that corresponds to the phase lag of 3.94 rad (1.254 π), where π radians is due to the signal inversion and the other 0.254π ≈ π/4 rad is the true dynamic phase lag.

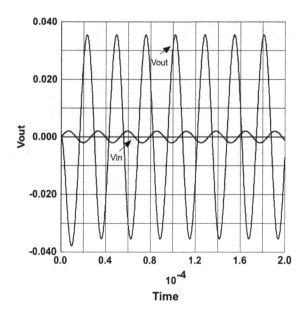

Fig. 7.93 Response of the amplifier to 2 mV at 38 kHz

7.5 Electromagnetic Systems

This section analyses applications of the bond graph approach to modelling electromechanical systems. The example chosen is an electromagnetic actuator. The permanent magnet data for the actuator were calculated by a magnetic field analysis program. These data are used as input to the BondSim program. More information on modelling magnetic systems by bond graphs can be found in [2].

7.5.1 Electromagnetic Actuator Problem

The schematic of an electromagnetic actuator is shown in Fig. 7.94. The actuator consists of a plunger, switching coils, and a permanent magnet for generating the holding force. The electrical driver circuit is represented by a capacitor, a resistor, and an ideal (lossless) switch. The plunger moves in the actuator assembly between hard stops from a value $x = x_{min}$ to a value $x = x_{max}$, where the displacement x is measured from the lower pole end. The electromagnet's behaviour is described by the tabulated values of flux in one turn (electrical side), and force on the plunger (mechanical side), for a range of the ampere-windings and plunger positions between x_{min} and x_{max}. These values were calculated using a magnetic field analysis program.

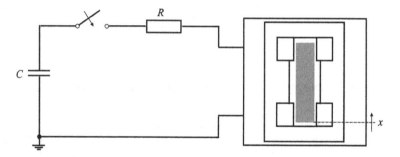

Fig. 7.94 Scheme of the electromagnetic actuator

The system parameters are:

1. The capacitor $C = 68$ mF, initially charged to 50 V (3.4 C)
2. The resistor $R = 0.2$ ohm
3. The switch starts closing at 10 μs, and closing takes 1 μs. The resistance changes linearly from 1×10^{9} (open) to 1×10^{-9} (closed) ohm
4. The plunger has mass 5.575 kg, its weight is neglected.
5. The hard stop positions are at $x_{max} = 0.0199$ m and $x_{min} = 0.0001$ m. The material poses a high stiffness coefficient (1×10^{9} N/m) and a high damping factor (1×10^{6} N s/m), corresponding to the impact without rebound.

7.5.2   System Bond Graph Model

We start by defining the basic structure of the *Electro-magnetic Actuator* system, as shown in Fig. 7.95. The model consists of an electrical part, a resistor R, capacitor C, and switch, and the `Magnetic Actuator` component. During the simulations

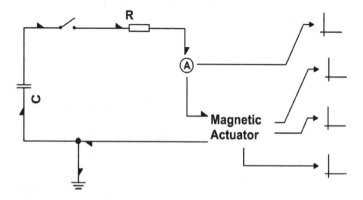

Fig. 7.95 Bond graph model of the electromagnetic actuator

the displayed variables are the current through electric circuit, the position and velocity of the plunger, and the total force on the plunger.

7.5.3   Electromagnetic Flux and Force Expressions

Before proceeding with the decomposition of the Magnetic Actuator component, we must define the *Flux(x,*i) and *EMForce(x,i)* functions. These interpolate data from tables of the flux and force, respectively. Each is a function of the plunger position x and ampere-winding i. In BondSim, one- and two-variate cubic B-spline interpolations are used for functions defined by one- and two-dimensional tables, respectively [15].

To define, for example, the *Flux* function, we select the *Function* command on the *Tools* menu, then *New* from the menu that drops down. In the *New Function Name* dialogue that opens, we define a unique name for the new function, e.g. *Flux*. This is the name that will be used when referring to the function in an algebraic expression. We accept this name by clicking the OK button. A dialog opens from which we can choose the type of the function we are defining (Fig. 7.96).

Currently we can choose between the *Mathematical expression* and *Table of values*. In the first case we can define a function symbolically in the form of a mathematical expression. We choose the second option, the *Table of values*. We chose next between *One* or *Two dimensional* functions. The *Flux*, as well as *EMForce*, has two arguments; thus, we select the *Two dimensional* and click OK.

A dialogue again opens, this time for defining the names and the values of the variables corresponding to the rows and columns of the function table (Fig. 7.97). The row variable represents the first argument, and the column variable the second argument of the function we are defining. Thus, in *Flux(x,i)* the row variable is *Position* (*x*) and column variable is *Ampere-Winding* (*i*). When we have added all values for both variables, we click the *Edit* button to edit the table of function values.

A grid opens, which in its leftmost column and topmost row, contains the values we have just defined (Fig. 7.98). We can type in the values of the function in the

Fig. 7.96 The function type dialog

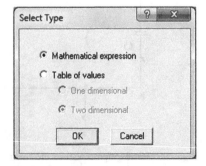

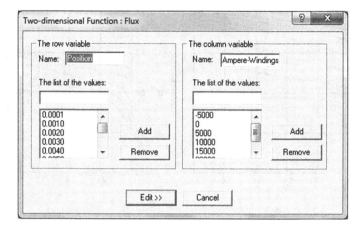

Fig. 7.97 Dialogue for defining values of variables

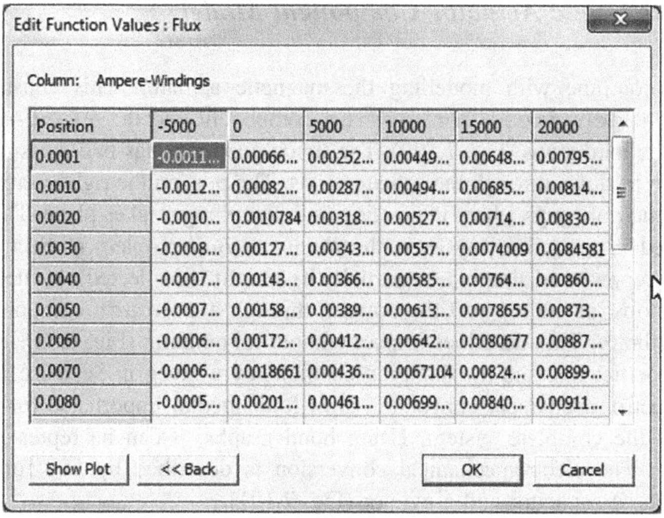

Fig. 7.98 Editing of the function values

cells of the grid, i.e. the flux value corresponding to a *Position* and an *Ampere-winding*. The function can be viewed using the *Show Plot* button. A *Plot Options* dialogue appears that offers several possibilities for the plotting. The *Flux* function appears as in Fig. 7.99 left. Similarly, we can define the *EMForce* function (Fig. 7.99 right).

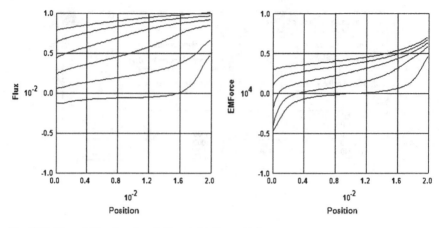

Fig. 7.99 Plot of *Flux* function (*left*) and *EMForce* (*right*)

7.5.4 Magnetic Actuator Component Model

We now continue with modelling the magnetic actuator. This consists of an Electro-Mechanical Conversion component and the Plunger moving between the hard stops (Fig. 7.100). The first component has two ports on the left where current flows through the actuator coils. The port on the right corresponds to the magnetic pole gaps where interaction with the plunger takes place. The plunger is modelled in a similar way as in the *Bouncing Ball* problem of Sect. 6.4. The plunger is represented simply as a particle (its weight is neglected). It interacts with the hard stops, at the top and the bottom, through a Contact component taken from the library. This describes impact without rebounding (Fig. 6.53). The mass, stiffness coefficients, and the damping coefficient are given in Sect. 7.5.1.

The electro-mechanical conversion is of fundamental importance for the functioning of the complete system. Using bond graphs, it can be represented very compactly. The electromechanical conversion is described by two fundamental components: a capacitor and a gyrator (Fig. 7.101).

The capacitor has two ports, magnetic and mechanical, and describes the storage of magnetic and mechanical energy in the actuator. The constitutive relations for these two ports are:

Magnetic port:

$$\left. \begin{array}{l} phi = Flux(x, M) \\ e = \frac{dphi}{dt} \end{array} \right\} \tag{7.65}$$

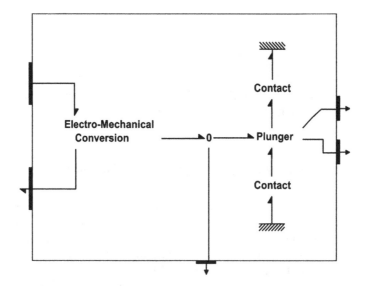

Fig. 7.100 Model of the magnetic actuator

Fig. 7.101 Modelling of
electromechanical conversion

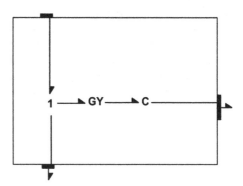

Mechanical port:

$$F = EMForce(x, M) \atop v = \frac{dx}{dt} \Big\} \qquad (7.66)$$

The gyrator describes the relations between the voltage V_d across the coil ports and the e.m.f. e, and between the magneto-motive force M and the current i through the coils:

Gyrator:

$$\left.\begin{array}{l} V_d = N \cdot e \\ M = N \cdot i \end{array}\right\}$$ (7.67)

where N is the number of coil turns. The (7.65)–(7.67) are familiar conversion relations from electromagnetics.

7.5.5 Simulation of Magnetic Actuator Behaviour

The simulation of the electromagnetic actuator was undertaken to determine how the system variables change during the switching of the actuator. The simulations were done using a simulation interval of 0.050 s. The output interval of 0.00001 s was chosen to determine the characteristic switching points accurately. Some of the results are shown in Figs. 7.102, 7.103 and 7.104.

Figures 7.102 and 7.103 show that the force on the plunger steadily increases until it crosses the zero value at 0.02253 s. Because the plunger is depressed to the hard stop, the real motion starts a little later, at 0.02358 s. The plunger hits the upper hard stop after 0.03753 s. The final value of the force at 0.04999 s is 5575 N.

The changes of the current through the coils are shown in Fig. 7.104. The current at first rises steeply, then reaches its maximum, 90.05 A, at 0.0258 s. It then drops down and reaches its minimum, 0.7481 A, when the plunger hits the hard stop. After that it rises somewhat to a final value of 20.45 A. This is the current that holds the plunger pressed to the upper hard stop.

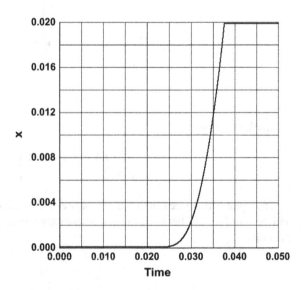

Fig. 7.102 The plunger position motion during switching on

Fig. 7.103 Total force on the plunger during switch-on

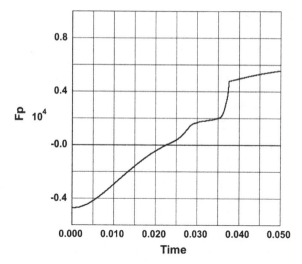

Fig. 7.104 Change of the current through the coils during the switch-on

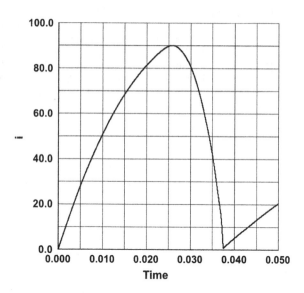

This section, as well as the previous sections of this chapter, shows the effectiveness of the bond graph approach in solving complex electronic and electro-mechanic systems.

References

1. Senturia SD (2001) Microsystems design. Kluwer Academic Publishers, Boston
2. Karnopp DC, Margolis DL, Rosenberg RC (2000) System dynamics: modeling and simulation of mechatronic systems, 3rd edn. Wiley, New York
3. Gawthrop P, Smith L (1996) Metamodelling: bond graphs and dynamic systems. Prentice Hall, Hemel
4. Vladimirescu A (1994) The SPICE book. Wiley, New York
5. Keown J (2001) OrCAD PSpice and circuit analysis. Prentice Hall, Upper Saddle River
6. Kielkowski R (1995) Spice practical device modeling, 2nd edn. McGraw-Hill, New York
7. Massobrio G, Antognetti P (1993) Semiconductor device modeling with SPICE, 2nd edn. McGraw-Hill, New York
8. Märtz R, Tischendorf C (1997) Recent results in solving index-2 differential-Algebraic equations in circuit simulation. SIAM J Sci Comput 18:139–159
9. Singh J (2001) Semiconductor devices, basic principles. Wiley, New York
10. Tieze T, Schenk C (1999) Halbleiter-Schaltungstechnik. Springer, Berlin-Heidelberg
11. Van Halen P (1994) A physical charged-based model for space charge region of abrupt and linear semiconductor junction. In: Proceedings of the 1994 international symposium on circuits and systems, London, pp 1.403–1.406
12. Christen E, Bakalar K, Dewey AM, Moser E (1999) Analog and mixed signal modeling using VHDL-AMS language (tutorial). In: The 36th design automation conference, New Orleans
13. Thoma J, Bousmsma BO (2000) Modelling and simulation in thermal and chemical engineering, a bond graph approach. Springer, Berlin-Heidelberg
14. Hafner AR, Blackburn DL (1993) Simulating the dynamic electrothermal behavior of power electronic circuits and systems. IEEE Trans Power Electron 8:376–385
15. de Boor C (1978) A Practical guide to splines. Springer, New York

Chapter 8
Control Systems

8.1 Introduction

Control system theory and practice play important roles in mechatronics. Their fundamental role is in the control of mechanical motion. Control systems are commonly described in terms of block diagrams, where input-output relations are represented by transfer functions. This offers a simplified picture of the system processes from a control point of view. The theory of control systems is well documented in numerous books, so we do not cover it here. What we wish to demonstrate is how the control actions taking place in mechatronic systems can be modelled using the time domain signal components developed here.

The bond graph elements that can serve as the mechatronic control interfaces are defined in Sect. 2.5.8. In Sect. 2.6.2 basic block diagram components are defined that can be used to synthesize the signal processing of the control loops. Control actions in real components can thus be simplified and represented by block diagram components. These component models control actions in the time domain. Modelling is not restricted to linear relations, but is applicable to non-linear relations as well including even those that feature discontinuities. Creation of such components, or development of more complex components, follows the philosophy of the component modelling approach discussed in previous chapters.

The next section explains the basic techniques of block diagram components as applied to a simple control system. Some modelling details specific to block diagram components are given. It will be seen that there is little difference from the power port component approach.

After that, a short overview of the modelling approach to control systems is given. This focuses mainly on modelling PID controllers in servo loops. Finally, modelling and simulation of a DC servomotor is presented.

© Springer-Verlag Berlin Heidelberg 2015
V. Damić and J. Montgomery, *Mechatronics by Bond Graphs*,
DOI 10.1007/978-3-662-49004-4_8

8.2   A Simple Control System

We start by modelling a typical feedback control system (Fig. 8.1). This system consists of a controller that regulates an object's motion. The motion in question is defined by a reference input and its output information is collected by a sensor. There are also disturbances acting on the object. In mechatronics, the object is typically a mechanical linkage actuated by a motor that executes the control action.

We begin the modelling process by creating a new project named *A Simple Control System*. Figure 8.2 shows the program screen with the empty project document window opened in the middle; the model structure tree is on the left and the editing tools on the right. The basic tools for editing control system models are contained in the *Editing Box*, under the *Continuous Signal Components* branch (Fig. 8.2). We also need the *Ports* branch when creating control signal ports. To

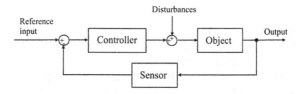

Fig. 8.1   A simple control system

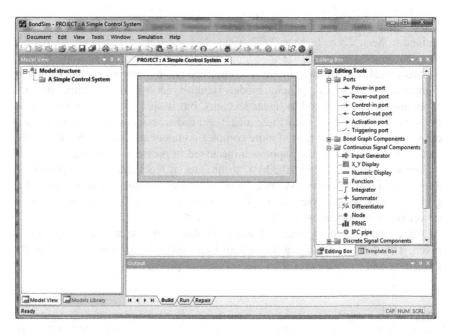

Fig. 8.2   A new control system project with editing tools shown

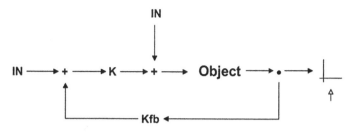

Fig. 8.3 The creation of the control system block diagram model

develop the model the same drag and drop technique is used as when editing the Bond Graph models. The basic difference is that we are now dealing with signals.

Starting from the left in the block diagram in Fig. 8.1, we first create the input component that generates the reference signal. This component is created by dragging the *Input Generator* tool from the *Continuous Signal Components*. We place the component near the left document edge (but not too close), Fig. 8.3. We may accept the default component name IN and click outside the component to finish its creation. The component has only one control-out port and by default generates a constant signal equal to one. We may change it as will be shown later.

We next create a component corresponding to the summation of the reference and feedback signals. Thus, we drag the *Summator* tool and create the summator to the right of the IN component. By default, the newly created component has a predefined name '+', which of course can be changed if we wish. We may move the *summator* by dragging its title ('+') using the mouse so that the left input port is on the same level as the output port of the IN component. We can move the component also by selecting it (clicking) and using the cursor arrows on the keyboard. We then connect the output port of IN component to the input port of the *summator* by a control line (active bond). We do it as we did when connecting the power ports by a bond line. We simply put the mouse cursor over the first port, press the left button, drag it to the next port, and then release the button.

By default the *summator* has two input ports and one output port. The input ports have predefined + signs, thus the component at the output generates the sum of the input signals. Because the feedback signal from the sensor in Fig. 8.1 is subtracted from the reference input, we have to change the sign of the bottom port of the *summator* to '−'. Thus, we double click the bottom port and in the dialogue window that opens (Fig. 8.4) we select the '−'sign in the *Summation Sign* combo box.

We can also change the name of the feedback signal to, e.g., *cfb*—short for "feedback control signal". In the same way we can change the name of the left port, e.g. to *cin*. Now, if we double-click the output port we can see that the output generated is *cin-cfb*.

Fig. 8.4 Dialogue for
defining the *summator* input
port sign and name

The controller in this example is implemented as a simple proportional
(*P*) control given by

$$cout = K \cdot cin \tag{8.1}$$

where *K* is the controller gain. This is accomplished by dragging the *Function* tool
and dropping it at the right of the *summator*. When the text *Fun* and a textual cursor
(carret) appears we delete this text and type in the K. By default, the constitutive
relation of the component is linear, and thus corresponds to (8.1). Thus, it is a gain
component. We can check this by double-clicking the output port. The value of the
gain defaults to 1 and we accept this for now. We move the gain component K so
that it's left port faces the output port of the *summator*, and then connect the ports.

The next step is to create another *summator* that will add disturbances to the
system. We create the *summator* to the right of the gain component, move the
component so that its left input port faces the output port of the gain component,
and connect these ports. The bottom port we drag around the summator periphery to
the top. We now create an IN component above it. This component will generate
the disturbances to the system. Thus, we drag its output port around its periphery to
face the other *summator* input port, and connect to it (Fig. 8.3).

The object of the control will be created as a word model component named
Object. To create this we need to expand the *Bond Graph Component* branch and
drag the *General Word Model Component* in the *Editing Box* and drop it at the right
of the last *summator*. When the component is created and a carret appears we type
in the Object as the name of the component and click outside to finish the editing.
By creation, however, the word model component has two power ports. We remove
these ports and at the left side drag and drop a control-in port, and at the right side a
control-out port. We move the component until its left control-in port directly faces
the *summator* output port, and connect them. Its model will be defined later.

Next, we create a node for the connection of the feedback signal (Fig. 8.1).
Hence, we drag the *Node* tool and drop it to the right of the Object with its left
input port facing the Object output port, and connect these ports.

To close the servo loop, the sensor component must be created. Continuing in a similar way as when dealing with the gain object we create a simple linear function component denoting it by Kfb. We must, however, exchange their ports by moving the right control-out port to the left side of the component, and control-in component from the left to the right side. We move the component to the mid position with respect to the left and right end of the block diagram and below the second summator. Connect the left port to the bottom input port of the left summator, and the right input port to the node (Fig. 8.3).

The final component that we create serves to display the control system's outputs. Note that every signal has to be fed to some destination; there cannot be 'dangling' signals. We do not have, however, enough space for this component and thus need to enlarge the document. To that end we put the mouse cursor on the right edge of the document rectangle, and when it changes shape into a double arrow, press it and drag the right edge of the document to the right and release the mouse. Now we may drag the *X-Y Display* tool to the right of the node (Fig. 8.3). Note that that the Display component is created with two input ports. We can drag the Display component or its left input port, so that it is in the line with the output port of the node, and then connect them by a signal line. This completes the generation of the block diagram component model of Fig. 8.1 (see Fig. 8.3).

At this point we introduce a slight modification to the model to permit display of the reference signal as well. To accomplish this we disconnect the signal line between the left IN component and the *summator*, and insert a node. Because there is not enough space for the node, we need to move the *summator* and all the connected components to the right. It is not necessary to redraw the complete block diagram: Instead we may use the mouse to draw a rectangle around all components we wish to move. Thus, we press and hold the mouse and drag it to the right. Now we may insert the node, connect it on the left to the IN component and on the right to the *summator*. We connect its bottom output port to the bottom input port of the Display. The final form of the block diagram is shown in Fig. 8.5.

The object of control in mechatronic applications is a mechanical object, such as a robot link or a headlight assembly in an automobile, driven by an actuator. The object's dynamics is commonly described by a transfer function relating the Laplace transform of the output to the input,

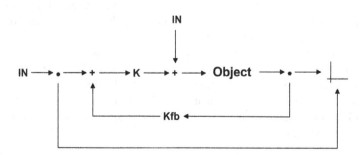

Fig. 8.5 The modified control system model

$$Cout(s) = G_{ob}(s)Cin(s) \tag{8.2}$$

where s is the complex variable. We use the following transfer function as a simple model of this object:

$$G_{ob}(s) = \frac{K_{ob}}{(T_{ob}^2 s^2 + 2\zeta_{ob}T_{ob}s + 1)s} \tag{8.3}$$

Here K_{ob} is the object's static gain, T_{ob} is the time constant of the object, and ζ_{ob} is the damping ratio. We cannot, however, model this transfer function directly; we must transfer it back to the time domain. The time model corresponding to (8.2) and (8.3) read:

$$T_{ob}^2 \frac{d^3 cout}{dt^3} + 2\zeta_{ob}T_{ob} \frac{d^2 cout}{dt^2} + \frac{dcout}{dt} = K_{ob}cin \tag{8.4}$$

To describe this equation in block diagram form using only elementary components—i.e. integrators, functions, summators, etc.—integrating (8.4) we obtain

$$\frac{d^2 cout}{dt^2} = \left(\frac{d^2 cout}{dt^2}\right)_0 + \int_0^t \frac{1}{T_{ob}^2}\left(K_{ob}cin - 2\zeta_{ob}T_{ob}\frac{d^2 cout}{dt^2} - \frac{dcout}{dt}\right)dt \tag{8.5}$$

with

$$\frac{dcout}{dt} = \left(\frac{dcout}{dt}\right)_0 + \int_0^t \frac{d^2 cout}{dt^2} dt \tag{8.6}$$

and

$$cout = (cout)_0 + \int_0^t \frac{dcout}{dt} dt \tag{8.7}$$

The first term on the right of each equation is the initial value of the variable in question. Written in this fashion, these equations can be described directly by the elementary block diagram components.

The input and output variables are the variables transferred to and from the control ports of the Object (Fig. 8.5). To describe its model according to (8.5)–(8.7), we create a new model document by double-clicking the Object. In the document window that opens, we create a model of the control system object in a way that is similar to the way in which we developed the basic model. The resulting model is shown in Fig. 8.6.

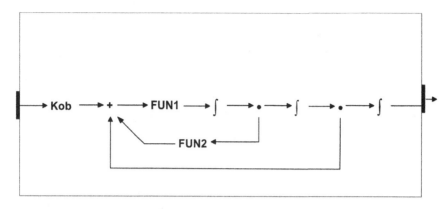

Fig. 8.6 The dynamics of the object

Note that the input port of the gain (function) component Kob and the output of the last integrator are connected to the document ports that serve as the internal connectors of the outside component ports. The integration operations in the above equations including the initial condition of the outputs are represented by the integrator components [see Sect. 2.6.2.4, Eq. (2.41)].

The function component FUN1 describes multiplication by $1/T_{ob}^2$ in (8.5), and the first integrator describes the corresponding integration. The FUN2 component inside the feedback loop describes the multiplication by $2\zeta_{ob}T_{ob}$. The other two integrators describe the integrations represented by (8.6) and (8.7). Finally, the summator describes the additions inside the parenthesis in the integrand of (8.5). Note that the signs of its bottom ports are set to minus.

The initial conditions that are required in (8.5)–(8.7) can be set by double-clicking the integrator's output ports. In the dialogue that follows (Fig. 8.7), the initial values of the output variables can be set in the *Initial Value* edit box, the default of which is zero.

Fig. 8.7 The integrator output dialogue

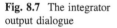

The control system model is now complete, but before simulating the system several parameters must be set and the input function must be defined.

We use the following parameter values:

Controller gain	$K = 30$
Object gain	$K_{ob} = 2.77$
Object time constant	$T_{ob} = 5.41 \times 10^{-3}$ s
Object damping ratio	$\zeta_{ob} = 1.35$
Feedback gain	$K_{fb} = 1$

The default values are accepted for all the other model parameters.

We first analyse the system response to a unit step input with no disturbances. Hence we set the output of the left IN object to 1, and of the upper (the disturbance) IN object to 0, and then build the model. The simulation interval was set to 0.2 s, the output interval and maximum step-size to 0.001 s, and defaults are accepted for all the other settings (error constants, method etc.).

The simulation results (Fig. 8.8) show that the settling time is about 0.12 s; and that there is an overshoot of about 26 %.

We next simulate the response to a unit disturbance with the reference input set to 0. As can be seen in the next graph (Fig. 8.9), the output does not return to zero, but settles at a value of 0.0333. Better response behaviour can be achieved using a better controller, e.g. a controller with proportional-integration and derivative action (PID) described in the next section.

Fig. 8.8 The response of the system to a unit step input

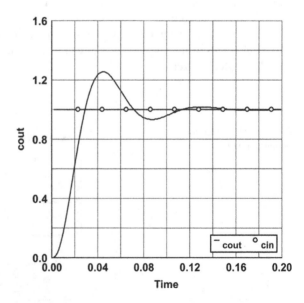

Fig. 8.9 The response to a
unit disturbance

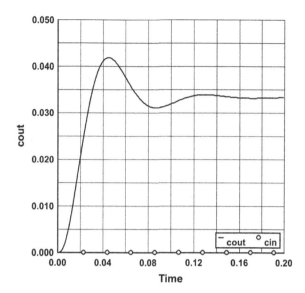

We end the discussion of this simple control system with the analysis of the response to a sinusoidal input. The reference signal generated by the IN component of Fig. 8.5 is now defined as a sine function of amplitude *cValue* and frequency *FREQ* Hz. The formula is:

$$cin = cValue * sin(2\pi \cdot FREQ \cdot t) \qquad (8.8)$$

We estimate the natural frequency of the oscillation from the step response in Fig. 8.8. The period of the oscillation, estimated as the difference between two successive crossings of the steady-state line at 1.0, is found to be 0.085 s. Hence the frequency is estimated as $1/0.085 = 11.76$ Hz $= 73.89$ rad/s. At a much lower frequency we can expect that output will follow the input closely. To check this, we simulate the response of the system to a unit sinusoid with the forcing frequency of 1.0 Hz. The simulation interval is taken 5 s and the output interval and the maximal step to 0.01 s. The result (Fig. 8.10) shows that the input and the output signals are nearly indistinguishable.

We repeat the simulation with an input frequency equal to the estimated natural frequency of 11.76 Hz. Because we are interested in the phase shift between the output and input, we select a simulation interval of 0.5 s and a rather short output interval of 0.0001 s.

The results of the simulation are shown in Fig. 8.11. We find that the difference between the points where the input and output sinusoids cross the time axis is about 0.0206 s. Hence, the phase lag is $2 \cdot \pi \cdot 11.76 \cdot 0.0206 = 0.4845\pi$, which is close to $\pi/2$. It was found numerically that the poles of the closed loop transfer function of the system in Fig. 8.5 are

Fig. 8.10 The frequency response at 1.0 Hz

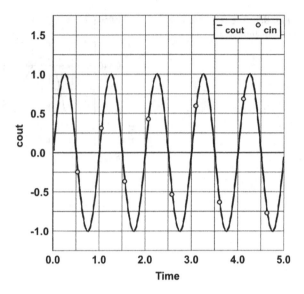

Fig. 8.11 The frequency response at 11.76 Hz (near the resonance)

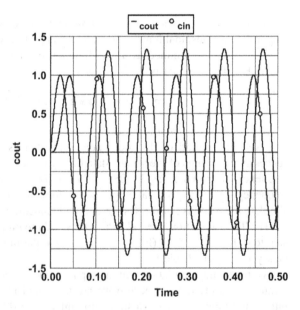

$$s_1 = -435.603$$
$$s_{2,3} = -31.7363 \pm i \cdot 74.2349$$

Hence, the system has a quick exponential rise with time constant $1/435.6 = 0.00230$ s, which appears in Fig. 8.8 as a short delay. The transient rises as a damped sinusoid of the frequency 74.23 rad/s. Thus, the estimation of the resonant frequency achieved above is a quite good.

8.3 PID Control System Modelling

Control systems described by transfer functions can readily be modelled by block diagram components. As shown in the previous section, it is necessary first to translate them into the time domain. Using a technique similar to analog-computer modelling, the time domain relations are described by block diagram components, such as the inputs, summators, functions, integrators and nodes. The output variables are fed to the components used to display the simulation results as x vs. t plots or x vs. y plots. The same technique can be used to process signals extracted from the bond graph components.

There is an elementary component that has not been used thus far: the *differentiator* (Sect. 2.6.2.5). One reason for introducing this component is to represent often-used controllers based on laws such as the *proportional-integration-derivative (PID)* law and their variants. The *PID* control law is commonly defined as:

$$cout = K_p \cdot cin + K_i \int_0^t cin \cdot dt + K_d \frac{dcin}{dt} \tag{8.9}$$

The K_p, K_i, and K_d are referred to as the *proportional, integrator,* and *derivative* constants (gains), respectively. There are other forms of PID controller laws [1]. The integrator and derivative constants are often expressed as:

$$\left.\begin{array}{l} K_i = K_p/T_i \\ K_d = K_p \cdot T_d \end{array}\right\} \tag{8.10}$$

where T_i and T_d are the *integrator* and *derivative time* constants, respectively.

The PID model (Fig. 8.12) has three branches. The middle branch with K_p gain defines the *P control*. The upper branch contains an *integrator* and K_i gain and defines the *I* control. Finally, the bottom branch contains the differentiator with K_d gain, and thus defines the *D control law*.

To evaluate the time derivative we do not use numerical approximations. Rather, the differentiation is applied directly. The equation is treated as a differential-algebraic equation and is solved jointly with the other equations (Chap. 5).

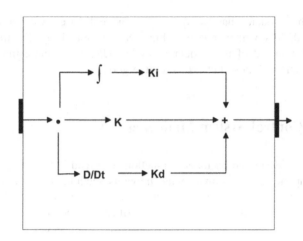

Fig. 8.12 The model of the PID controller

One of the central problems in the application of PID control is its *tuning*. The goal is to obtain good system response. A classical approach is to apply the well-known *Ziegler-Nichols tuning* [2, 3]. The application of *PID* control and the parameter tuning is illustrated using the example of the last section.

We begin by copying the *A Simple Control System* project to a new project named *A PID Control system*. In the new system we disconnect and delete the proportional controller K and insert a PID controller, whose model is described in Fig. 8.12. The resulting system is shown in Fig. 8.13.

There are two *Ziegler-Nichols* methods known as the *open loop* and *close loop* methods. We will apply the second, the close loop method.

To apply it is necessary to disable the *I* and *D* controllers, e.g. by setting $K_i = 0$ and $K_d = 0$. We starts then increasing the K_p gain and exciting the system by a step input until there are self-sustained oscillations in the system. This operation could be dangerous in real systems and there are some other more acceptable approaches. However, when simulating the system this is quite an acceptable operation. In fact the simulations can be used as a first step in setting the control of a real system. The value of gain $K = K_u$ where such oscillations appear is denoted as *ultimate* (or

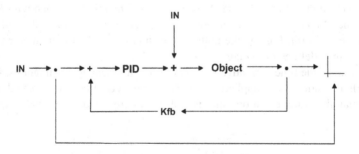

Fig. 8.13 The control system of Sect. 8.2 with PID controller

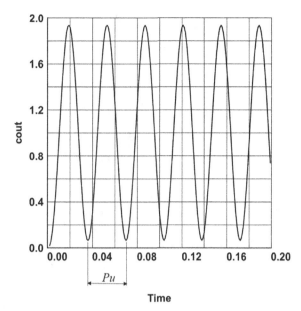

Fig. 8.14 Self-sustained oscillation in the system at $K_u = 180$

critical) gain. In our system it happens at $K_u = 180$. The resulting oscillations in the output are shown in Fig. 8.14. We find that the *ultimate* (or *critical*) *period* P_u of the self-sustained oscillation is $P_u = 0.034$ s.

According to the *Ziegler-Nichols* method, the PID constants are given by the following equations:

$$\left. \begin{array}{l} K = 0.6 \cdot K_u \\ T_i = 0.5 \cdot P_u \\ T_d = 0.125 \cdot P_u \end{array} \right\} \tag{8.11}$$

The oscillations in the system are acceptable if the ratio of the peaks in the same direction due to a step in the input or in the disturbance is approximately ¼.

By applying (8.11) we obtain the following values of the PID constants:

$$K = 0.6 \cdot 180 = 108$$
$$T_i = 0.5 \cdot 0.034 \text{ s} = 0.017 \text{ s}$$
$$T_d = 0.125 \cdot 0.034 = 0.00425 \text{ s}$$

The simulation of the system response to the step input using these values of PID constants is given in Fig. 8.15. The simulation parameters used are the same as used in the previous section, i.e. the simulation interval 0.2 s, the output interval and maximum step-size 0.001 s, and the defaults values for the other parameters (error constants, method etc.). The transient satisfies the requirements of the stability (¼ rule).

Fig. 8.15 The input step
response with $K_i = 108$,
$T_i = 0.017$ s and $T_d = 0.00425$

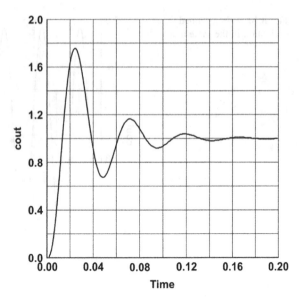

Fig. 8.16 The response to
step in the disturbance with
$K_i = 108$, $T_i = 0.017$ s and
$T_d = 0.00425$

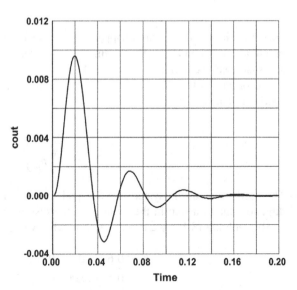

We can find the behavior of the system to a step in the disturbance (with the
reference input set to zero) as in Fig. 8.16. The controller successfully returned the
system output to zero. The stability satisfies the requirement of ¼ rule.

Ziegler-Nichols tuning is referred to as sub-optimal and is used only as the
starting point for a finer controller tuning. Several iterations typically are required to

get a satisfactory performance. It should be stressed that *PID* control tuning is not always easy to achieve; this is particularly true when the system includes nonlinearities, such as a dead zone or saturation which are often found in the servo-drives.

8.4   Permanent Magnet DC Servo System

The last section of this chapter deals with a position servo system. The next example shows the power of the combined bond graph-block diagram approach to modelling of mechatronics systems.

The scheme for the system for the control of the angular position of a body is given in Fig. 8.17. This is a simplified example of the control of a robotic arm. The servo loop consists of a servo driver and a permanent magnet DC motor that drives the body (arm) through a gearbox. The angular position of the body is measured by a sensor, the output of which is fed back to the servo-driver input. The servo driver consists of a *PID* controller as the input stage and an output stage that converts the low-power controller output to a higher power output needed to drive the motor.

The development of the system model starts by identifying the system components. The components are represented by the corresponding models and interconnected as in the real system. The example project is named *DC Servo System*. The system level model is shown in Fig. 8.18.

The `Servo Driver` has two control-input ports, one for the reference input and the other for the feedback signal. It also has a power port by which the transfer of electrical power to the motor takes place.

The permanent magnet motor is represented by the `DC Motor` component. It has two power ports, one electrical and one mechanical. The gearbox is represented by the transformer (`TF`) component (renamed `Gear`). The arm shaft rotates in a bearing fixed in the frame. The motor and gearbox body typically are fixed in the frame, and the gearbox shaft is connected to the arm shaft by a clutch.

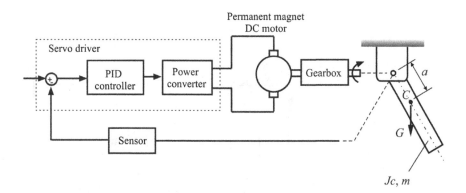

Fig. 8.17 Permanent magnet motor servo system

Fig. 8.18 Model of the permanent magnet motor servo system

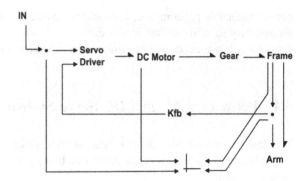

Following the approach of modelling mechanical systems in Sect. 2.7.3, we use the Frame object to model the connection of the arm shaft to the frame, and the Arm object to model the dynamics of the body. Thus, the power generated by the motor is transferred through the gear and the Frame to the Arm.

Information on the arm's angular velocity and position are measured at its shaft and fed out of the Frame. The position sensor is represented by a simple constant gain function component named Kfb. The reference signal of the position servo is generated by the IN component.

The variables that are of interest for observing the system behaviour, such as the reference input, arm angular velocity and position, and current drawn by the motor, are fed to an X-Y Display.

We next develop models of the main servo components, starting with the motor. The model of the DC Motor (Fig. 8.19) corresponds to the model of a permanent magnet DC motor often found in the literature [4, 5]. To show this we have added variables to the bond graph (normally stored in the ports).

Fig. 8.19 Model of the permanent magnet DC motor

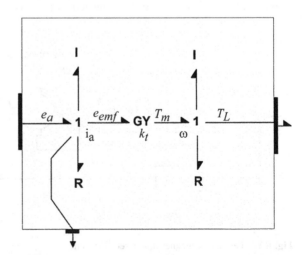

Gyrator GY describes the basic electromechanical conversion in the motor relating the *back emf* e_{emf} and the *armature current* i_a at the electrical side to the *torque* acting on the rotor T_m and its *angular velocity* ω at the mechanical side:

$$\left.\begin{array}{l} T_m = k_t \cdot i_a \\ e_{emf} = k_t \cdot \omega \end{array}\right\} \tag{8.14}$$

The coupling coefficient k_t is known as the *torque constant*. The coupling coefficient in the second equation usually denoted k_a and called the *back emf constant*, is, in fact, numerically the same coefficient. This is a consequence of the cross-coupling between variables in the electromechanical conversion and the conservation of the power in the conversion. The losses are taken care of at the electrical and mechanical sides of the motor.

The electrical process in the armature winding is commonly described in terms of the armature resistance R_a and the self-inductance L_a. In the bond graph model of Fig. 8.19 it is represented at the electrical side by the resistive and inertial elements, respectively, joined to a 1-junction. Thus, the relation between the *armature voltage* e_a across the electrical motor terminals and the armature current i_a through it reads:

$$e_a = L_a \frac{di_a}{dt} + R_a i_a + e_{emf} \tag{8.15}$$

Similarly, the process at the mechanical side is described by a resistive element that represents linear friction with coefficient B_m, and an inertial element that describes the rotation of the rotor of mass moment of inertia J_m. They are joined at the effort junction. T_L is the load torque. The corresponding relation reads

$$T_m = J_m \frac{d\omega}{dt} + B_m \omega + T_L \tag{8.16}$$

Equations (8.14)–(8.16) are the familiar equations used to describe motor dynamics [2, 4]. Applying the Laplace transformations and eliminating the torque and back emf by (8.14) yields

$$\left.\begin{array}{l} E_a = (L_a s + R_a) I_a + k_t \Omega \\ k_t I_a = (J_m s + B_m) \Omega + T_L \end{array}\right\} \tag{8.17}$$

If we eliminate the armature current from these equations, we obtain

$$E_a = \left[\left((L_a s + R_a)(J_m s + B_m) + k_t^2 \right) \Omega + (L_a s + R_a) T_L \right] / k_t \tag{8.18}$$

Neglecting the external (load) torque, we get the familiar motor transfer function

$$\frac{\Omega(s)}{E_a(s)} = \frac{k_t}{(L_a s + R_a)(J_m s + B_m) + k_t^2} \tag{8.19}$$

This is the transfer function used in Sect. 8.2. The time constant appearing in (8.3), denoted here as T_m, is given by

$$T_m = \sqrt{\frac{L_a J_m}{R_a B_m + k_t^2}} \tag{8.20}$$

Similarly, the motor damping ratio ζ_m is given by

$$\zeta_m = \frac{L_a B_m + R_a J_m}{2\sqrt{L_a J_m (R_a B_m + k_t^2)}} \tag{8.21}$$

and the static gain by

$$K_m = \frac{k_t}{R_a B_m + k_t^2} \tag{8.22}$$

We can also find an equation relating the load torque and the armature current by eliminating the motor angular velocity from (8.17):

$$T_L = \left[\left((L_a s + R_a)(J_m s + B_m) + k_t^2\right) I_a - (J_m s + B_m) E_a\right]/k_t \tag{8.23}$$

Equation (8.18) shows that the load torque reflects at the electrical side; its effect should be taken care of by the servo driver. In the same vein, the voltage across the motor terminals is reflected at the mechanical side and affects the torque delivered to the mechanical object that the motor drives. These two effects open various possibilities for the motor control. Thus, it is possible to control the angular velocity by controlling the armature voltage, treating the torque as a disturbance. Another possibility is to control the torque by regulating the armature current. Both approaches have advantages and disadvantages [5, 6].

The Servo Driver in Fig. 8.18 is modelled as in Fig. 8.20. It consists of a summator that outputs the difference of the reference and feedback signal at the

Fig. 8.20 Model of the Servo Driver

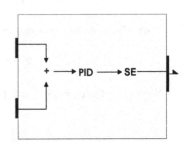

input ports, the PID controller already discussed in Sect. 8.3, and a controlled source effort that models the power output driver stage. Thus, the control of the DC motor is accomplished by controlling the armature voltage. The constitutive relation of the SE element is defined as

$$e_a = c < = \text{LIM? } (c < = - \text{LIM? } - \text{LIM:k} * c):\text{LIM} \qquad (8.24)$$

where e_a is the voltage generated by the driver and which is applied to DC motor input terminal; c is the control signal generated by the controller; k is a gain constant, and parameter *LIM* serves to limit the output voltage, and in this way the maximum current drawn by the motor.

The arm can be modelled as in Sect. 2.7.3 and Fig. 2.26 (Platform). The arm has a single port only, the shaft. We thus disconnect and remove the other two ports and all the components connected to them (Fig. 8.21). This model describes the arm as a rigid body, translating with its mass center CM, and rotating about it.

The model of the FRAME (Fig. 8.18) is a simplified variant of the FRAME object in Sect. 2.7.3; in the current model we are not interested in the reaction forces at the arm shaft. Its structure is shown Fig. 8.22 (left). The component Shaft Translation describes the effect of the bearing that fixes the arm shaft axis with respect to the frame. It consists of two source efforts that impose zero velocities of translation in two orthogonal directions. The model of Shaft Rotation is shown in Fig. 8.22 (right). The effort 1-junction, which represents the angular velocity node, is connected by the document port to the gearbox port (shaft). On the other side it is connected to the other external port connected to the arm. Thus, jointly with the Shaft translation it restricts the arm motion to the rotation about the shaft axis. The resistive element describes friction in the bearing. The component also supplies information on the angular velocity of the shaft, and by integrating it, on the angle of rotation.

Fig. 8.21 The model of the arm

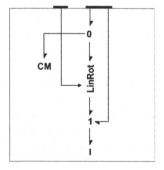

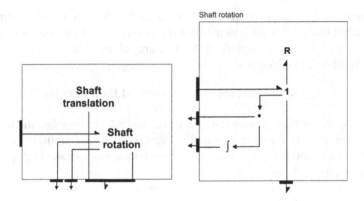

Fig. 8.22 The model of the frame: Structure of the model (*left*), Shaft rotation (*right*)

This completes the model of the servo system. The system parameters used are as follows:

Permanent magnet DC motor	
Armature resistance	R_a = 2 ohm
Armature inductance	L_a = 0.004 H
Torque constant	k_t = 0.360 N m/A
Rotor mass moment of inertia	J_m = 9.5 × 10^{-4} kg m^2
Rotor friction constant	B_m = 1.0 × 10^{-5} N m s
Peak current	i_{amax} = 50 A
Continuous current	i_a = 5 A
Gearbox	
Reduction ratio	1:20
Arm	
Mass	m = 10 kg
Mass moment of inertia about mass center	J_c = 0.4 kg m^2
Distance of mass center from the shaft	a = 0.1 m

Before starting with the simulations, several facts that are important for the interpretation of the results must be noted (Fig. 8.17). First, the arm is a non-linear object because the moment of its weight about the axis of rotation is equal to $a \cdot m \cdot g \cdot \sin\theta$, where θ is the angle of the arm longitudinal axis to the vertical axis of the co-ordinate system. Thus, changes in the moment of weight are adequately approximated by the linear relations only for small angle changes, e.g. $|\theta| \leq 0.1$. Because of this, the dynamic behaviour of the servo system differs for a small and large arm angles.

At any position—except the vertical—there is a moment of the arm weight that the motor must supply. Hence, even at a steady-state position of the arm, there is some current flowing through the armature winding of the motor. This effect is reduced by the action of the gear; its purpose is not only to reduce the angular velocity transmitted to the arm, but also to reduce the armature torque that the motor feels. This is a well-known problem that arises in robotics. It has motivated development of various weight-compensation schemes designed to minimise this effect. We will investigate how much this is a problem in the present example.

We first analyse the system response to a quarter-turn ($\pi/2$) step. The *PID* constants are set to K = 2160, K_d = 40, and K_i = 0. The system was simulated for 0.5 s with an output interval of 0.001 s. The arm rotation angle during the response is given in Fig. 8.23. It settles to a steady-state angle of 1.56954 rad. Thus the error is 0.00126112 rad, or 0.08 %. Hence, owing to a high P gain, the positional error is rather low.

Figure 8.24 shows transient in the current drawn by the motor. Initially the current drawn is very high, but when the system settles down it drops to 1.36203 A. It can be checked that this current corresponds to the moment of weight that the motor must supply in the steady-state. To limit the current drawn we can limit the voltage generated by the driver as given in (8.24). Setting LIMIT = 70 V we obtain the responses as given in Fig. 8.25 and 8.26. As Fig. 8.25 shows the settle time has now increased to about 0.2 s; the maximum current is now within the prescribed limits: the peak less than 50 A, and the continuous current less than 5 A (see the specifications above).

We also checked the response to a small input of 0.1 rad. This resulted in no apparent change in the behaviour (Fig. 8.27): The settling time is approximately the same as in the quarter turn step response without the limits (Fig. 8.23), but the motor now draws much less current.

Fig. 8.23 The rotation of the arm in response to quarter turn input (K = 2160, Kd = 40, Ki = 0)

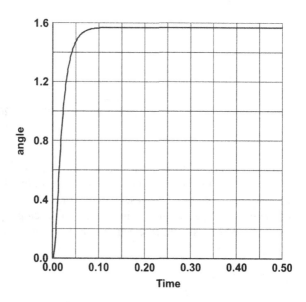

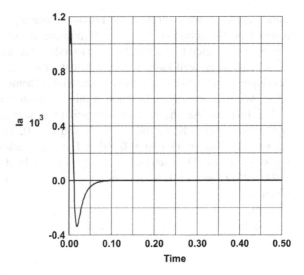

Fig. 8.24 The current drawn by the motor (K = 2160, Kd = 40, Ki = 0)

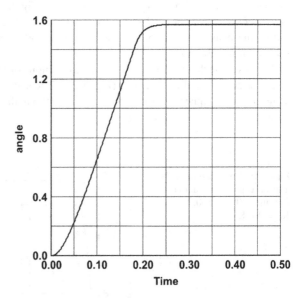

Fig. 8.25 The arm rotation to quarter turn input (K = 2160, Kd = 40, Ki = 0, LIMIT = 70 V)

One conclusion drawn from this simulation is that the simplified model of the controller, and in particular its power stage, can provide only a general impression of how such a system behaves. A much more thorough evaluation requires application of techniques presented in Chap. 7 to develop a physical model of the servo driver based on its detailed electronic design (e.g. as a MOS H-bridge).

Fig. 8.26 The current drawn by the motor (K = 2160, Kd = 40, Ki = 0, LIMIT = 70 V)

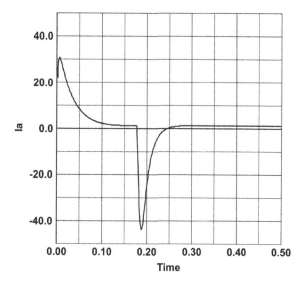

Fig. 8.27 The arm rotation to a small step input (K = 2160, Kd = 40, Ki = 0)

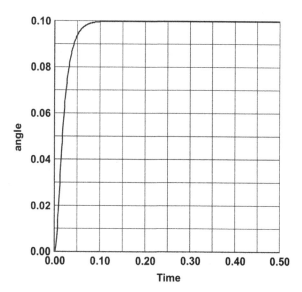

References

1. RM De Santis (1994) A Novel PID Configuration for speed and position control. Trans ASME J Dyn Syst Meas Cont 116:542–549
2. Vu HV, Esfandiary RS (1998) Dynamic systems: modeling and analysis. McGraw-Hill, New York
3. Peesen W (1994) A new look at PID-controller tuning. Trans ASME J Dyn Syst Meas Cont 116:553–557

4. Lyshevski SE (1999) Electromechanical systems, electric machines, and applied mechatronics. CRS Press, Boca Raton
5. Mohan N, Undeland TE, Robbins WP (1995) Power electronics: converters, applications, and design, 2nd edn. Wiley, New York
6. Sciavicco L, Siciliano B (1996) Modelling and control of robot manipulators. McGraw Hill, New York

Chapter 9
Multibody Dynamics

9.1 Introduction

There is an extremely large body of literature dealing with the modelling and simulation of multibody systems, e.g. [1–7]. The importance of multi-body systems is also recognized in robotics where different approaches have been developed taking into account the control aspect as well [8, 9]. The modelling of multibody systems has attracted attention in bond graph theory, too. The models are based on field multiport elements and multibonds [10–12].

In this section we describe the modelling and simulation of rigid multibody systems using the component model approach. In mechatronics, the problem is not only the mechanical part, but the complete system including the controls and the interaction with the environment as well. The general component model approach developed in this book can be applied readily to such complex systems.

The bond graph approach normally leads to the representation of the multibody system with system constraints described at the velocity, not positional level [13]. This is not specific to bond graphs, but is a characteristic property of the dynamics of systems that are described by the classical Newton-Euler approach. In this respect it corresponds more closely to the elegant approach of [7]. We will show that it is a viable approach not only from the modelling point of view, but also from the simulation aspect, as well. The component modelling approach enables the systematic development of the model, starting from the physical components and modelling the structure of the system. In this way the resulting model is more easily understood. Visual representation of the model helps this too.

We start with planar multibody systems first and develop a component model of body dynamics. Then the basic joints—such as revolute and prismatic joints—are analysed and the corresponding models developed. It is shown in an example of a quick return mechanism how the simulation model of mechanisms can be developed systematically. The system behaviour is analysed by simulation.

© Springer-Verlag Berlin Heidelberg 2015
V. Damić and J. Montgomery, *Mechatronics by Bond Graphs*,
DOI 10.1007/978-3-662-49004-4_9

To show the applicability of the approach to more complicated systems, the well-known Andrews' squeezer mechanism [13] is analysed. The accuracy of the simulation results is compared to the published results [13, 14]. It is shown that the simulation times and accuracy achieved are good, at least for engineering needs. As the last example of planar multibody dynamical systems, an engine torsional vibration problem is analysed.

The last section deals with modelling of space multibody systems. An approach to modelling of such systems is described and space component models of bodies and basic joints are developed. In the next section, the method and components developed are applied to the modelling and simulation of a complete robot system. A six-degree of freedom robotic manipulator the well-known Puma 560 is analysed. The last section deals with 3D visualization of robots.

9.2   Modelling of Rigid Multibody Systems in Plane

9.2.1   The Component Model of a Rigid Body in Planar Motion

Recall from engineering mechanics that the term *plane motion* denotes motion of a body in which all points move in parallel planes. Referring to Fig. 9.1, motion of the complete body can be represented by the motion of a body section in its plane.

The plane selected for representing the body motion is the one that contains the body's centre of mass. In mechatronic applications the bodies in question are usually the members of a mechanism assembled by connecting the bodies by suitable joints. The complete mechanism undertakes plane motion only if the joints allow motions in which all the members move in the same plane. The term

Fig. 9.1 Representation of a body motion in a plane

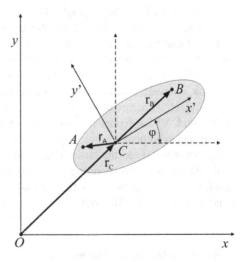

multibody system in plane refers to such a case. Otherwise the problem refers to motion in space. Plane motion of a rigid body has already been discussed in some detail in Sect. 2.7.3. For completeness we give here the essential points again.

According to the classical approach of engineering mechanics a plane body motion consists of a translation determined by the motion of the body's centre of mass and a rotation about an axis through the centre of mass that is orthogonal to the plane of motion. To describe the motion of the body, a base (inertial) frame Oxy is defined (Fig. 9.1). The translation part of the motion can be described by the position vector $\mathbf{r}_C$ of its centre of mass C in the base frame. Similarly, to describe the rotational part of the motion, a body frame $Ox'y'$ is defined, the origin of which is taken at the centre of mass, and which is moving with the body. The rotation of the body can be described by the angle ϕ, which the body frame makes with respect to the base frame.

The bond graph method uses velocities as fundamental quantities for the kinematic description of body motion. This is consistent with Newton's 2nd Law, as well as with other fundamental laws of body dynamics. Positional quantities are found from velocities by integration. Thus we take as the fundamental kinematical variables the vector of centre of mass velocity

$$\mathbf{v}_C = \begin{pmatrix} v_{Cx} \\ v_{Cy} \end{pmatrix} \tag{9.1}$$

and the angular velocity ω of the body.

The velocity of any other point P, such as A or B in Fig. 9.1, can be written as

$$\mathbf{v}_P = \mathbf{v}_C + \mathbf{v}_{CP} \tag{9.2}$$

i.e. as the sum of the velocity of the centre of mass and the relative velocity due to rotation of the body around the centre of mass. We are interested mostly in the points where the body is joined to other bodies. These points are normally defined in the body frame by the corresponding co-ordinates. The rotational part in (9.2) (from (2.87) to (2.91)) is given by

$$\mathbf{v}_{CP} = \mathbf{T}\omega \tag{9.3}$$

Here $\mathbf{T}$ is a matrix describing the transformation of rotational velocities to linear velocities and is defined by

$$\mathbf{T} = \begin{pmatrix} -x'_{CP}\sin\varphi - y'_{CP}\cos\varphi \\ x'_{CP}\cos\varphi - y'_{CP}\sin\varphi \end{pmatrix} \tag{9.4}$$

Assuming that there are two points in a body that serve for the connection to other bodies, a 2D body model can be represented by a component model (Fig. 9.2) having two power ports.

$$\longrightarrow\!\!\!\vartriangle \ \ \textbf{2DBody} \ \longrightarrow\!\!\!\vartriangle$$

Fig. 9.2 Bond graph representation of a 2DBody

The force that one body exerts on another can be represented by a resultant force vector at the connection point and a resultant moment about that point. To represent the forces and velocities at such points, the power ports of the body component model should be compounded. Power variables at the ports can be represented by 3D efforts

$$\mathbf{e}_P = \begin{pmatrix} \mathbf{F}_P \\ M_P \end{pmatrix} \tag{9.5}$$

and 3D flows

$$\mathbf{f}_P = \begin{pmatrix} \mathbf{v}_P \\ \omega \end{pmatrix} \tag{9.6}$$

The first part of such a port serves for the transfer of force $\mathbf{F}_P$ and linear velocity $\mathbf{v}_P$. The other part serves for the transfer of moment M_P and body angular velocity ω. The flow vector components in (9.6) are given by (9.2) and (9.3). Similar relations hold for the effort components of (9.5). They can be developed by applying the equivalent force and moment laws of engineering mechanics. We develop these by evaluating the power that is transferred to the body as a result of mechanical action as we did in Sect. 2.7.3. This alternative approach is convenient, for it simplifies the bond graph representation.

Using vector notation, the power transferred at the ports is given by

$$\mathbf{f}_P^T \mathbf{e}_P = \mathbf{v}_P^T \mathbf{F}_P + \omega M_P \tag{9.7}$$

By substituting from (9.2), and (9.3) we get

$$\mathbf{v}_P^T \mathbf{F}_P + M_P \omega = \mathbf{v}_C^T \mathbf{F}_P + (\mathbf{T}^T \mathbf{F}_P + M_P) \omega \tag{9.8}$$

Hence the force at a port not only tends to push the body's centre of mass, but also affects the rotation of the body. Term $\mathbf{T}^T \mathbf{F}_P$ represents the moment of the force $\mathbf{F}_P$ at the port P about the body's centre of mass. The matrix $\mathbf{T}^T$ that describes this transformation is the transposed matrix of (9.4). This term taken together with (9.3) describe the transformation between linear quantities at the port—the force and the linear velocity components—and the angular quantities referred to the body's centre of mass—moment about the centre of mass and the body angular velocity.

Using the foregoing equations the model of the body moving in a plane can be represented by the bond graph of Fig. 9.3. We can clearly see the compounded structure of the document ports (corresponding to the component ports of Fig. 9.2).

Fig. 9.3 Model of 2D body
motion

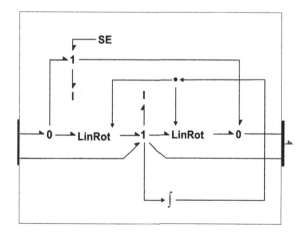

The 1 component at the top represents the body's centre of mass junctions and describes the summation of the forces acting at the centre of mass. The junctions' variables are x- and y- components of the body's centre of mass velocity. The components denoted by 0 describe the summation of the velocity vectors in (9.2). This component consists of two 0-junctions that describe the summation of the x- and y-components. These also serve as the junctions of the force components at the ports. The effort junction 1 in the middle of the model corresponds to the body angular velocity. It also describes the summation of the moments about the body's centre of mass. The components LinRot represent the transformations of the linear and angular quantities already discussed. The transformation matrix is given by (9.4). The matrix depends on the body rotation angle which is, in planar body motion, related to the body angular velocity by

$$\omega = \frac{d\varphi}{dt} \tag{9.9}$$

Hence, the rotation angle is given by

$$\varphi = \varphi_0 + \int_0^t \omega dt \tag{9.10}$$

where φ_0 is the initial value of the angle. The rotation angle is obtained in the model of Fig. 9.3 by an integrator component that, as input, takes a signal taken from the angular velocity junction. The integrator output branches through a node to the LinRot components. The rotation angle can be also extracted out of the body component if required.

The transformation of the LinRot component is represented as shown in Fig. 9.4. According to (9.3) and (9.4), and the $\mathbf{T}^T\mathbf{F}_P$ term in (9.8), it consists of two

Fig. 9.4 Transformation
between the linear and the
rotational quantities

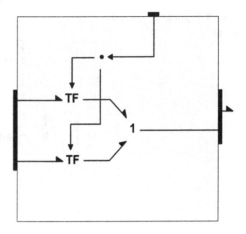

transformers and an effort junction. The transformation ratio is defined by the rows
of the transformation matrix of (9.4).

To complete the model, it is necessary to add the dynamics of body motion. This
consists of the translational dynamics governed by the body's centre of mass
motion and the dynamics of body rotation about it. The first is given by the
equations

$$\left.\begin{array}{l} \dfrac{d\mathbf{p}_C}{dt} = \mathbf{F}_C \\[2ex] \mathbf{p}_C = m\mathbf{I}\mathbf{v}_C \end{array}\right\} \tag{9.11}$$

$\mathbf{F}_c$ is the resultant force at the centre of mass, $\mathbf{v}_c$ its velocity, m is the body mass,
and $\mathbf{I}$ is the 2×2 identity matrix. The translational dynamics is represented in
Fig. 9.3 by an $\mathtt{I}$ component connected to the body's centre of mass junction $\mathtt{1}$. The
component consists of two inertial components that describe the momentum law of
(9.11). There is also a $\mathtt{SE}$ component connected to it. This is composed of the
source efforts representing the x- and y-components of the body weight.

The rotational dynamics is simpler and is described by the moment of
momentum law

$$\left.\begin{array}{l} \dfrac{dH_C}{dt} = M_C \\[2ex] H_C = J_C\omega \end{array}\right\} \tag{9.12}$$

In these equations M_c is the resultant moment about the centre of mass, ω is the
body's angular velocity and J_c is the centroidal mass moment of inertia of the body.
It is represented in Fig. 9.3 by an inertial element connected to the angular velocity
$\mathtt{1}$-junction. This junction is also connected to the lower part of the document ports
and, in this way, the moments at the body component ports are transferred directly
to the junction.

Fig. 9.5 The body frame
with the origin not at the
centre of mass

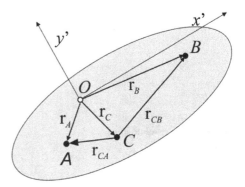

The component *2DBody* is stored in the *Model Library*, under the branch *Word Model Component Libr*ary (see Fig. 4.35), from which it can be used. It is described in terms of the absolute angular co-ordinate, i.e. the rotation angle with respect to the base (inertial) frame. It is also possible to develop a model based on the relative angular co-ordinates, i.e. of a body with respect to a second body. This approach is used in Sect. 9.6 when dealing with robot motion in space.

Another point we wish to stress is the selection of the body frame. We have assumed that its origin is at the body's centre of mass. Often, however, it is chosen with the origin at some other point in the body, e.g. at point O (Fig. 9.5). This selection affects the co-ordinates in the transformation matrix of (9.4) and, hence, the transformer ratios in the LinRot components (Fig. 9.4). These co-ordinates can be found from the vector relations

$$\mathbf{r}_{CA} = \mathbf{r}_A - \mathbf{r}_C \tag{9.13}$$

and

$$\mathbf{r}_{CB} = \mathbf{r}_B - \mathbf{r}_C \tag{9.14}$$

It is not necessary to change the transformer relations. The relations (9.13) and (9.14) can be defined by the parameter statements at the level of the LinRot components.

9.2.2   Joints

The bodies are connected by the *joints*. In multibody systems, such as machines and robot manipulators, the bodies are joined to permit some relative motions between them. The models of joints depend on their type, i.e. which motions are permitted and how the joint is physically designed. We look on a joint as a mass-less component, assuming that its mass has been included in the mass distribution of the

Fig. 9.6 A revolute joint

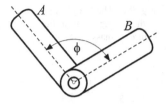

Fig. 9.7 The revolute joint
component

bodies that it connects. We develop here models of two basic joints: the revolute
and prismatic (translational). Others can be developed in a similar way.

9.2.2.1 The Revolute Joint

The revolute joint connects the bodies by a pin or a shaft (Fig. 9.6). We neglect
clearances between the pin (shaft) and the bearings. Their rotation axis is represented
in the plane of motion by the joint's coincidental central point. In the rigid revolute
joint, the only permitted motion is the relative rotation of the bodies about this point.

   We represent the joints by components that have two power ports to connect the
bodies (Fig. 9.7).

   A model of the joint is given in Fig. 9.8. The 1 component consists of the effort
junctions that represent the x- and y- velocity components of the joint centre. The
force is simply transmitted by the joint.

   The flow junction describes the relationship between the angular velocities ω_A
and ω_B of the bodies

$$\omega_A - \omega_B - \omega_{AB} = 0 \qquad (9.15)$$

where ω_{AB} is the relative velocity of body B with respect to body A.[1]

   This relative velocity is important if there is friction at the junction. This is
represented in Fig. 9.8 by a resistive element R. We also may use a SE element if
there is an actuator that drives the bodies about the junction axis, as is often the case
in robotics. The component *Joint* is stored in the *Model Library*, under the branch
*Word Model Component Libra*ry (see Fig. 4.35).

---

[1]This holds for planar motion of the bodies only, for in that case the rotation axis is orthogonal to
the plane of the motion.

Fig. 9.8 Model of the
revolute joint

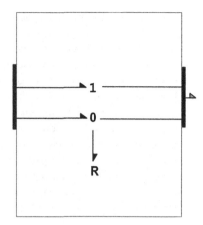

9.2.2.2 The Prismatic Joint

The prismatic or translational joint connects two bodies—one containing a straight
slot and the other having a part that fits precisely into the slot and can slide in it
without rotation (Fig. 9.9). The rotation is usually prevented by the form of the slot
and the body sliding in it, e.g. both having rectangular cross sections, or by the use
of a keyway. We can represent it by a Bond Graph component JointT (Fig. 9.10)
having two ports, as the revolute joint, but permitting the translational motion only.

Fig. 9.9 Prismatic joint

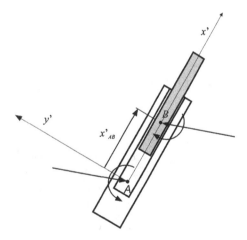

Fig. 9.10 The prismatic joint
component

⊾ **JointT** ⊾

The analysis of a prismatic joint is done in a similar way as the body motion in Sect. 9.2.1. We define a joint $Ax'y'$ co-ordinate frame that is fixed in one of the bodies, e.g. the one with the slot (Fig. 9.9). As the origin A of the frame, a convenient point on the centreline of the slot is chosen, for example, it can be the mid-point. The x-axis is directed along the slot axis and the y-axis is orthogonal to it. The other point, B, used for representing the joint is also chosen on the slot centreline, but belongs to the other body.

We assume that at these points the joint is connected to both bodies. They correspond to the ports of the prismatic joint component. Like other body connections, there is a force vector and moment acting on the joint at one port, and the reactions of other body at the other port. Likewise the ports flow consists of the velocity vectors of the corresponding junction points and the common angular velocity of the joined bodies.

The position vectors of point B and A, with respect to the base frame (not shown in Fig. 9.9), are related by

$$\mathbf{r}_B = \mathbf{r}_A + \mathbf{R}\mathbf{r}'_{AB} \tag{9.16}$$

where $\mathbf{r}'_{AB}$ is the relative position vector of B with respect to A, expressed in the frame of the joint, i.e.

$$\mathbf{r}'_{AB} = \begin{pmatrix} x'_{AB} \\ 0 \end{pmatrix} \tag{9.17}$$

The rotation matrix $\mathbf{R}$ of the joint frame with respect to the base frame (Sect. 2.7.3) is given by

$$\mathbf{R} = \begin{pmatrix} \cos\varphi & -\sin\varphi \\ \sin\varphi & \cos\varphi \end{pmatrix} \tag{9.18}$$

Thus (9.16) reads

$$\mathbf{r}_B = \mathbf{r}_A + \begin{pmatrix} x'_{AB}\cos\varphi \\ x'_{AB}\sin\varphi \end{pmatrix} \tag{9.19}$$

Here x'_{AB} represents the joint displacement co-ordinate and φ is the angle of the slot axis to the base x-axis. Both of these can change with time. Differentiation of (9.19) with respect to time gives

$$\mathbf{v}_B = \mathbf{v}_A + \begin{pmatrix} -x'_{AB}\sin\varphi \\ x'_{AB}\cos\varphi \end{pmatrix}\omega + \begin{pmatrix} \cos\varphi \\ \sin\varphi \end{pmatrix}v'_{ABx} \tag{9.20}$$

In the last equation

$$\omega = \frac{d\varphi}{dt} \tag{9.21}$$

is the angular velocity of the joint and

$$v'_{ABx} = \frac{dx'_{AB}}{dt} \tag{9.22}$$

is the velocity of the relative displacement of one of the junction parts with respect to the other along the joint axis. Note that the other velocity component

$$v'_{ABy} = 0 \tag{9.23}$$

Note that (9.20) is similar to (9.2) and (9.3), but has an additional term that comes from the relative translation of the bodies that make up the joint. The matrix of the angular velocity term in (9.20) can be also denoted as **T**, but is slightly simpler and is given by

$$\mathbf{T} = \begin{pmatrix} -x'_{AB} \sin \varphi \\ x'_{AB} \cos \varphi \end{pmatrix} \tag{9.24}$$

Thus (9.20) can be written as

$$v_B = v_A + \mathbf{T}\omega + \begin{pmatrix} \cos \varphi \\ \sin \varphi \end{pmatrix} v'_{ABx} \tag{9.25}$$

The expression for the power transferred at point (port) B is found as follows. From (9.25) we have

$$v_B^T \mathbf{F}_B + M_B \omega = v_A^T \mathbf{F}_B + (\mathbf{T}^T \mathbf{F}_B + M_B)\omega + v'_{ABx}(\cos \varphi \ \ \sin \varphi) \mathbf{F}_B \tag{9.26}$$

or

$$v_B^T \mathbf{F}_B + M_B \omega = v_A^T \mathbf{F}_B + (\mathbf{T}^T \mathbf{F}_B + M_B)\omega + v'_{AB} \mathbf{F}_B \tag{9.27}$$

Note that the power transferred at port A is $(v_A)^T \mathbf{F}_A + M_A \omega$. From this we cannot infer that the power transferred at point B is equal to the power input at point A because this would imply that $(v_{AB})^T \mathbf{F}_B = 0$. This holds, however, only for frictionless and unpowered prismatic joints. From the equilibrium equations of the mass-less joint we have $\mathbf{F}_A = \mathbf{F}_B$ and

$$M_A = M_B + \mathbf{T}^T \mathbf{F}_B \tag{9.28}$$

All these relations become clearer when we represent them by bond graphs!

We now formulate the bond graph model of the prismatic joint. It is given in Fig. 9.11.

Component 0 consists of flow junctions and describes the joint velocity relation given by (9.25). The junction variables are the components of the force $\mathbf{F}_A$ at the

Fig. 9.11 Model of the
prismatic joint

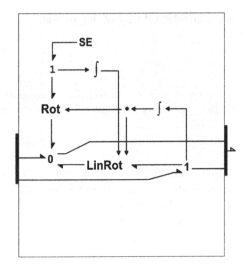

joint's left port, which is equal to the force transferred at other port, i.e. $\mathbf{F}_B$. The effort junction at the top corresponds to the relative velocity of the joint expressed in the joint frame. The corresponding position co-ordinate can be evaluated by the integrator that takes as its input the relative velocity of the joint. The component Rot describes the transformation of the x'-components of the relative velocity and the force at the joint to the base frame. It uses the first column of the matrix in (9.18) only. The source effort SE defines the force acting along the joint. It is simply zero if the friction is negligible, or can be replaced by a resistive element otherwise. This component can also be used to simulate an actuator driving the junction.

The effort junction 1 on the right corresponds to the joint angular velocity. It describes the balance of moments as implied by (9.28). The LinRot component describes the transformation between the linear and angular quantities. The corresponding transformation matrix, as given Eq. (9.24), needs information on the joint rotation angle and of the position co-ordinate of the joint. The angle of rotation is found from the integrator, which integrates the angular velocity taken from the corresponding junction.

9.2.3   Modelling and Simulation of a Planar Mechanism

We apply the modelling approach described to the quick-return mechanism of Fig. 9.12. The mechanism is relatively simple, but it contains all the elements that we have discussed so far—the bodies and the rotational and prismatic joints. More complex problems are analysed in the sections that follow.

The mechanism consists of a crank that rotates about a joint at O_1 with angular velocity ω_0. The end of the crank is connected by a revolute joint at O_2 to a block that can slide along the other member, which, in turn, can rotate about the joint at

Fig. 9.12 Quick return
mechanism

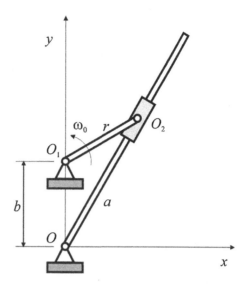

O. This simple mechanism generates an oscillatory motion of the driven member
with different forward and return times. We develop a model of the system and then
simulate its motion.

We start, as usual, by creating a project*Quick Return Mechanism*. We expand the
Model Library, the *Word Model Components and Mechanical Components*
(Fig. 9.13). We also expand the *Editing Tools* on the right, which we will need
occasionally. We will develop the model of the mechanism using the component
models we have developed in the previous section and stored in the library of models.
We will drag the library component models corresponding to the objects in Fig. 9.12
into the middle of the project window in. Going from the left we will first drag two
ground components, which will serve as the supports for joints (bearing). Next, we
drag two *Joint* components, then two *2DBody* components. Finally we drag another
Joint, which serves for connection to the prismatic joint, and then the *JointT* itself.

We can change the names (titles) of the components if we wish, as we will do later.
This operation is, however, somewhat involved. We cannot change the component
name if its ports are connected, outside or inside. Thus, we need to free it from other
components. We also need to open it and free its documents ports of the bonds. Only
then we can edit its name by selecting the component, activating its text editing mode
by clicking the bitmap icon in the form of "a page and pencil" (or using the *Edit Text*
command in the Edit menu), and typing in a new name. When we finish (by clicking
outside) we reconnect the component on the outside and inside.

We will adjust the components to correspond to the mechanism. We start with
the upper left `Joint` corresponding to $O1$ bearing (Fig. 9.13). Opening the
component a new tabbed window opens containing its Bond Graph model
(Fig. 9.14 left). The component `1` consists of two 1-junctions, which define the two
nodes representing the common x- and y-components of the velocity of the central
point. The corresponding external bonds transmit x- and y-components of the

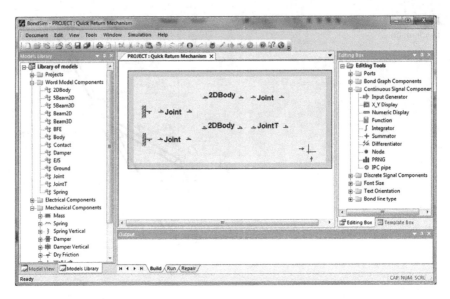

Fig. 9.13 The program window containing the project, library and tools windows

Fig. 9.14 The original model of the Joint (*left*) and its powered form (*right*)

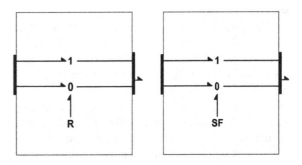

force/velocity pairs. The bottom bond line transmits the angular moment and velocity. We will thus have through the mechanism these two bond lines, one transmitting 2D vectors of force/velocity, and the other the scalar angular moment/velocity pair. This will affect also the structure of the connecting ports.

In the junction on the left appears a resistor R too. The connecting bond denotes the relative velocity between the left and the right end (inside and outside ring of the bearing). By default the resistor in library junction is zero, and hence the junction is frictionless. We can easily change this. We will retain this in the other two junctions (corresponding to O and $O1$), but this one is driven by the constant angular velocity $\omega_0 = 2\pi$ rad/s (1 c/s). Thus, we will break the bond connecting resistor the R to 0-junction, delete it; we drop from the *Edit Tools*, *Bond Graph Components* the *SF* tool in place of this component and connect it again (Fig. 9.14 right). By opening

Fig. 9.15 Model of the
supports

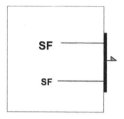

the port of SF component we can define its constitutive relation in the usual way. The flow of the *Flow Source* is omega, defined by the parameter expression *omega = 2*PI*RPS*. The *PI* is π defined at the program level, and the crank speed (cycles/s) is defined as *RPS* = 1 e.g. in the component.

Of the all components only the leftist two should be reworked. These are the two supports. Thus, we define that x- and y-components of translational velocity of the support, as well as its angular velocity, are zero. Figure 9.15 shows the model of these fixtures. We change the power sense of their ports so that in the upper component it is directed outward and in the bottom one inward. This follows from the assumed power flow along the mechanism: from the upper Joint (O1) through the upper body (the crank), then thorough the right Joint (O1) and then thorough the prismatic JointT and back by the bottom body (the swinging rod) to the lower Joint (O) and the support.

The final form of the system level model of the *Quick Return Mechanism* is shown in Fig. 9.16. The component titles are changed to better denote object in Fig. 9.13. The component ports are interconnected by the bond lines. In addition the signals are extracted from Crank and Swinging rod component models corresponding to the rotation angles with respect to the base co-ordinates. The Fun object is a function used to transform the rod rotation angle to the swing angle (see later).

To simulate the motion of the mechanism the following geometrical parameters are used: a = 0.8 m, b = 0.4 m and r = 0.2 m. The mass parameters are given in Table 9.1.

The geometry of the quick return mechanism is shown in Fig. 9.17. The maximum swinging angle is given by $\sin \alpha_{\max} = r/b$, or for the given lengths, as $\alpha_{\max} = \pi/6$. The limiting the position of the swinging bar are also shown. The upper part of the crank rotation angle is $\pi + 2\alpha$. This makes $0.5 + \alpha/\pi = 2/3$ of the full rotation. Thus

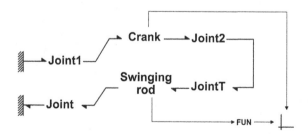

Fig. 9.16 The system level model of quick return mechanism

Table 9.1 Inertial parameters of the quick return mechanism

	Crank	Swinging rod
Mass (kg)	1.5	4.0
Moment of inertia (kg m^2)	0.005	0.25

The mass centers are located at the mid-points of the members

Fig. 9.17 Geometry of quick return mechanism

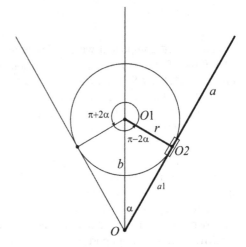

to return the crank needs to make another 0.333 part of the full rotation. Thus, for the given parameters the mechanism returns back two times quicker.

The length $a_1 = b \cos \alpha$. The centres of masses of the rods are at their mids, i.e. $0.5r$ from the crank ends, and $0.5a$ from the swinging rod ends. Figure 9.18 shows the initial position of the crank and swinging rod. This also determines the initial position of the prismatic joint on the swinging rod. This is one of characteristic points of the swinging rod. The other is its end where it is connected to the Joint. For the crank these are the end points. The values are summarised in Table 9.2. This is of course not the only possible selection of the initial position of the mechanism, but is a simple and logical one.

Fig. 9.18 The initial position of the mechanism and the local co-ordinates

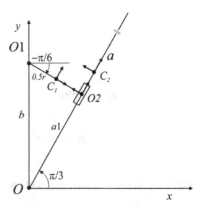

Table 9.2 The initial angles and co-ordinates of the end points

	Crank	Swinging rod
Initial angle	$-\pi/6$	$\pi/3$
x_{left}	$-0.5r$	$-0.5a$
x_{right}	$0.5r$	$a1-0.5a$

Fig. 9.19 Change of the swinging angle with the rotation of the crank

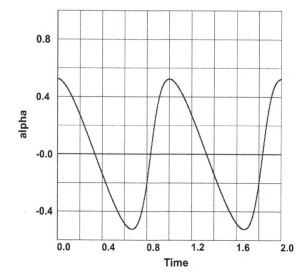

The models of `Crank` and `Swinging rod` are discussed in the previous section (see Fig. 9.3). To define these models we need the inertial parameters given in Table 9.1 and the co-ordinates of the characteristic points from Table 9.2 (see 9.4).

Now we can build the mathematical model and start the simulation. The simulation time is set to 2 s, corresponding to two revolutions of the crank. The output interval chosen is a relatively short (0.001 s) in order to obtain good resolution of the plot. The results are given in Fig. 9.19. The diagram shows the quick return behaviour of the mechanism. The results are in agreement with previous results obtained from the geometry of the problem.

9.3   Andrews' Squeezer Mechanism

We now apply the method developed in Sect. 9.2 to the well-known *Andrews' squeezer mechanism* problem. This problem has been promoted as a test of numerical codes [3, 13, 14]. We take the formulation of the problem as given in [3, 13] and compare the simulation results obtained by the BondSim program to the solution given in [14].

The mechanism (Fig. 9.20) consists of seven bodies that can move in a plane. The bodies are interconnected by revolute joints and also to the base. The arm $K1$

rotates about the fixed joint at O under the action of the driving torque M_d and this pushes, via body $K2$, the central revolute joint where three bodies—$K3$, $K4$ and $K6$ —are connected. Bodies $K4$ and $K6$ are further connected via bodies $K5$ and $K7$ to another revolute joint A that is fixed to the base. The third body, $K3$, can rotate about the fixed revolute joint at B. The end D of body $K3$ is connected to a spring that simulates the squeezer effect. The geometrical parameters of the mechanism are given in Table 9.3. The masses and mass moment of inertias with respect to the centre of mass of each body are given in Table 9.4. The spring stiffness and the driving torque are also given. The data were taken from [13].

To develop a simulation model using BondSim, we create a project called *Andrews Squeezer Mechanism*. All of the bodies—$K1$ to $K7$—will be represented by the standard plane motion body component model of Sect. 9.2.1. To create the model we will drag the corresponding *2DBody* components from the library at the appropriate places in the central window (Fig. 9.21), as we did in the previous example (Sect. 9.2.3). We will also change the component names from standard *2DBody* to the corresponding $K1$–$K7$. The weights of the bodies are neglected. The local co-ordinates of the connecting points of the bodies can be easily inferred from the data in Fig. 9.20 (right).

The revolute joints are created by copying (dragging) the standard revolute component model *Joint* from the library. The library also contains a joint for connecting more bodies *Joint3*, which represents a simple extension of the *Joint* component. This last component is used for connecting bodies $K3$, $K4$ and $K6$ to $K2$.

To save space the joints are reshaped by moving the ports from left—right to the bottom—up positions. This is achieved by opening the component and

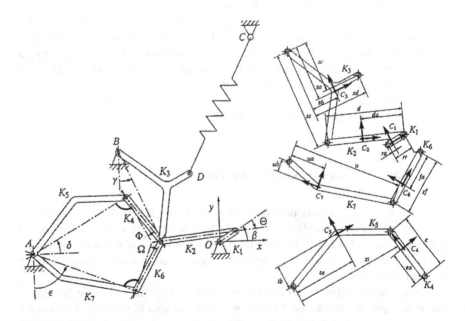

Fig. 9.20 The Andrews' squeezer mechanism ([3], used with permission)

Table 9.3 Geometrical parameters of the mechanism (Fig. 9.20)

Parameter	Value (m)	Parameter	Value (m)
ra	0.00092	tb	0.00916
rr	0.007	ss	0.035
d	0.028	sa	0.01874
da	0.0115	sb	0.01043
zf	0.02	sc	0.018
fa	0.01421	sd	0.02
el	0.02	lo	0.07785
ea	0.01421	xa	−0.06934
u	0.04	ya	−0.00227
ua	0.01228	xb	−0.03635
ub	0.00449	yb	0.03273
zt	0.04	xc	0.014
ta	0.02308	yc	0.072

Parameter *el* corresponds to the length *e* of body *K*4 in Fig. 9.20

Table 9.4 Mechanical parameters of the mechanism

	Mass (kg)	Inertia (kg m^2)	Other
K1	0.04325	2.194e−6	
K2	0.00365	4.410e−7	
K3	0.02373	5.255e−6	
K4	0.00706	5.667e−7	
K5	0.0705	1.169e−5	
K6	0.00706	5.667e−7	
K7	0.05498	1.912e−5	
Spring stiffness k			4530 N/m
Driver torque M_d			0.033 N m

disconnecting its ports on the inside off the bonds. The ports are then dragged on the front of the component to upper and bottom positions and then reconnected again on the inside. Some joints also have the control input ports and a summator to evaluate difference of the incoming signals (Fig. 9.22). This scheme is used mostly to find the difference in angular position of the bodies connected by the joint. This can be done also without the outsides signals and the summator by integrating the relative velocity flowing between 0-junction and the resistor R.

It was assumed that there is no friction in the joints. Thus, there are no reaction moments between the bodies. Hence, a joint restricts only the movement of the common points of the bodies allowing their free rotation about that point.

To simplify the model there are three components, which model the connection of the mechanism to the base. These are denoted as A, B and O. They consist of the supports and joints, as in the previous example (see Fig. 9.16). The component O also serves to drives the mechanism. Thus, as in Fig. 9.14, instead of the resistor a *Source Effort* component SE is used, which supplies the driving moment to the arm K1.

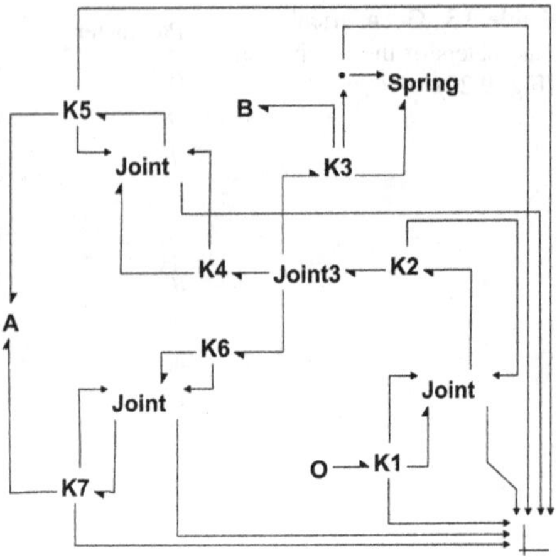

Fig. 9.21 The system level
model of the Andrews'
squeezer mechanism of
Fig. 9.20

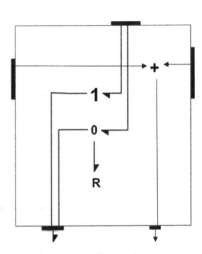

Fig. 9.22 The joint with
inserted summator of the
input signals

The body co-ordinate frames in the original Schiehlen scheme of Fig. 9.20
exactly correspond to the co-ordinate frames used in the formulation of the body
bond graph model in Sect. 9.2.1. We use as a base frame the co-ordinate frame Oxy
of Fig. 9.20. Many angles specified in the scheme, however, are not absolute, but
relative to the body frames of the connected bodies. Thus angle β of body K1 is
relative to the base frame, but the angle Θ of the next body K2 is given with respect
to the previous body K1 frame, etc. These angles are used as generalized
co-ordinates of the mechanism in [13, 14].

To compare the results we define a vector of generalized co-ordinates as in [13]

$$\mathbf{q} = (\beta \quad \Theta \quad \gamma \quad \Phi \quad \delta \quad \Omega \quad \varepsilon)^T \tag{9.29}$$

By inspection of Fig. 9.20, it is easy to find the relationships between these co-ordinates and the body rotation angles. The generalized co-ordinates of bodies K1, K3, K5 and K7 correspond to body rotation angles. Hence

$$q_1 = \varphi_1, q_3 = \varphi_3, q_5 = \varphi_5, q_7 = \varphi_7 \qquad (9.30)$$

For the others, these are the relative rotation angles

$$\left.\begin{array}{l} q_2 = \varphi_2 - \varphi_1 \\ q_4 = \varphi_4 - \varphi_5 \\ q_6 = \varphi_6 - \varphi_7 \end{array}\right\} \qquad (9.31)$$

These last relationships are evaluated by the summators inside the corresponding joint components (Fig. 9.22). This is the reason why the signals from some of the bodies in Fig. 9.21 are fed back to their common revolute joints. The output from the summators is taken out of the joint and connected to the display component at the bottom right corner. For the others, the outputs from the bodies are fed directly to the display component.

We now return to the problem of modelling the spring. The spring is attached between point D of the body and point C of the base. We assume that the attachments are such that the spring extends and contracts without bending. The co-ordinates of the attachment point D in the base frame are given by (Fig. 9.20)

$$\left.\begin{array}{l} xd = xb + sc \cdot \sin \varphi + sd \cdot \cos \varphi \\ yd = yb - sc \cdot \cos \varphi + sd \cdot \sin \varphi \end{array}\right\} \qquad (9.32)$$

The global co-ordinates xb and yb of point B, as well as the local co-ordinates sd and sc of point D with respect to B, are specified in Table 9.3. Angle φ in these equations is the rotation angle of body K3.

At point D there is a force acting on the spring, which is represented by two components F_x and F_y in the base co-ordinate frame (Fig. 9.23). F_s is the force acting along the spring.

The point moves with some velocity represented by the respective components. The movement of the point is opposed by the spring. To relate these quantities we introduce a co-ordinate frame attached to the spring. The rotation matrix of the spring local frame is given by

$$\mathbf{R} = \begin{pmatrix} \cos \varphi_s & -\sin \varphi_s \\ \sin \varphi_s & \cos \varphi_s \end{pmatrix} \qquad (9.33)$$

The rotation angle can be found from

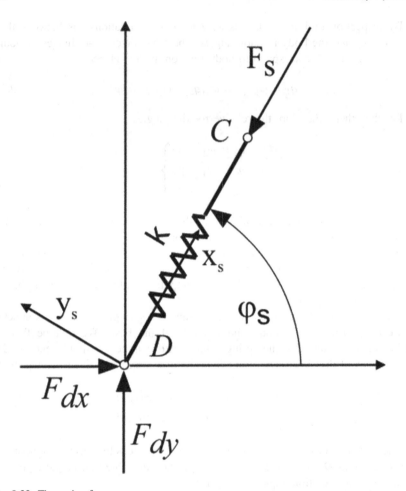

Fig. 9.23 The spring force

$$
\left.
\begin{aligned}
\cos \varphi_s &= \frac{xc - xd}{\sqrt{(xc - xd)^2 + (yc - yd)^2}} \\
\sin \varphi_s &= \frac{yc - yd}{\sqrt{(xc - xd)^2 + (yc - yd)^2}}
\end{aligned}
\right\}
\tag{9.34}
$$

The co-ordinates of point D in these relations are given by (9.32), and those of point C are the constants given in Table 9.3.

The force and velocity components of point D in the base and local spring frames are related by the transformations

Fig. 9.24 Model of the spring

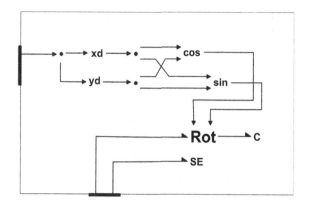

$$\left.\begin{array}{l} \mathbf{F}_d = \mathbf{R}\,\mathbf{F}'_s \\ \mathbf{v}'_s = \mathbf{R}^T\mathbf{v}_d \end{array}\right\} \tag{9.35}$$

From (9.33) and (9.35) follows

$$\begin{aligned} F_{dx} &= F_s \cos\varphi_s \\ F_{dy} &= F_s \sin\varphi_s \\ v_s &= v_{dx}\cos\varphi_s + v_{dy}\sin\varphi_s \end{aligned} \tag{9.36}$$

We can present the model of the spring as shown in Fig. 9.24. The functions in the first row represent the operations in (9.32) and (9.34). The transformations in (9.36) are represented by the Rot component, which has the usual structure of two transformers and a 0-junction. The spring is represented by the capacitive element C of stiffness k (Table 9.4). The SE component defines zero moment at the point of the connection of the spring to the body K3.

This concludes the model development of the Andrews' squeezer mechanism. The reader is advised to explore the project in the BondSim library for more details. The model developed is based on the physical modelling philosophy used in this book. The resulting model equations differ from those developed using other approaches. The model could be further simplified e.g. by removing the bonds transferring the moments between bodies. Owing to assumed frictionless joints the moments at the connection points are zero. The only exception is joint O where the driving moment acts. Therefore in the models of the bodies (see Fig. 9.3) we may remove the bottom two bonds transferring the moments from the external ports to the body angular velocity 1-junction in the middle. In the same vain we may remove the corresponding bonds, 0-junction and R element in the revolute joints (Fig. 9.8). However, as the simulations shows the improvements are only marginal, both in terms of the execution times and accuracy. Thus, even a relatively complicated multibody system can be modelled successfully in a systematically way

Table 9.5 The initial configuration of the mechanism

Parameter	Value
K1 angle (rad)	−6.17138900142764e−002
K2 angle (rad)	−6.17138900142764e−002
K3 angle (rad)	4.5527981916307e−001
K4 angle (rad)	7.10033369709727e−001
K5 angle (rad)	0.487364979543842
K6 angle (rad)	1.00787905438393
K7 angle (rad)	1.23054744454982
Spring displacement (m)	2.51774838892633e−002

using the standard component models from the BondSim program library leaving many points to the simulator.

Using the bond graph method the constraints on a body's motion are described at the velocity level. This generally leads to DAE models of index 2 (Chap. 5). In this case the model consists of 219 implicit equations. The equations are relatively simple, leading to a very sparse matrix of partial derivatives (Jacobian). It has 600 nonzero elements only, i.e. there is on average 2.7 variables per equation. In comparison the mathematical model of the mechanism based on the Lagrange multiplier form of constrained multibody mechanics [13, 14] consists of 21 differential and 6 rather complex algebraic equations. In spite of the great difference in the number of equations, it will be seen that the performance of the BondSim program is rather good.

To complete the model of the mechanism, it is necessary to define its initial configuration. This is defined by the initial values of the rotation angles of all bodies and the initial deformation of the spring. For the bodies we use the initial values of the generalized co-ordinates, as given in [13], and evaluate the corresponding initial rotational angles from (9.30) to (9.31). These are listed in Table 9.5 and correspond $\Theta = 0$ (Fig. 9.20). The initial spring displacement is calculated as

$$\Delta_s = l_0 - \sqrt{(xc - xd)^2 + (yc - yd)^2} \qquad (9.37)$$

where l_0 is the unstretched length of the spring (Table 9.3). Co-ordinates xd and yd are given by (9.32). This length is given in the last row of Table 9.5. The initial values of these angles are defined at the output port of the integrator evaluating these angles in the models of the corresponding bodies (see Fig. 9.3). Initial spring displacement is defined in the port of the capacitive element of Fig. 9.24.

The simulations were run under similar conditions as in [14]:

- Simulation interval 0.031 s
- Maximum step size and output interval 0.0001 s
- Absolute and relative error tolerance 1e−7

The reference solution in [14] was done on a CRAY computer using the PSIDE parallel software for DAEs. Experiments were also conducted using some other

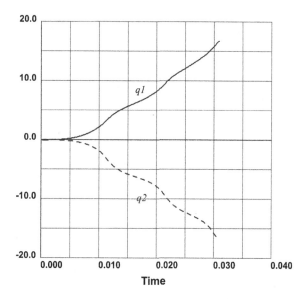

Fig. 9.25 Behaviour of q1 and q2 with time

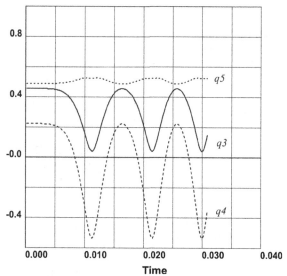

Fig. 9.26 The behaviour of q3, q4 and q5 over time

well-known codes including RADAUS [13]. These tests were run on a Silicon Graphics Indy workstation with 100 MHz R4000SC processor. The simulations with BondSim software on the other hand were done on a laptop with Intel Core I7-2630 QM CPU, 2 GHz, and 16 GB RAM.

Figures 9.25, 9.26 and 9.27 give time histories of the angles of the Andrews' squeezer mechanism as defined by Eq. (9.29). The diagrams closely correspond to those of [13, 14].

Fig. 9.27 The behaviour of q6 and q7 over time

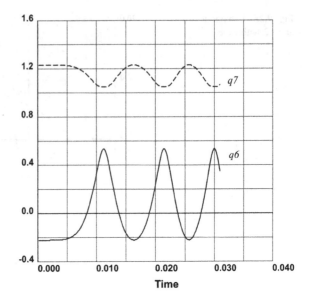

Table 9.6 Values at t = 0.03 s (default error control)

	BondSim	Referent solution [14]	−log (relerr)
q1	15.81163521	15.810771	4.3
q2	−15.7574488	−15.756371	4.2
q3	0.04082927883	0.040822240	3.8
q4	−0.534717726	−0.53473012	4.6
q5	0.5244112274	0.52440997	5.6
q6	0.5347175727	0.53473012	4.6
q7	1.048077964	1.0480807	5.6

Table 9.7 The values at t = 0.03 s (local error control)

	BondSim	Referent solution [14]	−log (relerr)
q1	15.81080552	15.810771	5.6
q2	−15.75641311	−15.756371	5.6
q3	0.0408227630	0.040822240	4.9
q4	−0.53472929241	−0.53473012	5.8
q5	0.5244103070	0.52440997	6.2
q6	0.5347292361	0.53473012	5.8
q7	1.048079695	1.0480807	6.0

Table 9.8 Simulation statistics

Parameter	Default error control	Local error control	RADAU5 [14]
Scd	3.8	4.9	4.46
Steps	397	489	117
#f	1107	1281	1321
#Jac	715	795	92
CPU (s)	0.13	0.17	0.83

Note scd is the minimum number of significant correct digits in the solution
Steps means the number of integration steps
#f is the number of function evaluations
#Jac is the number of Jacobian matrix evaluations

Table 9.6 gives the values at t = 0.03 s. The second column gives the values obtained by BondSim using the default error control method, i.e. by controlling the differentiated variables only. The third column lists the reference solution from [14]. To compare the values the number of significant correct digits are calculated as in [14] as $-\log|relerr|$. The simulation values agree with the reference solution to about 4 correct digits. Thus the simple error control used in BondSim by default gives relatively good accuracy.

The simulations were also repeated using the local error control of Sect. 5.3.3 (using the default differentiation weight of 1). The results are given in Table 9.7. The values obtained have at least one additional correct digit, that is, they are correct to a minimum of five digits.

Table 9.8 gives some of the simulation statistics. The second and third column gives the statistics when simulations are run using the default and local error control, respectively. The last column lists the values from [14] when solved numerically with the RADAU5 code.

The local error control is slightly more expensive in terms of CPU time. The accuracy of RADAU 5 lies between the BondSim default error control and the local error control. During the BondSim simulation there are also other operations besides the numerical solution.

9.4 Engine Torsional Vibrations

The determination of the torsional vibration characteristics of internal combustion engines plays an important role in the design of cars, ships and other vehicles. Traditionally, linear lumped-mass models are used in which the shafts, bearings, couplings and flywheels are modelled as concentrated disks, springs and dampers. The complete engine mechanism assembly of single and multi-cylinder engines is modelled as separate lumped inertias [15]. Such models are often supplied to shipbuilders for the torsional design of ships' propulsion systems. It is well known, however, that the reciprocating mechanisms of the engines are non-linear and, hence, their concentrated mass moments of inertias are not constant, but change

Fig. 9.28 Scheme of the
reciprocating mechanism

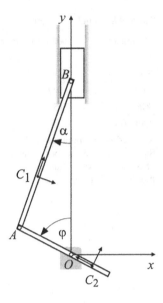

during the course of each revolution of the engine crankshaft. How important these variations are is difficult to comprehend a priori and requires detailed analysis of the complete system [16, 17]. There are also some non-linear effects that lumped parameter models cannot predict, e.g. the so-called secondary resonance [17]. Hence, a more detailed model of the engine in the time domain is important.

In this section we develop a model of a complete single-cylinder engine based on a multibody model of the reciprocating mechanism. The analysis is based on the plane body motion models of Sect. 9.2 [18, 19]. The scheme of the reciprocating mechanism is given in Fig. 9.28.

It consists of a crank that rotates about the bearing at O and which is connected to the engine piston by the connecting rod AB. All connections of reciprocating members are by bearings. We treat bearings as frictionless revolute joints. The piston slides in the cylinder bore and is acted upon by pressure forces developed by the combustion processes in the cylinder chamber. The piston friction is neglected.

There is also the friction in the mechanism, but this is quite difficult to predict. Thus its overall effect is represented by a linear resistive element (see later). The co-ordinate frames used are shown in the figure. The detailed geometry is generally not known, but it is reasonable to assume that the crank and the connecting rod mass centres are on their geometrical axes AO and AB, respectively, and the body-fixed axis is taken, in the each case, along these axes.

We develop a model to analyse the engine torsional vibrations using the BondSim. A project called *EngineTorsional Vibrations* is created. A model of the system is developed systematically by creating component models of all the main engine parts and then connecting them by bond lines (Fig. 9.29). The engine parameters are based on [17].

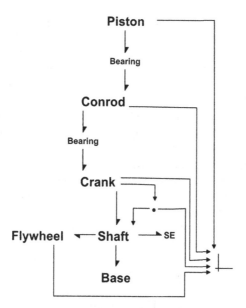

Fig. 9.29 Model for study of engine torsional vibration

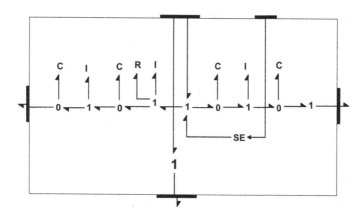

Fig. 9.30 Model of the engine shaft

The crank is fixed to the shaft. The shaft is connected on one side to the flywheel and on the other side (the output part) to the load. It is assumed that the engine is unloaded and, hence, the load is represented by a zero effort source SE. The shaft rotates in the bearings fixed in the engine body, which is here represented by the Base component.

The model of the shaft is given in Fig. 9.30. In the centre, the component 1 is connected to both power document ports. This component describes the translation of the shaft axis and is represented by an array of effort junctions corresponding to the x- and y-axis motions. It is connected to the Base by the lower power port. The shaft cannot translate in the engine body and hence the component Base is represented by two zero flow sources. The other effort junction connected to the shaft upper document port is the shaft angular velocity node.

On each side of the shaft angular velocity junction there are capacitive, inertial and resistive elements that model the shaft torsional dynamics. The model used corresponds to the lumped mass model of [17]. Thus, the first inertial element on the left is the angular inertia of the reduction gear and the camshaft. The resistive element describes the overall friction effect in the engine reduced to the shaft.

A pair of capacitive elements on each side of the centre in Fig. 9.30 represents the stiffness of the shaft parts between the crank and the bearings. The bearings are represented by the inertial elements. The last capacitive element on the left of the centreline represents the stiffness of the shaft between the left bearing and the flywheel; that on the right side represents the stiffness of the output portion of the shaft. The controlled source effort models the effect of the combustion forces on the piston, reduced to the shaft. The element is controlled by the angle of shaft rotation. It simulates the load torque by [16, 17]

$$M = M_m(1 + B \cdot \sin(n\varphi)) \tag{9.38}$$

where M_m is the average indicated torque and B and n are appropriate constants.

The Flywheel (Fig. 9.29) is modelled simply by an inertial element connected to an effort junction, from which information on the flywheel angular velocity is obtained. This is fed to the display component (at the right bottom in the document).

The Crank and Conrod components are modelled as rigid bodies in plane motion using the general model for plane motion bodies of Sect. 9.2.1. The components are created simply by dragging the 2DBody component from the library. The main difference is the change of name. The ports are moved to the upper and lower parts of the components in order to represent the connections as depicted in Fig. 9.28. Thus, the model of the Crank is as depicted in Fig. 9.31.

Fig. 9.31 The model of the Crank

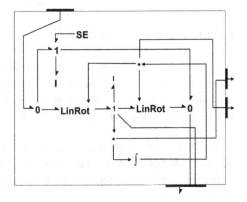

Fig. 9.32 Simplified model
of the Piston

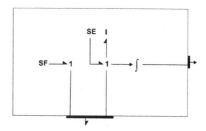

Note that, because of the connection through the frictionless bearings, there is no transfer of external moments to the crank. The SE connected to the mass centre joint describes the weight of the component. It can be seen in the lower part of the model that signals of the crank rotation angle and the angular velocity are obtained. These are used to display the crank motion, as well as for simulating the shaft load torque (Fig. 9.29). The model of the Conrod is almost the same, except that only the component rotation angle is picked up. The bearings are modelled by the revolute joint component of Sect. 9.2.2. The Bearing components imply the common velocity condition of the bodies at the connection points (the bearing axes).

The engine piston slides inside the cylinder. This is described by a translation joint. The model is simplified by assuming that the engine body is fixed in the base frame (Fig. 9.32). The left effort junction represents the x-axis piston velocity, which is zero.

Table 9.9 The engine model parameters

Parameter		Value
Crank	Length OA	0.02491 m
	Ratio OC/OA	0.143
	Mass	0.557 kg
	Inertia	3.21×10^{-4} kg m^2
Conrod	Length AB	0.09847 m
	Ratio AC/AB	0.164
	Mass	0.1040 kg
	Inertia	1.53×10^{-4} kg m^2
Piston mass		0.1649 kg
Shaft stiffness	The left part	146,000 N m
	The right part	146,000 N m
	The flywheel part	11,090 N m
	The output part	7960 N m
Shaft inertia	Gear and camshaft	7.99×10^{-5} kg m^2
	Bearing left	5.92×10^{-6} kg m^2
	Bearing right	5.92×10^{-6} kg m^2
Flywheel inertia		9.35×10^{-3} kg m^2
Engine friction constant		0.01 N m s
Average torque		1.86 N m

This condition is implied by a zero flow source. The piston can move in the y-direction and is affected by its inertia and gravity. Note also that the position of the piston is evaluated by integration and the signal is fed out for display.

The effect of the combustion pressure is not represented here but is taken into account by the equivalent load torque reduced to the Shaft as has already been described (see 9.38). This analysis of engine torsional vibrations is a common one, for it is well known that the pressure forces can be described by a Fourier series [15]. In (9.38) the parameter n takes into account different harmonics contained in the piston force and, thus, the effect on the torsional vibration characteristic can be examined by varying it. Alternatively, a detailed model of the combustion process in the cylinder can be developed and included in the model.

The model parameters are summarized in Table 9.9. These are based on the measured data of a small four-stroke single cylinder engine [17]. The last two parameters were chosen in such a way as to achieve, on the one hand, the free running engine velocity of about 1800 rpm, as in the experiments of [17], and on the other hand a relatively good mechanical efficiency (light damping) of the system. The parameter B of (9.38) was set to 1.

After the model is completed, it is possible to analyse the dynamic behaviour of the engine by simulation. We are interested mainly in the behaviour of the system around the first natural frequency. Thus the time-response of the system to the load torque is found first, and then the frequency spectrum of the engine angular velocity is generated. Because the time constant of the system is about 1 s, we need about 5 s to be sure that the velocity settles down to a steady state value. After that, we run the system for an additional 5 s.

Thus the simulation interval is taken to be 10 s. The range of frequency we are interested in is 0–1000 Hz. Thus the output interval should be at least 1/2000 s or 5×10^{-4} s. We choose the default error tolerance of 10^{-6} and the default method.

Fig. 9.33 Response of the engine near the resonant frequency ($n = 25$)

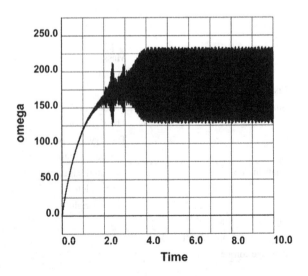

Fig. 9.34 Frequency spectrum of engine angular velocity

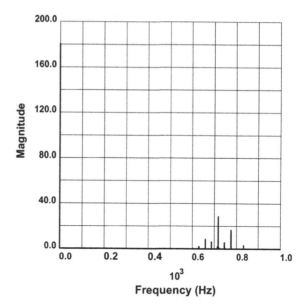

Fig. 9.35 The lower part of the spectrum (1–100 Hz)

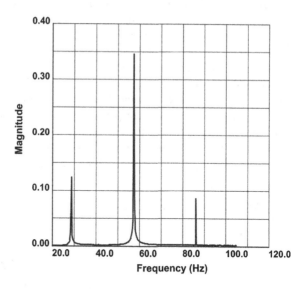

The first resonance appears when parameter n of (9.38) is about 25. Figure 9.33 shows transients in the angular velocity for $n = 25$. The output is very noisy because it is on the border of instability. The first signs of instability appears after $t = 2$ s.

We get the frequency spectrum by right-clicking the time plot and select *Frequency spectrum* from a drop-down menu. In the dialogue that appears we select a time window of 5–10 s and, and retain the default *Rectangular* window. The resulting spectrum calculated by Fast Fourier Transform is shown in Fig. 9.34. As can be seen, there is a large DC component of magnitude 180.570 rad/s (1724 rpm).

Fig. 9.36 High frequency range of the spectrum (500–900 Hz)

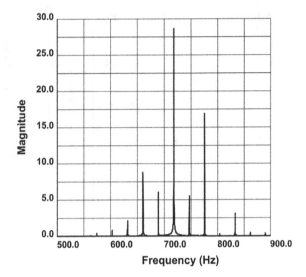

We expand the low frequency region up to 100 Hz, Fig. 9.35, by clicking with the right mouse button on the frequency plot, and then selecting the *Expand* command from a drop-down menu. In a similar way the high frequency region can be expanded. This is shown in Fig. 9.36.

The low frequency region shows characteristic harmonics that appear at multiples of the fundamental frequency of 28.7 Hz ($\sim$1724 rpm) because of the inertial effects in the engine. It can be seen that the second natural frequency is the largest, then the first, and followed by the third. The higher order harmonics are much lower in amplitude and there is no response until frequencies in the range 600–900 Hz. The higher natural frequency is estimated at 718.5 Hz with a magnitude of 28.7 rad/s. There are also amplitude peaks on both sides of the resonant frequency, which are displaced by twice the fundamental frequency, i.e. 57.5 Hz. This is

Fig. 9.37 The base and body frames

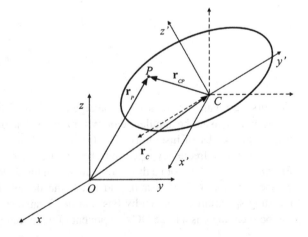

termed the *secondary resonance* and is the result of non-linear inter-coupling in the engine's reciprocating mechanism [17]. The spectrum diagrams, as well as the characteristic frequencies, agree well with the experimental results reported in [17].

9.5   Motion of Constrained Rigid Bodies in Space

9.5.1   Basic Kinematics

To describe motion of a body in space we use two fundamental co-ordinate frames (Fig. 9.37)—a base frame $Oxyz$ and a body frame $Cx'y'z$ moving with it, as we did in Sect. 9.2. There can be a number of bodies and, hence, a number of body frames that are used for their description. On the other hand, there is a single base (inertial) frame. In robotics it is often convenient to introduce other frames, as well [9]. All of the frames are *3D Cartesianco-ordinate* frames.

We assume that the *position* of a body frame with respect to the base is defined by a position vector $\mathbf{r}_C$ of the origin C of the body frame. The point C is the reference point for describing the body position—a body centre (pole). As in the planar case, it is the body mass centre, but it could be also some other point. The body *orientation* is defined by the rotation matrix $\mathbf{R}$ composed of the direction cosines of the body axes with respect to the base axes (see e.g. [2]). The position of any point P fixed in the body with respect to the base is given by

$$\mathbf{r}_P = \mathbf{r}_C + \mathbf{R}\mathbf{r}'_{CP} \qquad (9.39)$$

where $\mathbf{r}'_{CP}$ is the vector of its *material* co-ordinates, i.e. the co-ordinates with respect to the body frame.

During motion, the position of the body centre $\mathbf{r}_C$ and its orientation—represented by matrix $\mathbf{R}$—changes with time. The material co-ordinates of a point fixed in the body do not change, but the co-ordinates change with respect to the base frame. The velocity of the point P, as seen from the base frame, can be found by differentiating (9.39) with respect to time, i.e.

$$\mathbf{v}_P = \mathbf{v}_C + \frac{d\mathbf{R}}{dt}\mathbf{r}'_{CP} \qquad (9.40)$$

The relative velocity of the point P with respect to the origin of the body frame C is

$$\mathbf{v}_{CP} = \mathbf{v}_P - \mathbf{v}_C \qquad (9.41)$$

and thus the velocity of point P is

$$\mathbf{v}_P = \mathbf{v}_C + \mathbf{v}_{CP} \qquad (9.42)$$

From (9.40) the relative velocity depends on the body orientation change

$$\mathbf{v}_{CP} = \frac{d\mathbf{R}}{dt}\mathbf{r}'_{CP} \qquad (9.43)$$

The velocity can be represented as a vector in any frame by its rectangular co-ordinates. The relative velocity representations in the body frame $\mathbf{v}'_{CP}$ and in the base frame are related by the co-ordinate transformation

$$\mathbf{v}_{CP} = \mathbf{R}\mathbf{v}'_{CP} \qquad (9.44)$$

The rotation matrix is *orthogonal* and thus we have for its inverse

$$\mathbf{R}^{-1} = \mathbf{R}^T \qquad (9.45)$$

where the superscript T is the transposition operator. From (9.44) to (9.45),

$$\mathbf{v}'_{CP} = \mathbf{R}^T \mathbf{v}_{CP} \qquad (9.46)$$

Substituting from (9.43) we get

$$\mathbf{v}'_{CP} = \mathbf{R}^T \frac{d\mathbf{R}}{dt}\mathbf{r}'_{CP} \qquad (9.47)$$

But,

$$\mathbf{r}_{CP} = \mathbf{R}\mathbf{r}'_{CP} \qquad (9.48)$$

Thus, we have

$$\mathbf{v}_{CP} = \frac{d\mathbf{R}}{dt}\mathbf{R}^T \mathbf{r}_{CP} \qquad (9.49)$$

It is a simple matter to show that the factors of the position vector on the right side of (9.47) and (9.49) are skew-symmetric. From (9.45) it follows that

$$\mathbf{R}^T\mathbf{R} = \mathbf{R}\mathbf{R}^T = \mathbf{I} \qquad (9.50)$$

where $\mathbf{I}$ is the identity matrix. Hence, by differentiating with respect to time we get

$$\frac{d\mathbf{R}^T}{dt}\mathbf{R} + \mathbf{R}^T\frac{d\mathbf{R}}{dt} = 0 \qquad (9.51)$$

or

$$\mathbf{R}^T \frac{d\mathbf{R}}{dt} = -\frac{d\mathbf{R}^T}{dt}\mathbf{R} = -\left(\mathbf{R}^T \frac{d\mathbf{R}}{dt}\right)^T \tag{9.52}$$

And, similarly

$$\frac{d\mathbf{R}}{dt}\mathbf{R}^T = -\mathbf{R}\frac{d\mathbf{R}^T}{dt} = -\left(\frac{d\mathbf{R}}{dt}\mathbf{R}^T\right)^T \tag{9.53}$$

The vector product of two vectors $\boldsymbol{\alpha}$ and $\boldsymbol{\beta}$ defined by their components in a Cartesian frame

$$\boldsymbol{\alpha} = \left(\alpha_x\ \alpha_y\ \alpha_z\right)^T \tag{9.54}$$

and

$$\boldsymbol{\beta} = \left(\beta_x\ \beta_y\ \beta_z\right)^T \tag{9.55}$$

reads

$$\boldsymbol{\alpha} \times \boldsymbol{\beta} = \begin{pmatrix} \alpha_y\beta_z - \alpha_z\beta_y \\ \alpha_z\beta_x - \alpha_x\beta_z \\ \alpha_x\beta_y - \alpha_y\beta_x \end{pmatrix} = \begin{pmatrix} 0 & -\alpha_z & \alpha_y \\ \alpha_z & 0 & -\alpha_x \\ -\alpha_y & \alpha_x & 0 \end{pmatrix} \begin{pmatrix} \beta_x \\ \beta_y \\ \beta_z \end{pmatrix} \tag{9.56}$$

Following [7] we denote by $\boldsymbol{\alpha}\times$ a skew-symmetric tensor that is, through the ordinary vector product operation, associated with vector $\boldsymbol{\alpha}$. Vector $\boldsymbol{\alpha}$ is the *axial vector* associated with a skew-symmetric tensor. We can also write (9.56) as

$$\boldsymbol{\alpha} \times \boldsymbol{\beta} = \begin{pmatrix} \alpha_y\beta_z - \alpha_z\beta_y \\ \alpha_z\beta_x - \alpha_x\beta_z \\ \alpha_x\beta_y - \alpha_y\beta_x \end{pmatrix} = -\begin{pmatrix} 0 & -\beta_z & \beta_y \\ \beta_z & 0 & -\beta_x \\ -\beta_y & \beta_x & 0 \end{pmatrix} \begin{pmatrix} \alpha_x \\ \alpha_y \\ \alpha_z \end{pmatrix} \tag{9.57}$$

and, hence,

$$\boldsymbol{\alpha} \times \boldsymbol{\beta} = -\boldsymbol{\beta} \times \boldsymbol{\alpha} \tag{9.58}$$

Thus, we have from (9.49)

$$\boldsymbol{\omega}\times = \frac{d\mathbf{R}}{dt}\mathbf{R}^T \tag{9.59}$$

and

$$\boldsymbol{\omega} = axial\left(\frac{d\mathbf{R}}{dt}\mathbf{R}^T\right) \tag{9.60}$$

Similarly, from (9.47), we have

$$\boldsymbol{\omega}' \times = \mathbf{R}^T \frac{d\mathbf{R}}{dt} \tag{9.61}$$

and

$$\boldsymbol{\omega}' = axial\left(\mathbf{R}^T \frac{d\mathbf{R}}{dt}\right) \tag{9.62}$$

From (9.59) and (9.61) follows

$$\frac{d\mathbf{R}}{dt} = \boldsymbol{\omega} \times \mathbf{R} \tag{9.63}$$

and

$$\frac{d\mathbf{R}}{dt} = \mathbf{R}\boldsymbol{\omega}' \times \tag{9.64}$$

Vectors $\boldsymbol{\omega}$ and $\boldsymbol{\omega}'$ are the base frame and body frame representations of the body angular velocity. These representations are related by

$$\boldsymbol{\omega} = \mathbf{R}\boldsymbol{\omega}' \tag{9.65}$$

Thus, the relative velocity expressions of (9.47) and (9.49) read

$$\mathbf{v}'_{CP} = \boldsymbol{\omega}' \times \mathbf{r}'_{CP} \tag{9.66}$$

and

$$\mathbf{v}_{CP} = \boldsymbol{\omega} \times \mathbf{r}_{CP} \tag{9.67}$$

The velocity equations of (9.42) can now be written as

$$\mathbf{v}_P = \mathbf{v}_C + \boldsymbol{\omega} \times \mathbf{r}_{CP} \tag{9.68}$$

and

$$\mathbf{v}'_P = \mathbf{v}'_C + \boldsymbol{\omega}' \times \mathbf{r}'_{CP} \tag{9.69}$$

We transform these equations a little more using (9.58). Thus, (9.68) can be written as

$$\mathbf{v}_P = \mathbf{v}_C + (-\mathbf{r}_{CP} \times \boldsymbol{\omega}) \tag{9.70}$$

Similarly, from Eq. (9.69), we have

$$\mathbf{v}_P' = \mathbf{v}_C' + (-\mathbf{r}_{CP}' \times \boldsymbol{\omega}') \tag{9.71}$$

To represent these relations compactly we introduce a generalized vector of 6D vector composed of the linear velocity vector of the body centre and the body angular velocity. In the base frame this vector is given by

$$\mathbf{f}_C = \begin{pmatrix} \mathbf{v}_C \\ \boldsymbol{\omega} \end{pmatrix} \tag{9.72}$$

and similarly in the body frame as

$$\mathbf{f}_C' = \begin{pmatrix} \mathbf{v}_C' \\ \boldsymbol{\omega}' \end{pmatrix} \tag{9.73}$$

We denote such vectors by $\mathbf{f}$ because, as will be shown later, these are simply the flows of the bond graph component ports.

In a similar way, a generalized velocity of any other point P fixed in the body can be represented in the base frame by

$$\mathbf{f}_P = \begin{pmatrix} \mathbf{v}_P \\ \boldsymbol{\omega} \end{pmatrix} \tag{9.74}$$

and in the body frame as

$$\mathbf{f}_P' = \begin{pmatrix} \mathbf{v}_P' \\ \boldsymbol{\omega}' \end{pmatrix} \tag{9.75}$$

From (9.70) we have

$$\mathbf{f}_P = \begin{pmatrix} \mathbf{I} & -\mathbf{r}_{CP}\times \\ \mathbf{0} & \mathbf{I} \end{pmatrix} \mathbf{f}_C \tag{9.76}$$

Also, we have

$$\mathbf{f}_C = \begin{pmatrix} \mathbf{R} & \mathbf{0} \\ \mathbf{0} & \mathbf{R} \end{pmatrix} \mathbf{f}_C' \tag{9.77}$$

From (9.76) and (9.77) we have

$$\mathbf{f}_P = \begin{pmatrix} \mathbf{I} & -\mathbf{r}_{CP}\times \\ \mathbf{0} & \mathbf{I} \end{pmatrix} \begin{pmatrix} \mathbf{R} & \mathbf{0} \\ \mathbf{0} & \mathbf{R} \end{pmatrix} \mathbf{f}_{Ci}' \tag{9.78}$$

or

$$\mathbf{f}_P = \begin{pmatrix} \mathbf{R} & -\mathbf{r}_{CP} \times \mathbf{R} \\ \mathbf{0} & \mathbf{R} \end{pmatrix} \mathbf{f}'_C \tag{9.79}$$

Matrix

$$\mathbf{C} = \begin{pmatrix} \mathbf{R} & -\mathbf{r}_{CP} \times \mathbf{R} \\ \mathbf{0} & \mathbf{R} \end{pmatrix} \tag{9.80}$$

can be termed a representation of the body frame *configuration tensor* [7]. Thus, we have

$$\mathbf{f}_P = \mathbf{C}\mathbf{f}'_C \tag{9.81}$$

We can also express (9.71) as

$$\mathbf{f}'_P = \begin{pmatrix} \mathbf{I} & -\mathbf{r}'_{CP} \times \\ \mathbf{0} & \mathbf{I} \end{pmatrix} \mathbf{f}'_C \tag{9.82}$$

From (9.77) we find

$$\mathbf{f}'_C = \begin{pmatrix} \mathbf{R}^T & \mathbf{0} \\ \mathbf{0} & \mathbf{R}^T \end{pmatrix} \mathbf{f}_C \tag{9.83}$$

After substituting in (9.82) we find the body frame configuration representation in the same frame

$$\mathbf{C}' = \begin{pmatrix} \mathbf{R}^T & -\mathbf{r}'_{CP} \times \mathbf{R}^T \\ \mathbf{0} & \mathbf{R}^T \end{pmatrix} \tag{9.84}$$

Thus, we have the inverse transformation

$$\mathbf{f}'_P = \mathbf{C}'\mathbf{f}_C \tag{9.85}$$

In this section some of the basic kinematic quantities have been developed. They are generalizations of the corresponding expressions for planar body motion developed in Sect. 9.2.1.

9.5.2   Bond Graph Representation of a Body Moving in Space

Now we can proceed with the development of a bond graph model of a body moving in space. It is similar to the one used for bodies in plane motion in

Fig. 9.38 Representation of a body moving in space

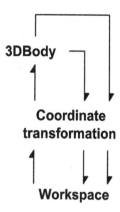

Sec. 9.2.1. The main difference lies in the more complex relationship that governs motion of bodies in space.

To develop a fundamental bond graph representation we consider a single body interacting with its environment. The environment consists of the bodies to which the analysed body is connected. We model the body and its environment by two component objects—3DBody and Workspace (Fig. 9.38).

We assume that the body is connected to other bodies at two points. We also assume that a body in the environment acts on *the considered body* by a force and a moment. Likewise *the body* acts to another body in the environment. In this way there is a transfer of power from the environment to the body and from the body back to the environment. In order to represent these interactions we assume that the 3DBody component has *three* ports. Two of these, the bottom and the upper ones in Fig. 9.38, correspond to the power interactions just described. We have added the third one (the side port) to correspond to the body centre of mass. It is possible to develop a model without the explicit use of the centre of mass port, but its use simplifies the description of the body dynamics.

As explained previously, velocities seen at a port can be represented by the 6D *flow* vectors of (9.74) and (9.75). Similarly we can use 6D *efforts* to represent the resultant force and moment at a port. Such efforts can be represented in the base frame by

$$\mathbf{e}_P = \begin{pmatrix} \mathbf{F}_P \\ \mathbf{M}_P \end{pmatrix} \tag{9.86}$$

and, similarly, in the body frame as

$$\mathbf{e}'_P = \begin{pmatrix} \mathbf{F}'_P \\ \mathbf{M}'_P \end{pmatrix} \tag{9.87}$$

Here $\mathbf{F}_P$ and $\mathbf{F}'_P$ are 3D representations of the resultant force at connection point P and $\mathbf{M}_P$ and $\mathbf{M}'_P$ the corresponding representations of the resultant moment.

To represent these 6D vector quantities, the ports of the Base and Body components are assumed to be compounded (Sect. 2.4). The first sub-port corresponds to the linear velocity and resultant force pair, and the other to the body angular velocity and resultant moment. This way the ports can be used to access efforts and flows at the port by two ordered 6D efforts and flows. At the mass centre ports however, as will be seen later, a 3D representation is sufficient. The flow is the mass centre velocity and the effort is the resultant of the forces acting on the body at other ports (interconnection points) reduced to the mass centre. The gravity force (the body weight) acting there is not accounted for and is taken into account when dealing with the body dynamics. The reason for this is that body weight can be easily described in the base frame. However, this is not so in a body frame, because it is generally rotated with respect to the first.

We represent effort and flow vectors at body ports as seen in the body, i.e. with respect to a body fixed frame. Similarly at the base ports the corresponding quantities are expressed with respect to the base frame. Transformations between these two representations are given in Fig. 9.38 by the Coordinate transformation component. This component transforms effort and flow at a body port to the corresponding base representation and vice versa. The transformation between efforts is given by

$$\mathbf{e}_P = \begin{pmatrix} \mathbf{R} & \mathbf{0} \\ \mathbf{0} & \mathbf{R} \end{pmatrix} \mathbf{e}_P' \tag{9.88}$$

and similarly for the flows

$$\mathbf{f}_P = \begin{pmatrix} \mathbf{R} & \mathbf{0} \\ \mathbf{0} & \mathbf{R} \end{pmatrix} \mathbf{f}_P' \tag{9.89}$$

where $\mathbf{R}$ is the rotation matrix of the body frame with respect to the base.

We use Cartesian 3D frames only. Thus the transformation matrix is orthogonal as shown in the first part of this section. Thus from (9.89) we get the inverse transformation

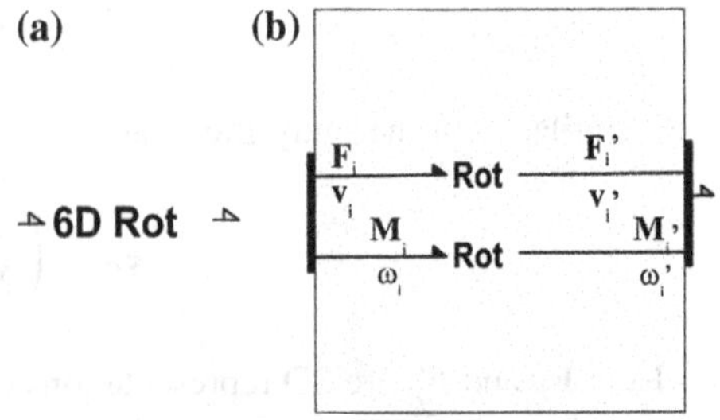

Fig. 9.39 6D transformation: **a** the component representation, **b** the structure

$$\mathbf{f}'_P = \begin{pmatrix} \mathbf{R}^T & \mathbf{0} \\ \mathbf{0} & \mathbf{R}^T \end{pmatrix} \mathbf{f}_P \tag{9.90}$$

It follows, then, that the transformation does not change the power transfer between the Body and the Base

$$\mathbf{f}'^T_P \mathbf{e}'_P = \mathbf{f}^T_P \mathbf{e}_P \tag{9.91}$$

The transformation of port quantities can be represented by the 6D transformation component of Fig. 9.39a. This component transforms both 3D parts of the 6D effort-flow pairs of the corresponding ports, according to (9.88) and (9.89), as shown in Fig. 9.39b. The transformations are represented by the Rot components that describe the transformation of 3D effort-flow vectors by the rotation matrix $\mathbf{R}$. The transformation can be represented using transformer elements TF (see Sect. 9.6).

The velocities at a port are related to the velocity of the body mass centre by (9.82). To find the relation between forces and moments, we evaluate the power transfer between the ports as in Sect. 9.2.1. From (9.82) we get

$$\mathbf{f}'^T_P \mathbf{e}'_P = \mathbf{f}'^T_C \begin{pmatrix} \mathbf{I} & \mathbf{0} \\ \mathbf{r}'_{CP}\times & \mathbf{I} \end{pmatrix} \mathbf{e}'_P \tag{9.92}$$

After substituting from (9.73), (9.75) and (9.87), and evaluating, we get

$$\mathbf{v}'^T_P \mathbf{F}'_P + \boldsymbol{\omega}'^T \mathbf{M}'_P = \begin{pmatrix} \mathbf{v}'^T_C & \boldsymbol{\omega}'^T \end{pmatrix} \begin{pmatrix} \mathbf{F}'_P \\ \mathbf{r}'_{CP} \times \mathbf{F}'_P + \mathbf{M}'_P \end{pmatrix} \tag{9.93}$$

or

$$\mathbf{v}'^T_P \mathbf{F}'_P + \boldsymbol{\omega}'^T \mathbf{M}'_P = \mathbf{v}'^T_C \mathbf{F}'_P + \boldsymbol{\omega}'^T (\mathbf{r}'_{CP} \times \mathbf{F}'_P + \mathbf{M}'_P) \tag{9.94}$$

Fig. 9.40 The model of 3D body motion

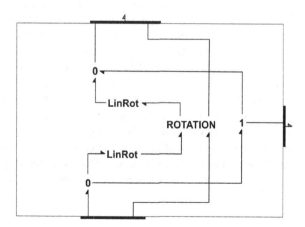

Fig. 9.41 The representation of LinRot transformation

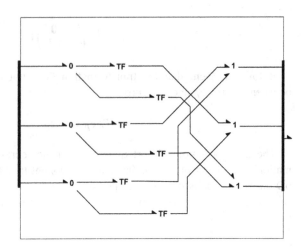

The last equation is the generalization of the planar case (9.8). It describes, jointly with (9.82), the basic velocity and force relationships for rigid bodies.

The basic structure of the Body component, defined by (9.82) and (9.94), is given in Fig. 9.40. It is very similar to that of the planar body model in Fig. 9.3, but involves three-dimensional quantities.

Here the component 1 is also an array of the effort junctions corresponding to the centre of mass velocity. Instead of a simple effort junction for the body angular velocity, the component Rotation is used. This is because the rotational dynamics in 3D is much more involved than in the planar case. (We discuss this component later.). The components 0 represent the array of flow junctions that describe the port velocities relationship given by the first 3D row of (9.82). It gives the linear velocity at a port as the sum of the velocity of the body centre of mass and the relative velocity of the port due to the rotation of the body about the centre of mass. The joint variable is the force at the port. The body angular velocity, however, is a property of the body; as such—according to the bottom row of (9.82)—it is directly transferred to the corresponding Rotation component port.

The LinRot components represent transformations between linear and angular quantities. That is, between the relative velocity at the port with respect to the centre of mass of the body and the body angular velocity on the one hand, and of the force at the port and its moment about the mass centre, on the other. These transformations are defined by the skew-symmetric tensor operations in (9.82) and (9.94) and can be represented by transformers, as in Fig. 9.41.

Every transformer in Fig. 9.41 corresponds to a nonzero entry of tensor $-\mathbf{r}'_{CP}\times$ or its transpose. Thus, the flow junctions on the left evaluate the relative velocities at the left port as a result of the multiplication of the angular velocity at the right port by matrix $-\mathbf{r}'_{CP}\times$. Likewise, the effort junctions on the right give the moment at the right port of the force at the left port by multiplying it by the $\mathbf{r}'_{CP}\times$ matrix. The ratios of the transformers are the material co-ordinates of a body point

Fig. 9.42 Motion of the body
in space

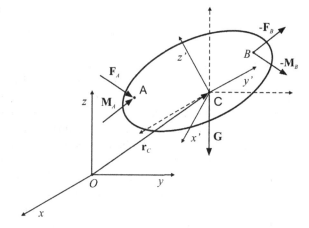

corresponding to the port with respect to the mass centre. Thus, they are parameters
that depend on the geometry of the rigid body and don't change with its motion.

9.5.3 Rigid Body Dynamics

To complete the model, we need a dynamic equation governing rigid body motion.
The simplest form of such an equation is given with respect to axes translating with
the body mass centre (Fig. 9.42). We assume that the base frame is an inertial
frame.

The translational part of the motion can be described by

$$\left.\begin{array}{c} \mathbf{p} = m\mathbf{I}\mathbf{v}_C \\ \dfrac{d\mathbf{p}}{dt} = \mathbf{F} \end{array}\right\} \tag{9.95}$$

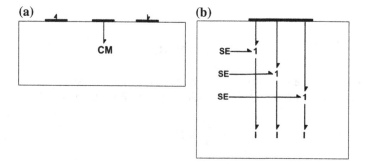

Fig. 9.43 Rigid body translation. **a** Representation in the Base. **b** The CM component

Here, m is the body mass and $\mathbf{F}$ is the resultant of the forces reduced to the mass centre. We represent the dynamics of body translation in the Workspace (Fig. 9.43a) by the CM component that describes the motion of the centre of mass of the body. This last component (Fig. 9.43b) consists of effort junctions corresponding to the x, y and z components of the velocity of the mass centre of the body with respect to the base frame. These junctions are connected to the Workspace port to which the resultant of the forces acting on the body is transferred.

Note that the weight of the body is added here. This force is represented by source efforts defining the weight components with respect to the base frame axes. The momentum law in (9.95) is represented by the inertial elements I with the body mass as a parameter. Because it is used both in the SE and I components, it can be defined just once, at the level of the CM document (Fig. 9.43b).

The rotational part of the body motion is commonly described in a frame translating with the mass centre of the body and that is parallel to the base frame. The moment of momentum law with respect to such a frame reads

$$\left.\begin{aligned} \mathbf{H} &= \mathbf{J}\omega \\ \frac{d\mathbf{H}}{dt} &= \mathbf{M} \end{aligned}\right\} \tag{9.96}$$

where $\mathbf{J}$ is the mass inertia matrix and $\mathbf{M}$ is the resultant moment about the centre of mass of the body. Due to rotation of the body, the inertia matrix changes during the motion. Because of this the equations of the rotational dynamics are as a rule (at least for rigid bodies) represented with respect to a frame fixed to the body. It is not difficult to transform Eq. (9.96) to this form.

In the body frame the body moment of momentum is given by

$$\mathbf{H} = \mathbf{R}\mathbf{H}' \tag{9.97}$$

Substituting from (9.96) we obtain

$$\mathbf{H}' = \mathbf{R}^T\mathbf{H} = \mathbf{R}^T\mathbf{J}\omega = \mathbf{R}^T\mathbf{J}\mathbf{R}\omega' = \mathbf{J}'\omega' \tag{9.98}$$

where

$$\mathbf{J}' = \mathbf{R}^T\mathbf{J}\mathbf{R} \tag{9.99}$$

is the body mass inertia matrix with respect to the centroidal body axes. Multiplying the second Eq. (9.96) by $\mathbf{R}^T$ and substituting from Eq. (9.97), we obtain

$$\mathbf{R}^T\frac{d}{dt}(\mathbf{R}\mathbf{H}') = \mathbf{M}' \tag{9.100}$$

or

Fig. 9.44 The structure of the ROTATION component of Fig. 9.40

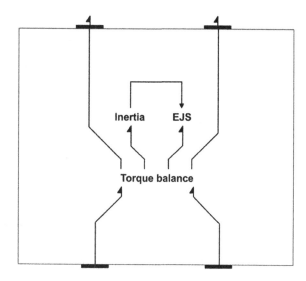

$$\frac{d\mathbf{H}'}{dt} + \mathbf{R}^T \frac{d\mathbf{R}}{dt} \mathbf{H}' = \mathbf{M}' \tag{9.101}$$

Using Eq. (9.61), the last expression can be written as

$$\frac{d\mathbf{H}'}{dt} + \boldsymbol{\omega}' \times \mathbf{H}' = \mathbf{M}' \tag{9.102}$$

These are the famous Euler equations of body rotation in which the rate of change of the moment of momentum is represented by its local change and a part generated by the body rotation. This other part has a very elegant representation in a bond graph setting. This is the celebrated *Euler Junction Structure (EJS)* [10, 11].

Now we can describe the ROTATION component of Fig. 9.40, which represents the body rotation about the centre of mass with respect to the body frame (Fig. 9.44). The Torque balance component in the middle consists simply of an array of three effort junctions that corresponds to the x, y and z components of the body angular velocity with respect to the body frame. The left and right ports transfer the moment of the forces and the moments acting on the body. In the centre it is connected to Inertia and EJS components. The first of these consists of an array of inertial elements that describes the local rate of change of the moment of momentum. There is a control-out port that serves for the transfer of information on the moment of momentum vector $\mathbf{H}'$ that the EJS component needs. This can be seen in (9.102). The EJS component is shown in Fig. 9.45.

It consists of three gyrators connected in a ring. Note that the power circulates inside the structure, thus there is no net power generation or dissipation. The gyrators are modulated by the body moment of momentum rectangular components. The EJS can be slightly simplified if the body axes correspond to the

Fig. 9.45 The Euler Junction
Structure (EJS)

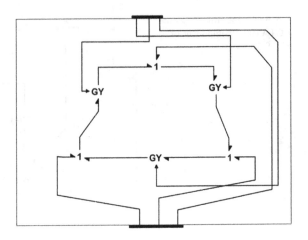

principal axes of the inertia tensor. We do not assume this, as selection of the body
axes is usually based on how the body is connected. Thus, we assume that the body
inertia matrix is fully populated. We do not recalculate the moment of momentum
in the EJS component, but rather take it from the inertial component.

This completes development of the component model of a body moving in 3D
space. The model is stored in the BondSim *Models Library*, under *Word Model
Components*, from which it can be easily used by dragging into the working
window where the models are developed. Now we turn our attention to the modelling of interconnections between the bodies in space.

9.5.4 Modelling of Body Interconnections in Space

Typical mechatronics systems, such as robots, consist of manipulators guided by
controllers. The manipulators are multibody systems consisting of several members
(links) interconnected by suitable joints. They are powered by servo-actuators. In
the previous sections we developed a fairly general component model of bodies that
can be used for the representation of manipulator links. Now we develop models of
the joints. Two types of joint are considered—revolute and prismatic. Based on
these component models, a typical robotic manipulator can be represented in a way
similar to which real manipulators are assembled. In the next section we apply the
methods of this section to the modelling of a complete robot.

The approach is applicable to other multibody systems as well. Robot manipulators are chosen for several reasons. These are fascinating systems that have
influenced development in many fields such as multibody mechanics, conventional
and intelligent control, sensors and actuator technology, and have promoted
mechatronics as a design philosophy.

Fig. 9.46 Revolute joint in space

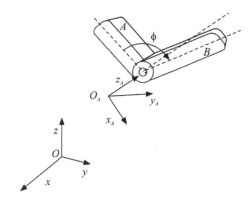

The modelling and simulation of such complex systems is not an easy task. The problem of modelling manipulators as multibody systems is only one part of it; there is also the problem of the control of such complex space systems, particularly when there are interactions with the environment. The bond graph method is a good candidate for solving such multi-disciplinary problems. We can go a step further and show in the last section how the motions of such complex mechanical structures can be visualized in a virtual 3D scene.

9.5.4.1 Revolute Joints

Revolute joints have already been discussed in Sect. 9.2.2. The basic difference with those discussed earlier is that, because bodies connected by a revolute joint can move in three-dimensional space, the axis of the joint is not confined to specific motions but can move freely. To describe this effect, we analyse the bodies A and B connected by a revolute joint, as shown in Fig. 9.46. The bodies could, for example, be two links of a robot manipulator joined by a revolute joint. We assume that the z-axis of the co-ordinate frame $O_A x_A y_A z_A$ of body A is directed along the joint axis. We assume further that there is a body B frame $O_B x_B y_B z_B$. The precise position of the frame is not prescribed and in a specific multibody system can be defined as is most convenient, e.g. using the Denavit-Hartenberg convention [9]. The frame $Oxyz$ is the base frame.

Let P be the centre point of the revolute joint used as the reference connection point. The joint is represented by word model components as in Sect. 9.2.2 (Fig. 9.7). The ports are assumed compounded such that the port variables are 6D flows and efforts at the connection of the joint to body A and B These are expressed in their respective frames. At A (the power-in) these are

$$\mathbf{f}_P^A = \begin{pmatrix} \mathbf{v}_P^A \\ \boldsymbol{\omega}_A^A \end{pmatrix} \tag{9.103}$$

and

$$\mathbf{e}_P^A = \begin{pmatrix} \mathbf{F}_P^A \\ \mathbf{M}_P^A \end{pmatrix} \tag{9.104}$$

Similarly at B (the power-out) we have

$$\mathbf{f}_P^B = \begin{pmatrix} \mathbf{v}_P^B \\ \boldsymbol{\omega}_B^B \end{pmatrix} \tag{9.105}$$

and

$$\mathbf{e}_P^B = \begin{pmatrix} \mathbf{F}_P^B \\ \mathbf{M}_P^B \end{pmatrix} \tag{9.106}$$

The linear velocities of bodies A and B at the connection point are common to both of the bodies. Hence

$$\mathbf{v}_P^A = \mathbf{R}_B^A \mathbf{v}_P^B \tag{9.107}$$

where $\mathbf{R}_B^A$ is the rotation matrix of the body B frame with respect to the body A frame.

A similar relation holds for the forces. Because the linear velocities at the common point are equal, the same is true for the power transferred across the joint during its translation, i.e.

$$(\mathbf{v}_P^A)^T \mathbf{F}_P^A = (\mathbf{v}_P^B)^T \mathbf{F}_P^B \tag{9.108}$$

By substituting from (9.107) we obtain

$$\mathbf{F}_P^B = \mathbf{R}_A^B \mathbf{F}_P^A \tag{9.109}$$

Equations (9.107) and (9.109) describe the relationships between the linear parts of the flow-effort 3D vectors at the joint's ports. We develop next the relationships between the angular parts, i.e. the angular velocities and moments.

Rotation matrices of the bodies A and B frames with respect to the base frame are related by composition of the rotations, i.e.

$$\mathbf{R}_B = \mathbf{R}_A \mathbf{R}_B^A \tag{9.110}$$

Differentiating it with respect to time we get

$$\frac{d\mathbf{R}_B}{dt} = \frac{d\mathbf{R}_A}{dt}\mathbf{R}_B^A + \mathbf{R}_A\frac{d\mathbf{R}_B^A}{dt} \tag{9.111}$$

From (9.63) and (9.64) we obtain

$$\boldsymbol{\omega}_B \times \mathbf{R}_B = \mathbf{R}_A\boldsymbol{\omega}_A^A \times \mathbf{R}_B^A + \mathbf{R}_A\boldsymbol{\omega}_{AB}^A \times \mathbf{R}_B^A \tag{9.112}$$

where

$$\boldsymbol{\omega}_{AB}^A = axial\left(\frac{d\mathbf{R}_B^A}{dt}\mathbf{R}_A^B\right) \tag{9.113}$$

is the relative angular velocity of body B with respect to body A expressed in body A frame. Simplifying (9.112) we get

$$(\mathbf{R}_A)^T\boldsymbol{\omega}_B \times \mathbf{R}_A = \boldsymbol{\omega}_A^A\times + \boldsymbol{\omega}_{AB}^A\times \tag{9.114}$$

Hence (see Eqs. 9.63 and 9.61)

$$\boldsymbol{\omega}_B^A\times = \boldsymbol{\omega}_A^A\times + \boldsymbol{\omega}_{AB}^A\times \tag{9.115}$$

The last equations imply that

$$\boldsymbol{\omega}_B^A = \boldsymbol{\omega}_A^A + \boldsymbol{\omega}_{AB}^A \tag{9.116}$$

In addition we have

$$\boldsymbol{\omega}_B^A = \mathbf{R}_B^A\boldsymbol{\omega}_B^B \tag{9.117}$$

Note also that

$$\boldsymbol{\omega}_{AB}^A = \begin{pmatrix} 0 & 0 & \dot{\varphi} \end{pmatrix}^T \tag{9.118}$$

where is φ is the joint's angle of rotation.

We use the component in Fig. 9.47 to represent a revolute joint. The component has an additional port that corresponds to the relative rotation of the joint. This port can be used for actuation of the joint. There are also a signal port for the joint's rotation angle output.

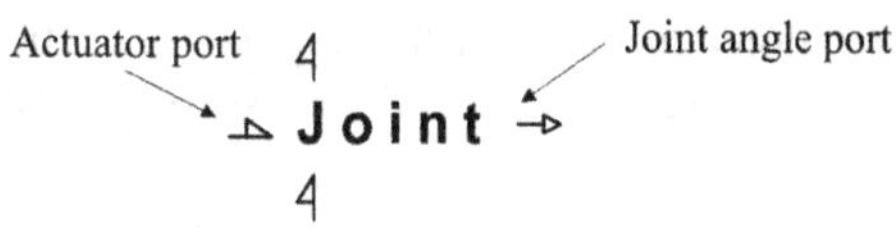

Fig. 9.47 Revolute joint component model

Actuator port   Joint angle port

Joint

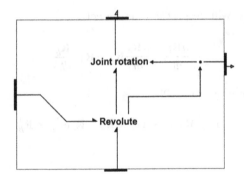

Fig. 9.48 Structure of the revolute joint 3D component model

(a)   **(b)**

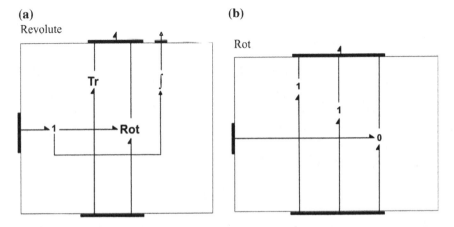

Fig. 9.49 The structure of: Revolute component, Rot subcomponent

We can now represent the model of a revolute joint. It consists of two main parts (Fig. 9.48): Revolute and Joint Rotation. The Revolute represents the basic relations between the port variables in the frame of lower joint part (body A). The other component, Joint Rotation, transforms the port variables from the frame of the other body (body B) to the frame of the lower (body A).

The structure of the Revolute component is shown in Fig. 9.49a. It consists of two components. The Tr component represents the translation part of the joint model and expresses the fact that the joint centre point is common to the both bodies. Thus it consists solely of three effort junctions for the translation in the direction of x_A, y_A and z_A axes. This ensures that the joined ends of the bodies move with a common velocity and that the corresponding forces are the same.

The other component, Rot, whose structure is shown in Fig. 9.49b, represents the relationship between the angular velocities as given by (9.116) and (9.118). The model is very simple and consists of two effort junctions in the x_A- and y_A- directions

Fig. 9.50 The rotational
transformations by the joint

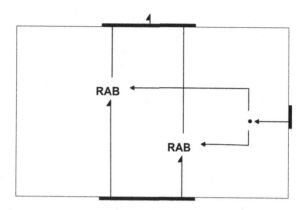

in which the angular components are the same for both of the joined bodies. There is
also a flow junction that corresponds to the relative rotation about the z_A-axis.

The 1-junction on the left in Fig. 9.49a is inserted to extract the joint relative
angular velocity and integrate it to obtain the joint rotation angle. It is used in the
Joint rotation component, but is also transmitted to the output port
(Fig. 9.48).

An important function of the joint is the rotation transformation between the two
link frames (Fig. 9.48). This is depicted in Fig. 9.50. The transformations are
applied to the linear effort-flow parts, and separately to the angular. These are
represented by components RAB. The components represent transformations, as
given by (9.107) and (9.108) for linear variables. The same transformations hold for
the angular variables.

The transformations are usually expressed by transformers TF, which transform
flow and effort components. These are then summed as given in (9.107) and
(9.109), and similarly for the angular quantities. We do not give here the general

Fig. 9.51 Prismatic joint in
space

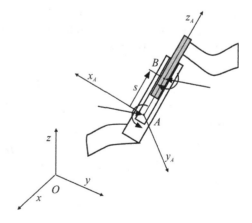

structure of the component. Later in Sect. 9.6, we meet some specific examples of such transformations.

Revolute joints play an important role in the design of robotic manipulators. They offer the simplest way to change the orientation of the robot links. The component model introduced here gives the main functionality of such joints. They are used later for the building of manipulator models. This is illustrated in Sect. 9.6.

9.5.4.2 Prismatic Joints

Prismatic joints have already been described in Sect. 9.2.2. The basic difference here is that the axis of the joint can be anywhere in space (Fig. 9.51). To describe the effects of prismatic joints on bodies connected by such a joint, we define a body frame $Ax_Ay_Az_A$ attached to one body at point A. The z-axis is directed along the relative displacement of the joint (joint axis). There is also a second body B and a frame attached to it.

The precise positions of the frames are not specified and can be defined as is the most convenient, e.g. by using the Denavit-Hartenberg convention [9]. The frame $Oxyz$ is the base frame.

Let B be a point on the second body. This body can move along the joint slot. The joint can be represented by a word model component as in Fig. 9.47, but of course its model will be different. The ports are assumed compounded such that the port variables are 6D flows and efforts at the connection of the joint to the bodies (the upper and lower in Fig. 9.47). The side port is used for actuation of the joint. We develop the governing equations first. They are generalizations of the corresponding equations for the planar case in Sect. 9.2.2, but are expressed with respect to the body frame, not the base frame, as we did for plane revolute joints.

The position vector of point B (Fig. 9.51), the reference point of the body that can slide along the joint, is given by

$$\mathbf{r}_B = \mathbf{r}_A + \mathbf{R}_A \mathbf{r}_{AB}^A \tag{9.119}$$

where $\mathbf{R}_A$ is the rotation matrix of frame A with respect to the base and $\mathbf{r}_{AB}^A$ is its relative position with respect to the body frame A. By differentiation with respect to time, we get for its velocity

$$\mathbf{v}_B = \mathbf{v}_A + \mathbf{R}_A \mathbf{v}_{AB}^A + \frac{d\mathbf{R}_A}{dt} \mathbf{r}_{AB}^A \tag{9.120}$$

where

$$\mathbf{v}_{AB}^A = \frac{d\mathbf{r}_{AB}^A}{dt} \tag{9.121}$$

is the relative velocity of the junction. Substituting from (9.64) in Eq. (9.120) we get

$$\mathbf{v}_B = \mathbf{v}_A + \mathbf{R}_A \mathbf{v}_{AB}^A + \mathbf{R}_A \boldsymbol{\omega}_A^A \times \mathbf{r}_{AB}^A \qquad (9.122)$$

Multiplying from the left by the transposed rotation matrix $\mathbf{R}_A^T$, we find

$$\mathbf{v}_B^A = \mathbf{v}_A^A + \mathbf{v}_{AB}^A + \boldsymbol{\omega}_A^A \times \mathbf{r}_{AB}^A$$

or

$$\mathbf{v}_B^A = \mathbf{v}_A^A + \mathbf{v}_{AB}^A + (-\mathbf{r}_{AB}^A \times \boldsymbol{\omega}_A^A) \qquad (9.123)$$

In addition, we have

$$\mathbf{v}_B^A = \mathbf{R}_B^A \mathbf{v}_B^B \qquad (9.124)$$

Here, $\mathbf{R}_B^A$ is the rotation matrix of the second body frame with respect to the body frame of A. Note that in prismatic joints this is a constant matrix.

In the body frame A, the relative position vector is very simple

$$\mathbf{r}_{AB}^A = (0 \quad 0 \quad s)^T \qquad (9.125)$$

and hence the relative velocity of the joint is

$$\mathbf{v}_{AB}^A = (0 \quad 0 \quad \dot{s})^T \qquad (9.126)$$

We also introduce a matrix $\mathbf{T}$ as in Sect. 9.2.2 defined by

$$\mathbf{T}_A^A = -\mathbf{r}_{AB}^A \times \qquad (9.127)$$

From (9.56) and (9.125) we have

Fig. 9.52 The structure of the prismatic joint component model

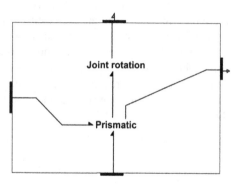

Fig. 9.53 The representation
of the model of the space
prismatic joint

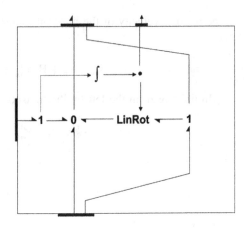

$$\mathbf{T}_A^A = \begin{pmatrix} 0 & s & 0 \\ -s & 0 & 0 \\ 0 & 0 & 0 \end{pmatrix} \qquad (9.128)$$

Now (9.123) reads

$$\mathbf{v}_B^A = \mathbf{v}_A^A + \mathbf{v}_{AB}^A + \mathbf{T}_A^A \boldsymbol{\omega}_A^A \qquad (9.129)$$

This is similar to (9.25) for the planar case. We can expect that the model of the
space prismatic joint is just a generalization of the model of Fig. 9.10.

The structure of the prismatic joint component model is similar to the revolute
joint of Fig. 9.48 and is shown in Fig. 9.52. The `Joint Rotation` component
transforms efforts and flows between the two body frames. This is similar to the
revolute joint (Fig. 9.50), but here the matrix is constant.

The model of the `Prismatic` component is given in Fig. 9.53. The 0
component-junction describes the prismatic joint velocity relationship given by
Eq. (9.129). It consists of flow junctions that describe the velocity relationships in
the direction of the body axes.

Component 1 on the right corresponds to the joint's angular velocity vector. The
`LinRot` component describes the transformation between linear and angular
quantities given by the last term, which is defined by matrix (9.128). It depends on
the joint's relative position, evaluated by the integrator that uses the joint sliding
velocity from the corresponding junction.

Note that because of the very simple form of the matrix in (9.128), there are no z-
components generated by the linear-to-rotational transformation. Summation of the
velocity components represented by the flow junction that corresponds to (9.129) is
simplified, too. Thus, we see that the space prismatic junction is even simpler than
the planar one.

9.6 Dynamics of Puma 560 Robot

9.6.1 Problem Formulation

In this section we apply the component-modelling technique to a problem of dynamic modelling of robots. The term robot was coined to science fiction play "Rosum's Universal Robots" written 1921 by Czech writer Karl Čapek. In the period from 1950 to the 1980s robots were the subject of many science fiction stories written by Isaac Asimov. The first robot patent was filled by George Devol in 1954 and issued in 1961. The Unimation founded by G. Devol and Joseph Engelberger was the first robot manufacturing company, which started producing robots in 1956. From that time robotics really started. Millions of robots of very different designs are produced and have found application in very diverse fields from industry to personal use.

The PUMA 560 robot, which is short of *Programmable Universal Manipulator for Assembly*, was released in 1978 by Unimation and shortly was accepted worldwide by industry, but also in the research field. It is perhaps one of the best known robot manipulators and was the subject of much research both in academic institutions and industry. Its characteristics are well documented ([21, 22] etc.). In this section we will deal with the systematic approach to modelling of this well-known robot by Bond Graphs.

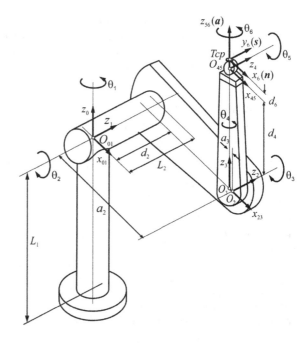

Fig. 9.54 PUMA 560 with attached co-ordinate frames

Table 9.10 DH parameters

i	θ_i	α_i	a_i	d_i
1	θ_1	$-90°$	0	0
2	θ_2	0	a_2	d_2
3	θ_3	$90°$	$-a_3$	0
4	θ_4	$-90°$	0	d_4
5	θ_5	$90°$	0	0
6	θ_6	0	0	d_6

The control of robots is a field in its own. Originally the PUMA was controlled by PID controllers. Even nowadays the PID is commonly used for robot control. We will present here a complete PUMA 560 system based on a PID controller.

Figure 9.54 shows a 3D view of PUMA 560 robot with the attached co-ordinate frames. The PUMA robot was the first anthropomorphic robot designed and was the leader of many robots that follows. It has a human form. Even its size is chosen to correspond to a typical man or women. It is customary to describe its parts by human terms such as a body, shoulder, arm, elbow, wrist, joints, hand, etc.

To model a robot two kinds of 3D Cartesian co-ordinate systems are typically used: the basic and links systems. The basic is an inertial co-ordinates system, and represent the basic reference system for the description of the robot motion.

Each robot link has a separate body fixed co-ordinate system. Figure 9.54 shows these co-ordinate systems. They are defined following the well-known standard Denavit-Hartenberg (DH) scheme [22].[2] In this scheme, the co-ordinate frame i is attached to joint $i + 1$ and z_i axis lies along the axis of the rotation. The origins of the link 4 and 5 frames coincide with the central point of the spherical wrist, often used in robots. The origin of the last frame 6 is at the *TCP* (tool central point). The corresponding DH parameters are shown in Table 9.10.

The assumed values of the parameters in meters following [21, 22] are given below. The length L_1 is the height of the basic co-ordinate frame from the floor, and L_2 defines the position of the mean plane of the link 2 orthogonal to z-axis:

- $a_2 = 0.4318, a_3 = 0.0203$
- $d_2 = 0.1501, d_4 = 0.4331, d_6 = 0.056$
- $L_1 = 0.672, L_2 = 0.2435$

The joint angles are limited to the values as given below:

- $-160° \leq \theta_1 \leq 160°$
- $-215° \leq \theta_2 \leq 35°$
- $-45° \leq \theta_3 \leq 225°$
- $-140° \leq \theta_4 \leq 170°$

---

[2]There is also a DH modified scheme introduced by Craig [8]. In this scheme the frame i is attached to joint i. It is perhaps even simpler to use, but it is not used here.

Fig. 9.55 The positions of a
point with respect to two
co-ordinate frames

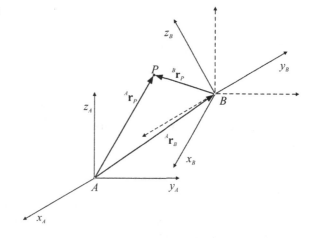

- $-100° \leq \theta_5 \leq 100°$
- $-266° \leq \theta_6 \leq 266°$

Using the link frames we can find the relative homogenous transformation matrices from one link frame to the previous one going from the first link to the last, Fig. 9.54.

The homogenous transformation is a compact way to represent the position and orientation of one co-ordinate frame with respect to the other (Fig. 9.55). It also allows projection and scaling, but is not used here.

Figure 9.55 shows two co-ordinates frames A and B and a point P. The frame B can be obtained by translation of the frame initially coincident with the frame A by a vector $\mathbf{r}_B^A$ with respect to frame A (indicated by the superscript). The translated frame axes are shown by the broken lines. Next, the translated frame is rotated by a 3×3 matrix $\mathbf{R}_B^A$. This is a matrix whose columns are the co-ordinates of the unit vectors of axes x_B, y_B, z_B in the co-ordinate frame A (or the translated frame). It is an orthogonal matrix we spoke about in Sect. 9.5.1.

These transformations can be represented compactly by a 4×4 matrix of homogenous co-ordinates, which has the following structure

$$\mathbf{T}_B^A = \left(\begin{array}{c|c} \mathbf{R}_B^A & \mathbf{r}_B^A \\ \hline \mathbf{0} & 1 \end{array} \right) \qquad (9.130)$$

Note that the last row consists of a 3D row-vector of projective transformations, which is assumed zero, and a scaling factor assumed 1.

The position of a point P in frame A, which is defined in the frame B by the vector $\mathbf{r}_P^B$, as discussed in Sect. 9.5.1, (9.39), can be written as

$$\mathbf{r}_P^A = \mathbf{r}_B^A + \mathbf{R}_B^A \mathbf{r}_P^B \tag{9.131}$$

Introducing the vector of the homogenous 4×1 co-ordinates by

$$\hat{\mathbf{r}} = \begin{pmatrix} \mathbf{r} \\ 1 \end{pmatrix} \tag{9.132}$$

and using the homogenous transformation matrix (9.130) we can write (9.131) compactly as

$$\hat{\mathbf{r}}_P^A = \mathbf{T}_B^A \hat{\mathbf{r}}_P^B \tag{9.133}$$

We now return to the Denavit-Hartenberg co-ordinate frames of Fig. 9.54 and Table 9.10. Following Corke [22] we describe the homogeneous transformation between frame i and frame $i - 1$ as a composition of the following elementary transformations:

$$\mathbf{T}_i^{i-1} = \mathbf{R}_z(\theta_i)\mathbf{D}_z(d_i)\mathbf{R}_x(\alpha_i)\mathbf{D}_x(a_i) \tag{9.134}$$

where $\mathbf{R}$ is a rotation transformation about an axis by an angle, and $\mathbf{D}$ is a translation transformation along an axis by a value. The product of the first two matrices may be denoted as

$$\mathbf{Screw}_z(d_i, \theta_i) = \mathbf{R}_z(\theta_i)\mathbf{D}_z(d_i) = \begin{pmatrix} c\theta_i & -s\theta_i & 0 & 0 \\ s\theta_i & c\theta_i & 0 & 0 \\ 0 & 0 & 1 & d_i \\ 0 & 0 & 0 & 1 \end{pmatrix} \tag{9.135}$$

and represents a *screw* transformation consisting of rotation about z-axis by angle θ_i and translation along the same axis by d_i. Similarly, the product of the two last matrices gives the screw about the x axis, i.e.

$$\mathbf{Screw}_x(a_i, \alpha_i) = \mathbf{R}_x(\alpha_i)\mathbf{D}_x(a_i) = \begin{pmatrix} 1 & 0 & 0 & a_i \\ 0 & c\alpha_i & -s\alpha_i & 0 \\ 0 & s\alpha_i & c\alpha_i & 0 \\ 0 & 0 & 0 & 1 \end{pmatrix} \tag{9.136}$$

In the above expressions we used the shorthand notation $c\alpha_i = \cos\alpha_i, s\alpha_i = \sin\alpha_i, c\theta_i = \cos\theta_i, s\theta_i = \sin\theta_i$.

DH transformation matrix, thus, can be written as product of these two screw matrices about z and x axes respectively:

$$\mathbf{T}_i^{i-1} = \mathbf{Screw}_z(d_i, \theta_i)\mathbf{Screw}_x(a_i, \alpha_i) \tag{9.137}$$

It can be shown that modified DH scheme applies these operations in the opposite order, i.e. first the screw about x-axis and then screw about the rotation axis z.

By evaluating the last expression we obtain DH transformation matrix as

$$\mathbf{T}_i^{i-1} = \begin{pmatrix} c\theta_i & -s\theta_i c\alpha_i & s\theta_i s\alpha_i & a_i c\theta_i \\ s\theta_i & c\theta_i c\alpha_i & -c\theta_i s\alpha_i & a_i s\theta_i \\ 0 & s\alpha_i & c\alpha_i & d_i \\ 0 & 0 & 0 & 1 \end{pmatrix} \tag{9.138}$$

Now we can find transformation matrices between the frames of Puma 560

$$\mathbf{T}_1^0 = \begin{pmatrix} c\theta_1 & 0 & -s\theta_1 & 0 \\ s\theta_1 & 0 & c\theta_1 & 0 \\ 0 & -1 & 0 & 0 \\ 0 & 0 & 0 & 1 \end{pmatrix} \tag{9.139a}$$

$$\mathbf{T}_2^1 = \begin{pmatrix} c\theta_2 & -s\theta_2 & 0 & a_2 c\theta_2 \\ s\theta_2 & c\theta_2 & 0 & a_2 s\theta_2 \\ 0 & 0 & 1 & d_2 \\ 0 & 0 & 0 & 1 \end{pmatrix} \tag{9.139b}$$

$$\mathbf{T}_3^2 = \begin{pmatrix} c\theta_3 & 0 & s\theta_3 & -a_3 c\theta_3 \\ s\theta_3 & 0 & -c\theta_3 & -a_3 s\theta_3 \\ 0 & 1 & 0 & 0 \\ 0 & 0 & 0 & 1 \end{pmatrix} \tag{9.139c}$$

$$\mathbf{T}_4^3 = \begin{pmatrix} c\theta_4 & 0 & -s\theta_4 & 0_3 \\ s\theta_4 & 0 & c\theta_4 & 0 \\ 0 & -1 & 0 & d_4 \\ 0 & 0 & 0 & 1 \end{pmatrix} \tag{9.139d}$$

$$\mathbf{T}_5^4 = \begin{pmatrix} c\theta_5 & 0 & s\theta_5 & 0 \\ s\theta_5 & 0 & -c\theta_5 & 0 \\ 0 & 1 & 0 & 0 \\ 0 & 0 & 0 & 1 \end{pmatrix} \tag{9.139e}$$

$$\mathbf{T}_6^5 = \begin{pmatrix} c\theta_6 & -s\theta_6 & 0 & 0 \\ s\theta_6 & c\theta_6 & 0 & 0 \\ 0 & 0 & 1 & d_6 \\ 0 & 0 & 0 & 1 \end{pmatrix} \tag{9.139f}$$

Note that modeling approach developed here is based on velocities as explained before. Therefore we do not need the complete DH matrices given in (9.139a), only their rotation parts. However we will use these matrices to calculate expressions for the position of *TCP* in the basic frame, to check the accuracy of evaluations based on the Bond Graph method.

We determine the complete transformation between the tool frame and the robot base by multiplying these six transformation matrices:

$$\mathbf{T}_6^0 = \mathbf{T}_1^0\mathbf{T}_2^1\mathbf{T}_3^2\mathbf{T}_4^3\mathbf{T}_5^4\mathbf{T}_6^5 \tag{9.140}$$

The position of the *TCP* in the base frame can be found as the last column of this transformation matrix. Thus, we find

$$
\begin{aligned}
x_{tcp}^0 &= c\theta_1(-a_3c\theta_{23} + d_4s\theta_{23} + a_2c\theta_2) - d_2s\theta_1 \\
&\quad d_6[c\theta_1(c\theta_{23}c\theta_4s\theta_5 + s\theta_{23}c\theta_5) - s\theta_1s\theta_4s\theta_5] \\
y_{tcp}^0 &= s\theta_1(-a_3c\theta_{23} + d_4s\theta_{23} + a_2c\theta_2) + d_2c\theta_1 \\
&\quad + d_6[s\theta_1(c\theta_{23}c\theta_4s\theta_5 + s\theta_{23}c\theta_5) + c\theta_1s\theta_4s\theta_5] \\
z_{tcp}^0 &= a_3s\theta_{23} + d_4c\theta_{23} - a_2s\theta_2 + d_6(-s\theta_{23}c\theta_4s\theta_5 + c\theta_{23}c\theta_5)
\end{aligned}
\tag{9.141}
$$

Note that as earlier $c\theta_{23} = \cos(\theta_2 + \theta_3)$ and $s\theta_{23} = \sin(\theta_2 + \theta_3)$.

The dynamical parameters of Puma 560 robots are taken from [21, 22], basically according to Armstrong et al. In [21, 22] data are given with respect to the modified

Table 9.11 Masses (kg) and COG (m) of the links

Link	Mass	x_C	y_C	z_C
Link 1[a]	13.0	0	0.309	0.004
Link 2	17.40	−0.3638	0.006	0.0774
Link 3	4.80	0.0203	0.014	0.070
Link 4	0.82	0	0.019	
Link 5	0.34	0	0	0
Link 6	0.09	0	0	0.032

[a]According to Tarn from [22]

Table 9.12 Moments of inertia about COG kg m^2

Link	I_{xx}	I_{yy}	I_{zz}
Link 1[a]	1.100	0.177	1.100
Link 2	0.130	0.524	0.539
Link 3	0.066	0.086	0.0125
Link 4	1.80×10^{-3}	1.30×10^{-3}	1.80×10^{-3}
Link 5	0.30×10^{-3}	0.40×10^{-3}	0.30×10^{-3}
Link 6	0.15×10^{-3}	0.15×10^{-3}	0.04×10^{-3}

[a]According to Tarn [22]

Table 9.13 Motor and drive parameters

	Joint 1	Joint 2	Joint 3	Joint 4	Joint 5	Joint 6
Resistance R (Ohm)	1.6	1.6	1.6	3.9	3.9	3.9
Inductivity L (Henry)	0.0048	0.0048	0.0048	0.0039	0.0039	0.0039
Back e.m.f. constant Ke (V s/rad)	0.19	0.19	0.19	0.12	0.12	0.12
Torque constant K_m (N m/A)	0.260	0.260	0.260	0.090	0.090	0.090
Viscous friction B (N m s/rad)	4.20	8.1	3.15			
Armature inertia Im (kg m^2)	200×10^{-6}	200×10^{-6}	200×10^{-6}	18×10^{-6}	18×10^{-6}	18×10^{-6}
Maximum torque (N m)	97.6	186.4	89.4	24.2	20.1	21.3
Break away torque (N m)	6.3	5.5	2.6	1.3	1.0	1.2
Gear ratio	62.61	107.36	53.69	76.01	71.91	76.63

DH co-ordinates, which slightly differs from the standard DH co-ordinates used here. Therefore the dynamic parameters are adjusted to the assumed link co-ordinates. In addition Armstrong et al. didn't give the dynamic parameters for link 1, hence the parameters for this link are based on Tarn data [22].

The masses of the links and the positions of their respective centres of gravity (COG) are given in Table 9.11.

Table 9.12 lists the diagonal components of the inertia tensor of the links. The moments of inertias are expressed in a co-ordinate frame obtained by translation of the corresponding link frame to its centre of the mass (gravity).

The parameters for the motors and drives are given in Table 9.13.

9.6.2 Model of the Robot

9.6.2.1 System Level Model

We next give a brief description of the model. The complete model is held in the BondSim program library under the project name *PUMA 560*. The system level Bond Graph model is shown in Fig. 9.56. It consists of three interconnected main components: Controller, Manipulator and Workspace. To the right of

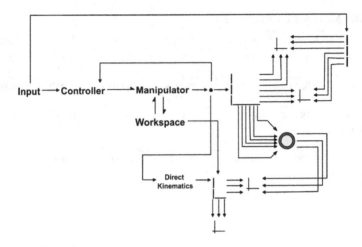

Fig. 9.56 System level model of Puma 560

these components are $x - y$ displays, which monitor how the system behaves during the simulation.

Below these displays a component in the form of a ring is inserted, which will be used for inter-process communication (*IPC*) between the PUMA simulation and its 3D visualization. *BondSim* supports building and running of the *named pipe* connections with a visualization application *BondSimVisual*, running on the same or a remote computer connected by a local net. Using this it is possible to visualize motion of the PUMA robot during the simulation in a 3D virtual scene. The signals connected on the left side of the *IPC* supplies the joint angles calculated by the model during the simulation. These angles are regularly (e.g. every 40–50 ms) picked and packed into a message, which is then sent by the pipe to the visualization application. After redrawing the scene it responds by sending back information e.g. on the new position of *TCP*. These data are picked by the signal lines shown on the right side of *IPC* component and sent to a display. We will discuss this in more detail in the next section on visualization.

Below the robot model there is the Direct Kinematics component. It is not part of the model but is used for the model testing. It evaluates the co-ordinates of *TCP* given by (9.141) using for the joint angles the values generated by the Manipulator. The calculated co-ordinates are compared with the corresponding values generated by the robot model in the Workspace component. The generated values of the co-ordinates and differences from the values evaluated by the direct kinematics are displayed in the corresponding plots.

The Input component on the left generates the joint angles input trajectories that robot arm should follow. We will discuss it when analyzing the design of the Controller which we will use to drive the PUMA 560 model.

Fig. 9.57 Schematic of a spherical joint composed of three revolute joints

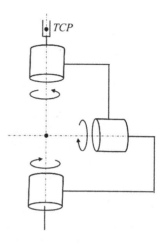

9.6.2.2 Model of PUMA 560 Manipulator

PUMA 560 (Fig. 9.54) is a six degree of freedom (DOF) robot designed as an anthropomorphic (human like) structure. It consists of an arm and wrist with hand. The arm is formed of three links: the body, upper arm and forearm. The body (Link 1) can rotate about a revolute joint in the immobile base. The upper arm (Link 2) is connected to the body by the shoulder revolute joint, and the forearm (link 3) to the upper arm by the elbow revolute joint. These last two joints have parallel rotation axes, orthogonal to the body axis.

The forearm at its end carries a wrist with hand. The wrist is a spherical joint realized as three links which can rotate about revolute joints with orthogonal axes intersecting at a point (Fig. 9.57). This arrangement is used also in many other industrial robots. The last link plays the role of a hand and enables connecting different tools, which robot uses in different applications. Its central point is known usually as the *TCP* (tool central point).

Fig. 9.58 The structure of the model of manipulator

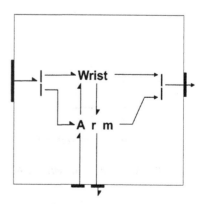

The links are rotated by the servo drives acting at the joints. The arm is responsible for positioning the wrist in space, and the last three links comprising the wrist orient the hand in space. For more details see e.g. [9].

The model of the `Manipulator` is shown in Fig. 9.58. It consists of two components: the `Arm` and `Wrist`. The first models the robot hand described above and the other the spherical wrist with hand. There are several bond power and signal lines between these components and outside components. Their connection points —the ports—are compounded. We used this approach in the previous models too, but in the multibody models it is very important. Otherwise, we will have to deal with forest of connecting bond and signal lines. It is not only very difficult to represent this in a tidy and clear way, but it is also prone to errors.

Using the component model approach and compounded ports and bonds we hope that rather complex multibody models of the robot can be represented in a clear and understandable way. The bonds on the left side in Fig. 9.58 represent the efforts (moments) and flows (angular velocities) of the servo drives (motors and gears) to the joints. The signal on the right transmit out the collected joint angles. The power transfer in the vertical direction shows the power transfer from the workspace thorough the links to the robot tip and again back to the workspace.

Figure 9.59 shows models of the `Arm` and `Wrist` components. The models follow the structure of the corresponding robot components in Fig. 9.54. They

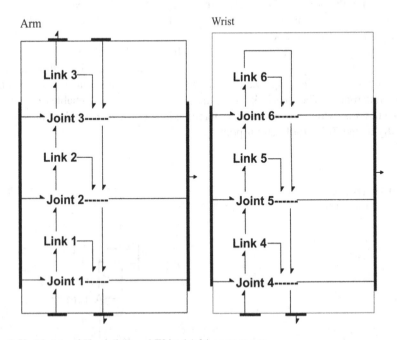

Fig. 9.59 Models of Hand (*left*) and Wrist (*right*) components

consist of series connections of the joints and links. The rotation of the links about the joints is represented as a series of the rotation transformations.

The left vertical bonds describe the efforts (forces and moment vectors) and the flows (liner and angular velocities) on the links at the joints. They are expressed in the link body frames. Therefore it is necessary to transform these quantities from one link frame to the next one using the corresponding rotation matrices as discussed in Sect. 9.5.2.

There is also power transfer in the opposite direction. These bonds represent the transformations of the effort/flow quantities at the robot free end and at the center of mass of the links across the joints to the workspace. For instance to find the position of the TCP in the working (operating) space we must know its velocity in this space. The weights of the links are well defined in the workspace. Thus we need to transform them along with the velocities of the centers of masses of the links between the workspace and the links frames. The force on the robot tip is also defined in the workspace (here it is zero) and should be transformed into the last link frame as well.

To achieve these transformations we use an approach that has the root in Damic [20]. Thus we apply transformation in such a way to support the object oriented paradigm. A link does not know anything about the other links. Therefore we cannot apply all rotation transforms needed to evaluate e.g. the weight of a link, because it is necessary to supply the information on other rotation matrices and the values of the rotation angles as well. Thus we apply only the transformations in a joint that the joint knows about, i.e. its own rotation. Therefore we systematically apply the rotation transformation going from the last joint to the first one. Gradually we build the compound bonds that define quantities transformed by the joint.

Thus starting from the last link of the Wrist (Fig. 9.59 right) we draw a bond from the free end of the link (*TCP*) and connect it to the Joint 6. We connect at the left the bond connecting the Link 6 center of mass port. Inside the Joint 6 component these bonds are connected to the components that apply the corresponding rotation transformations. At the other side of these components the bonds are connected to the corresponding joint document port. At this port, on the outside of the Joint 6 we get the compounded bond that is composed of these bonds. It is farther connected to the Joint 5 component. Again a bond from the centre of the mass port of the Link 5 is connected to Joint 5 at the left of the previous bond. These operations are continued systematically until the workspace is reached.

Note that bonds do not carry anything with them. They only define the ports that are connected and in that way the quantities that are transformed. The compounded bonds define in a concise way only the component ports that participate in these operations.

To illustrate how this is coded in Bond Graphs we consider Joint 2 object of Fig. 9.59 left. It has a general structure of the revolute joints discussed in Sect. 9.5.4 and shown in Fig. 9.48. The Joint rotation component changes from joint to joint. For Joint 2 it has the form shown in Fig. 9.60.

Fig. 9.60 Structure of Joint
rotation component of Joint 2

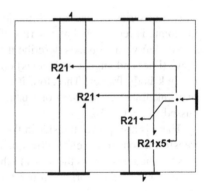

In this joint the rotation transformations are applied by component R21 discussed shortly. On the left the joint's efforts and flows from the previous link frame are transformed to the next link frame. On the right are the transformations we are speaking about. The first one from the left are the transformation of the efforts and flows from the center of mass of the previous Link 2. The component denoted as R21 × 5 contains the transformations connected to the bonds contained in the compound bond from the previous Joint 3 (see Fig. 9.59 left). Number 5 in the component name denotes that there are five transformations applied to the efforts and flows of the links 3–6 centres of masses ports, and of the TCP in that order. Later, in the Workspace component, this compounded bond is unpacked by extracting the bonds in just opposite order as they were packed in. Thus all this is relatively simple, only some discipline is necessary!

Model of R21 is shown in Fig. 9.61. It describes the transformation given by

$$\left.\begin{array}{l} \mathbf{f}_1 = \mathbf{R}_{21}\mathbf{f}_2 \\ \mathbf{e}_2 = \mathbf{R}_{21}^T\mathbf{e}_1 \end{array}\right\} \tag{9.142}$$

Fig. 9.61 Rotation
transformation by **R21**

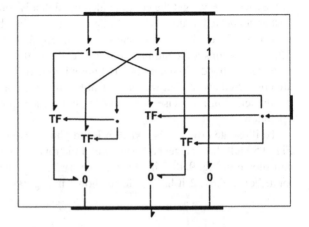

The rotation matrix is given by (9.139b) and has the following form

$$\mathbf{R}_{21} = \begin{pmatrix} \cos\theta_2 & -\sin\theta_2 & 0 \\ \sin\theta_2 & \cos\theta_2 & 0 \\ 0 & 0 & 1 \end{pmatrix} \tag{9.143}$$

There are four controlled transformers corresponding to non-zero elements of the matrix. They multiply the corresponding effort or flow components by $\sin\theta_2$ and $\cos\theta_2$, where θ_2 is the rotation angle of the joint, which is evaluated in the joint (see Fig. 9.49a) and is supplied by the port on the right side of the component.

The model of the link has been discussed already in Sect. 9.5.3.

9.6.2.3 Model of Workspace

We will now direct our attention the Workspace component. This component models the dynamics of the robot as seen in the basic frame. Its structure is shown in Fig. 9.62 left.

The component Robot Base consists simply of *Source Flow* (*SF*) components which confirm that the linear and angular velocity of the base is zero. Thus, the robot base is fixed in its initial position.

The next port is on the outside connected to the manipulator by the compounded bond we spoke about. Thus, inside the Workspace component we unpack it by defining two components.

The first one is denoted as CM1 and represents the dynamics of motion of the centre of the mass of the first link (see Sect. 9.5.3 and Fig. 9.43). It is connected by a bond to the right Workspace port. To establish the correct connections we create a new component and connect it to the right the CM1 connection. It is denoted simply as Other Part 6 indicating that it must contain six other components—five components representing the dynamics of the centres of the mass of the other links, and a *TCP* object. We add these components gradually one by one. The structure of the last component Other Part 2 is shown in Fig. 9.62 right. It contains CM6 component describing the dynamics of the centre of the mass of the last link and the TCP component. This last component consists simply of three 1-junstions

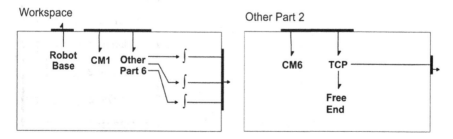

Fig. 9.62 The structure of the workspace component

connected to the component port. These junctions represent the nodes of x, y and z velocity components of the *TCP* in the basic frame. The information on these quantities is transferred out and connected to corresponding integrators as shown in *Workspace* document in Fig. 9.62. The Free End component defines that the forces on the *TCP* are zero, i.e. the robot is free of contacts with other objects.

This completes description of the model of PUMA 560 robot. We will direct our attention now to the problem of its control.

9.6.2.4   The Robot Controller

In order to drive the robot to follow the required joint angles trajectories we need a controller. We develop here a simple controller based on the independent joint control concept. This is a very common control concept in robotics. A good review of robotic control approaches can be found elsewhere (see e.g. [9]).

The model of the Controllercomponent is shown in Fig. 9.63. It consists of six identical drivers, which as input receive the values of the joint angles references generated by the Input component (Fig. 9.56). The drivers power ports are connected internally to the controller power port, which is further directly connected to the joint ports of the manipulator (see Figs. 9.56 and 9.58).

The Controller upper port is connected to the Manipulator output. The connecting signal is the compounded one consisting of the ordered signals connecting the joint rotation angles from the first to the last. On inside the port is connected to the upper input signal ports of the drivers. The connections are ordered from the left to the right corresponding to the structure of the signal line. This insures that Driver 1 receives information on the first joint angle (θ_1), Driver 2 from the second (θ_2), and so on.

The drivers have identical models. We designed them similarly to the independent joint controller described in Corke [23]. Figure 9.64 shows the driver position loop. It consists of the Velocity Loop and the gain (function)

Fig. 9.63 Independent joints controller for PUMA 650

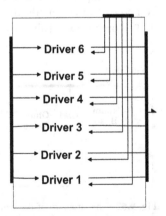

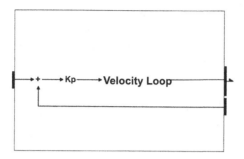

Fig. 9.64 Position loop of the drivers

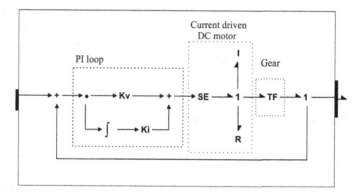

Fig. 9.65 The internal velocity loop

component Kp defining a *Pcontroller* with proportional gain *Kp*. Note that the feedback signal is really subtracted from the reference input. This is not clearly seen in the figure. (The signs of a summator input ports are set by double-clicking the ports and selecting the corresponding sign + or −.)

The Velocity Loop is shown in Fig. 9.65. The input to the loop is the angular velocity required by the position loop (Fig. 9.64). The feedback signal is the joint angular velocity taken at the driver power port. It is again subtracted.

The loop uses a *PI* controller to control the joint servo driver. The *Kv* and *Ki* are corresponding proportional (velocity) and integration gains. The PUMA 560 servo driver consists of the controlled DC motors, which through the gears, drive the robot joints. In Sect. 8.4 we analyzed a voltage controlled DC motor. Here, we will use the current controlled motor. The current control is implemented by an ideal current source

$$i_a = K_a u_c. \tag{9.144}$$

In this relationship i_a is the armature current, u_c voltage control input and K_a transconductance of the motor driver (in A/V). We do not need the complete model of the DC motor as given in (8.14), but only the torque-current relationship,

$$T_m = K_m i_a. \tag{9.145}$$

where T_m is developed motor toque and K_m the motor (torque) constant. Combining (9.144) and (9.145) we obtain

$$T_m = K_t K_a u_c \tag{9.146}$$

This relationship is modeled in Fig. 9.65 by a SE (*Source Effort*) component, having the constitutive relation given by (9.146).

Motor inertia and viscous friction are modeled in standard way by inertial I and resistive R elementary components. We didn't implement dry friction in the model. The reducing gear is used to drive the joints and is represented by a transformer with transformer (gear) ratio of 1:N.

9.6.3   Simulation of PUMA 560

We will now conduct the simulations of the developed PUMA 560 model. The robot parameters used in the model are defined in Sect. 9.6.1. We assume that initially all the joint angles are zero (the zero pose). In addition we need the initial position of *TCP* which is used in the integrators in the Workspace (Fig. 9.62). In the zero pose from Fig. 9.54 we find the following values:

$$x_{TCP}(0) = a_2 - a_3, \quad y_{TCP}(0) = d_2, \quad z_{TCP}(0) = d_1 + d_6$$

These values agree with the values obtained from (9.141) for zero values of the joint angles.

Finally we need to define the trajectories of the joint angles, which we wish the robot to follow. We will define these trajectories using the trapezoidal velocity profile with constant acceleration and deceleration at the beginning and at the end (Fig. 9.66). During the acceleration period the velocity changes linearly and at end of this interval t_c it reaches the value

$$v_c = a_c t_c. \tag{9.147}$$

The angle changes in this interval parabolic and reaches the value

$$q_c = \frac{1}{2} a_c t_c^2 = \frac{1}{2} v_c t_c. \tag{9.148}$$

Fig. 9.66 Angle trajectory
based on trapezoidal velocity
profile

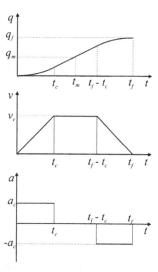

In the mid interval from t_c to $t_f - t_c$ the angle changes with a constant velocity, usually termed the *cruising* velocity. This velocity is equal to value reached at the end of acceleration period. Finally there is deceleration interval at end t_f of which the angle reaches the final value q_f. The velocity profile is symmetric about mean time $t_m = t_f/2$, and the corresponding value of the angle $q_m = q_f/2$. This value of the angle can be expressed using the velocities as

$$\frac{1}{2}q_f = \frac{1}{2}v_c t_c + v_c\left(\frac{1}{2}t_f - t_c\right).$$
(9.149)

or

$$q_f = v_c\left(t_f - t_c\right)$$
(9.150)

Introducing the mean velocity of the motion as

$$v_m = \frac{q_f}{t_f}.$$
(9.151)

We can write (9.150) as

$$v_m = v_c\left(1 - \frac{t_c}{t_f}\right)$$

From this we obtain the expression for the length of the acceleration and deceleration intervals as

$$\frac{t_c}{t_f} = \frac{v_c - v_m}{v_c} = \left(\frac{v_c}{v_m} - 1\right)\bigg/\frac{v_c}{v_m} \tag{9.152}$$

The change of the angle with time can be expressed as

$$q(t) = \begin{cases} \frac{1}{2}a_c t^2, 0 \leq t < t_c \\ \frac{q_f}{2} - \left(\frac{t_f}{2} - t\right), t_c \leq t < t_f - t_c \\ q_f - \frac{1}{2}a_c\left(t_f - t\right)^2, t_f - t_c \leq t \leq t_f \end{cases} \tag{9.153}$$

We assume that $v_c = 1.5 \, v_m$. From (9.152) follows that $t_c = t_f/3$. As final values of the joint angles the following are taken:

$$\theta_1 = 100°, \theta_2 = -100°, \theta_3 = 80°, \theta_4 = 120°, \theta_5 = 60°, \theta_6 = 150°$$

After some experimenting with controller drives with the robot and without it the following values of the controller gains are accepted: Kp = 50, Kv = 50, and Ki = 10. These constants give accepted results as will be shown shortly. The controller can be improved further by adding feed forward actions, gravity compensations or other changes. We leave it to the reader.

We will analyze two cases. In the first we will simulate motion of PUMA 560 under quick input trajectories with final time of 1 s. Before running the simulation we need to build the model. Below are given the basic information on the characteristics of mathematical model generated (see Chap. 5 for details):

- Number of equations generated 1047
- Number of switch elements 12
- Number of partial derivative matrix entries 2843
- Number of leading coefficient matrix entries 57

Hence, the model contains rather a large number of equations. This is expected because the robot model developed is a multi-body model in a descriptive form. It is based on velocity formulation and this as is well known leads to index 2 differential-algebraic model, a pretty challenging task. The partial derivative matric (Jacobian) is very sparse containing on average 2.8 elements per row. Thus, the most of equations of the model are very simple; some are not of course. Also the leading coefficient matrix is rather small, which indicates that we mainly deal with the algebraic equations. We can examine the model in more detail by opening the project PUMA 560 from library and applying methods described in Chap. 5.

After the model is successfully built, we start simulation. We use the following parameters: simulation time 1 s, maximum and output step 0.001 s, default error tolerance (1e−6) and select as method BDF method with the local error control. The simulation runs is rather quick. Below is given the simulations statistics:

- Number of integration steps 1087
- Number of function (equations) evaluations 2207

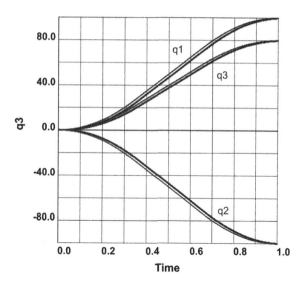

Fig. 9.67 The response curves for the first three joint angles

- Number of partial derivative matrix evaluations 1120
- Elapsed time[3] 1.25 s

We selected at the start of simulation rather short maximum step size and output interval of 0.001 s, or 1000 steps per simulation interval. The above data shows that there was not much looping during the execution; practically every step was accepted. Note that the matrix of partials was evaluated every step. As explained in Chap. 5 we use variable coefficient form of BDF, which generally requires more frequent evaluations of functions and matrix of partials. Thus it is more expensive in the processing terms than more common constant coefficient form. On other side, it is more robust, and as this not so simple example shows rather efficient.

The response curves for the first and the last three joint angles are shown in Figs. 9.67 and 9.68. On the same plot are drawn both the reference inputs (the tin line) and responses (the thick line). It is apparent the responses lag the reference input trajectory. The displacement is not large, but is visible. Similar curves under the similar conditions were reported by Corke [23].

The projections of *TCP* trajectory on $x - y$ and $x - z$ planes in the global frame are shown in Fig. 9.69. The errors in TCP position in comparison to values evaluated by the direct kinematics are shown in Fig. 9.70. The accuracy is really excellent.

---

[3]Simulation was conducted on a Toshiba laptop with Intel i7 (four core) 2.0 GHz processor, and 16 GB RAM, under Windows 7.1 operating system.

Fig. 9.68 The response
curves for the last three joint
angles

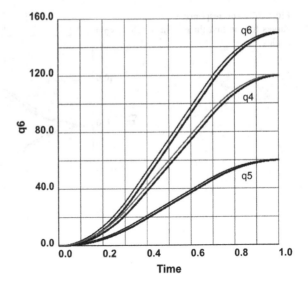

Fig. 9.69 Trajectory of TCP
in the global x − y and x − z
plane

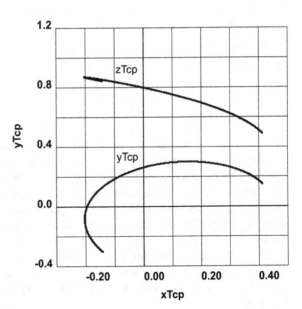

Fig. 9.70 Errors in positions of TCP

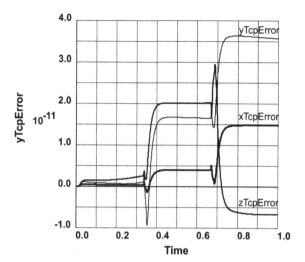

Fig. 9.71 The responses of the first three joint angles when the final time is 5 s

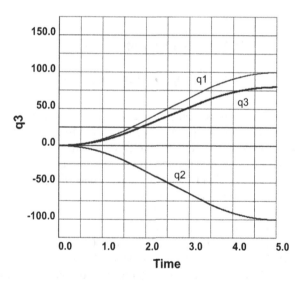

The simulations are repeated with larger input trajectory final time of 5 s. We set simulation time to 5 s, and maximum and output intervals to 0.005 s. The new response curves are shown in Figs. 9.71 and 9.72.

We can see that the joint angles now much better follow the input trajectories and both curves in the plot appear as a single curve. The other results are similar to that of the previous simulation. The elapsed time of the simulation increases now from 1.25 to 3.39 s.

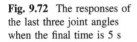

Fig. 9.72 The responses of the last three joint angles when the final time is 5 s

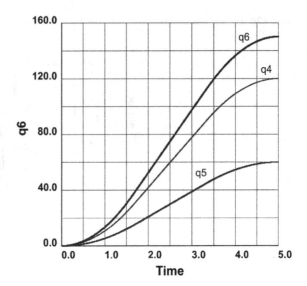

9.7   3D Visualization of Robots

9.7.1   Concept of 3D Visualization

In the previous Sect. 9.6 we dealt with the dynamic of a PUMA robot and its control. It gives a deep insight into the behavior of the robot. It could be further improved by adding flexibility of the links, the robot behavior at contact with surrounding objects and similar. The results are presented in the form of different plots and qualitative and quantitative data. For people not interested too much in the mathematic details this is rather esoteric. The visualization of the robot motion in space is perhaps more informative.

Many robot manufacturers use 3D robot models for off-line programming, e.g. ABB Robot studio [24], Fanuc RoboGuide [25], KUKA KukaSim [26] and others. Perhaps the first software of this kind was Grasp, developed by BYG Systems Ltd. [27]. These software applications were designed with the specific goals in mind and are based on geometric and kinematical models.

The geometrical models driven kinematically are useful for solving some problems, such as off line programming, but cannot simulate the real behaviour of the robots. It is quite complicated, however, to have all in one application, both the geometry and dynamics. Thus, we developed a separate application *BondSimVisual*, which supports developing of 3D geometric models of robots and similar mechanical systems and their rendering on the computer screen in a virtual scene. The simulation of robot motion based on Bond Graph models can be visualized by the motion of a virtual robot on the computer screen.

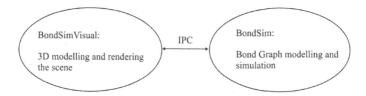

Fig. 9.73 Concept of virtual and dynamic modeling and IPC

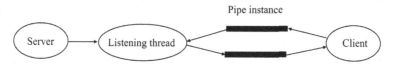

Fig. 9.74 The configuration of IPC by named pipe

The communication between these two model spaces is established through the support of the corresponding modelling programs—*BondSimVisual* and *BondSim*. This is realized in the form of *IPC* (Inter Process Communication) between these two applications, which is based on the named pipes [28], Fig. 9.73.

The configuration of the named pipe IPC used is shown in Fig. 9.74. The server (the *BondSimVisual*) is responsible for creating the pipe having a specified name. It also creates a special processing thread, which enables that the program simultaneously with its other task listens for the message from the client. The server also asks the client to connect. When the client (the *BondSim*) connects the two-way communication is established and simulation can start.

As already discussed in Sect. 9.6.2 *BondSim* supports exchange of data by using a special *IPC interface* component (Fig. 9.56). Its task is to collect the data from the Bond Graph model, pack it and send it with a message to the server. In the same vein after receiving the messages from the server it unpacks it and makes it available to the dynamic model. During the simulation the dynamic model regularly sends the messages containing information e.g. on the current values of the joint angular or linear displacements to the server. After receiving them the server extracts the content of the pack, reevaluates the 3D virtual model and renders it to the screen. If asked it can calculate the position of some point in a robot link and return its co-ordinates in the form of the message packs to the client. The IPC communication exists so long as the simulation session is active. Thus, it is possible to repeat different simulation runs. When the session is closed, the IPC communication and the server's listening tread are closed as well.

The 3D virtual and Bond Graph dynamic models represent two different, but closely related views of the same problem. As discussed above they are generated by two different processes that communicate one with the other. However, these processes are running in different time domains. There is also a third process, that of the user. When there is the simulation of the dynamic model only the problem of the time scales usually is not a severe one. The term simulation time denotes the real time of the physical process. The time that is necessary to execute simulation is basically different and can be much shorter or much longer. But, there is no confusion.

However, when there is visualization the situation is quite different. The user expects that the time that the objects move over the screen corresponds to the user time scale. To try to solve this problem the currently adopted approach is described below.

We do not allow the simulation runs as quickly as the processor can execute it, i.e. we will slow it down. Therefore, in the *Simulation Option* dialog (see e.g. Fig. 6.21) there is a simulation parameter *Output delay* (*ms*). It is by default equal to zero. But, when there is the visualization and we wish to slow down the simulation, we set it to some value, e.g. 50 (ms). This means that the simulation, after it generates the outputs, packs the respective data into a message and updates the virtual scene every 50 ms of the real-world time. Thus, the simulation shoots a new picture every 50 ms. Note that shooting the picture every 50 ms corresponds to frequency of $1/0.050 = 20$ pictures per second. Thus the virtual scene is refreshed 20 times in second. We are free to use a different delay time and/or different simulation output interval. Note that the film makers in order to generate a continuous motion picture for human eyes usually shoot 20–25 picture frames per second. It is usually explained by the persistence of human eyes. The computer games screens are updated typically 30 times per second.

9.7.2 Generating 3D Virtual Scene

Generatinga virtual 3D scene in the program *BondSimVisual* is based on a powerful Visual Toolkit (VTK) hy Kitware [29]. It is a large open source C++ library. The central structure of VTK is the pipeline of data, from a source of information to an image rendered to the screen. A typical application can have several such pipelines, corresponding to different objects appearing on the screen.

The typical robot consists of several links connected by the joints to the common base. The base itself can move over the workspace floor. There can be several such robots or other similar mechanical objects. Thus, as a basic item of the visualization we take a link visualization pipeline shown in Fig. 9.75.

The *Source* is a component that initializes the visualization pipeline. It is used to generate the link geometry. There are several types of the sources depending on type of geometric object that they generate. The basic forms of sources are shown in Fig. 9.76. Thus it is possible to generate simple objects such as the *Cylinders*, *Cubes* and *Spheres*. A more complex is *PolyPrism*, which represents a prism having as the base an arbitrary polygon composed of the straight lines and arcs. Finally,

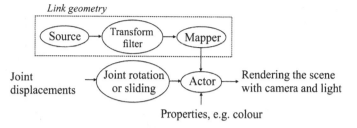

Fig. 9.75   Visual representation of a robot link

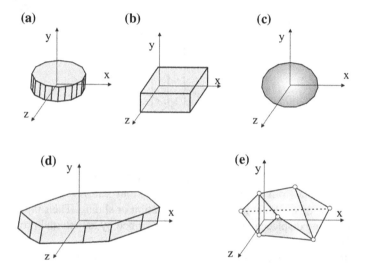

Fig. 9.76   The basic forms: **a**_Cylinder_. **b**_Cube_. **c**_Sphere_. **d**_PolyPrism_. **e**_PolyDataObject_

there is a general object in form of *PolyDataObject*. It is composed of *points*, and *polygons*. In addition to these objects there are the general objects termed the *Parts*. They represent the geometric object generated by 3D CAD systems such as Catia, Solid Works and others. These components are imported into *BondSimVisual* in the form of stl (stereo-lithographic) files.

Before using the data they must be transformed by *Transform filters*. The most important transformations that we need are the translating (shifting) of bodies along and their rotation about the coordinate axes.

The *Mappers* are VTK components that receive data from the sources and filters and maps them into a form that can be used for the rendering. Thus the mappers define the geometry of a link. The Actors are responsible for rendering geometric objects into the rendering window. We assumed that the rendering window contains

Fig. 9.77 PUMA 560 with
alternative co-ordinates
frames

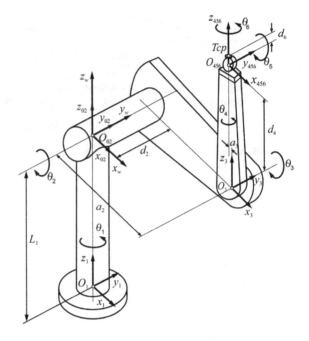

the complete computer screen. Their properties can be defined, such as colours. The
actors receive data from the mappers. However, they can receive the user inputs
from the transforms, which define the joint rotation or sliding. Their input data, the
angular or sliding displacements are, however, defined by an external program such
as *BondSim*.

To define the configuration of the system for visualization we use *scripts*.
A script consists of commands that *BondSimVisual* uses to create the visualization
pipeline for the problem. We will illustrate this on the problem of visualization of
Puma 560 robot analyzed in the Sect. 9.6. The script is shown in Textbox 9.1 on the
following two pages.

We define the coordinate frames differently from we did in Sect. 9.6.1, where we
used the Denavit-Hartenberg scheme. Here we will use an alternative scheme,
which is similar to that of Cork [23], and is based on a series of elementary
transforms. We start from the world coordinate frame $Ox_wy_wz_w$, which defines the
3D virtual space (Fig. 9.77). It has the origin in the middle of the scene, z-axis
directed out of the screen, x-axis to the right, and y-axis upward. We can define the
robot base coordinate frame $x_0y_0z_0$ as usual, with x-axis out of the screen, y-axis to
the right and z-axis upward. It is defined relative to the world frame.

Box 9.1 Visualization script for Puma 560

```
!-------------------- 3D model of robot Puma 560-------------------------------------------------------
! The world co-ordinates are:  x right, y - upward, and z - out of the screen
! The robot base co-ordinates: x - out, y - right, and z - upward

robot Puma (euler -90.0 -90.0 0.0)
    joint 1 ( shift z   -672.000 ) revolute Z
    joint 2 ( shift z   672.000) revolute Y
    joint 3 ( shift x   432.000 y   150.000) revolute Y
    joint 4 ( shift x   -20.330 z   433.000) revolute Z
    joint 5 revolute Y
    joint 6 revolute Z
    tcp( shift z 56.0)
    initial(   0.0 -90.0  90.0   0.0   0.0   0.0 );


cylinder Puma560-0cyl00  diameter 150.000000  resolution 28  length 570.000000;
cylinder Puma560-0cyl01  diameter 10.000000  resolution 6  length 570.000000;
cylinder Puma560-0cyl02 diameter 10.000000  resolution 6  length 570.000000;
cylinder Puma560-0cyl03  diameter 10.000000  resolution 6  length 570.000000;
cylinder Puma560-0cyl04  diameter 10.000000  resolution 6  length 570.000000;
cube Puma560-0cu01 70.000000 10.000000 70.000000;
cube Puma560-0cu02 70.000000 10.000000 70.000000;
cube Puma560-0cu03 70.000000 10.000000 70.000000;
cube Puma560-0cu04 70.000000 10.000000 70.000000;


set Puma560-0set00 add Puma560-0cyl00
            Puma560-0cu01 ( shift X 30  Y -5  Z 30  euler 0 45 0 )
            Puma560-0cyl01 ( shift X -78 )  Puma560-0cyl02 (shift X 78 )
            Puma560-0cyl03 (shift Y -78 ) Puma560-0cyl04 (shift Y 78 )
            Puma560-0cu02 (shift -5  Y -30  Z 30  euler -90 45 0 )
            Puma560-0cu03 (shift t X 5  Y 30  Z 30  euler 90 45 0 )
            Puma560-0cu04 (shift X -30  Y 5  Z 30  euler 180 45 0 ) ;
set Puma add Puma560-0set00 (shift z -650);
cube Puma560-0cu05 60.000000 60.000000 60.000000;
cylinder Puma560-0cyl05  diameter 90.000000  resolution 14  length 250.000000;
cylinder Puma560-0cyl06 diameter 380.000000  resolution 30  length 10.000000;
cylinder Puma560-0cyl07  diameter 400.000000  resolution 28  length 50.000000;
set Puma560-0set07 add Puma560-0cyl07 Puma560-0cyl06 ( shift Z 50 ) ;
cube Puma560-0cu06 108.000000 180.000000 102.000000;
set Puma560-0set06 add Puma560-0cu06 Puma560-0cu05 ( shift X 25  Y -20  Z 20 )
            Puma560-0cyl05 ( shift X 54  Y 68.7385  Z 22  euler 0 0 90 )
            Puma560-0set07 ( shift X 54  Y 235  euler 0 0 90 ) ;
set Puma add Puma560-0set06 ( shift X -235  Y 54  Z -672  euler 0 0 -90 );
render Puma color 0.8 0.8 0.8;


cylinder Puma560-1cyl00  diameter 160.000000  resolution 20  length 320.000000;
cylinder Puma560-1cyl01  diameter 150.000000  resolution 28  length 60.000000;
set Puma#1 add Puma560-1cyl00 ( shift Y -105  Z 672  euler 90 90 -90 )
            Puma560-1cyl01 ( shift Z 592 ) ;
render Puma#1 color 0.90 0.90 0.90;
```

```
polyprism Puma560-2poly01
height 110.000000 axis Y
 0.000000 0.000000
arc to 149.000000 -57.000000 radius -223.246063 tolerance 1.000000
arc to 298.000000 0.000000 radius -223.246033 tolerance 1.000000
 298.000000 196.000000
 224.000000 619.000000
arc to 182.425995 671.353027 radius -76.875259 tolerance 1.000000
arc to 115.573997 671.353027 radius -76.875053 tolerance 1.000000
arc to 74.000000 619.000000 radius -76.874718 tolerance 1.000000
 0.000000 196.000000
 0.000000 0.000000;

set Puma#2 add Puma560-2poly01 ( shift X -172  Y 200  Z 149  euler 0 90 0 ) ;

render Puma#2 color 0.9 0.9 0.9 ;

polyprism Puma560-3poly01
height 88.000000 axis Y
 0.000000 0.000000
arc to 76.000000 -32.000000 radius -132.999954 tolerance 1.000000
arc to 152.000000 0.000000 radius -132.999954 tolerance 1.000000
 99.190002 437.000000
 12.190000 437.000000
 0.000000 0.000000;

set Puma#3 add Puma560-3poly01 ( shift X -76  Y -44  Z -85.5 ) ;
render Puma#3  color 0.90 0.90 0.90 ;

polyprism Puma560-4poly00
height 88.000000 axis Y
 0.000000 0.000000
 57.000000 0.000000
 57.000000 10.000000
arc to 69.594002 45.959000 radius -41.500034 tolerance 1.000000
arc to 47.549999 77.035004 radius -41.499989 tolerance 1.000000
arc to 9.450000 77.035004 radius -41.500027 tolerance 1.000000
arc to -12.594000 45.959000 radius -41.499992 tolerance 1.000000
arc to 0.000000 10.000000 radius -41.500000 tolerance 1.000000
 0.000000 0.000000 ;

cube Puma560-4cu00 87.000000 88.000000 40.000000;
set Puma#4 add Puma560-4poly00 ( shift X -28.5  Y -44  Z -42.5 )
           Puma560-4cu00 ( shift X -43.4  Y -44  Z -81.5 ) ;
render Puma#4 color 0.4 0.4 0.4;

cylinder Puma560-6cyl00  diameter 53.000000  resolution 8  length 20.000000;
cube Puma560-6cu00 12 13.717000 10.000000 ;
set Puma560-6set00_set add Puma560-6cu00
                           Puma560-6cyl00 ( shift Z -15 ) ;
set Puma560#6 add Puma560-6set00 ( shift Z 46 );
render Puma560#6 color 0.25 0.25 0.25;

Probe Point1 Puma560#6( shift z 56) refer Puma560 ;
!------------------------End of Robot Puma 560 ------------------------------------------------
```

We can obtain the base frame from the world frame by applying the rotations. We use *Euler zyz* transform. Assuming the base frame is coincident with the world frame, we rotate it by −90° about *z*-axis, and by −90° about new *y*-axis. The joints axes are defined similarly by a series of the elementary transformations. This is defined in the script in Box 9.1 by the *robot* command. This command defines first the name of the robot simply as *Puma*, and then the base coordinate frame using *euler* zyz transform. The *joint* 1 frame is defined with respect to the base by a translation (*shift*) along z-axis, and then applying *revolute* Z transform. The first is a constant transform defined by the height of the robot shoulder above the floor (672 mm). The other transform defines that the joint is a revolute, with the rotation axis along *z*-axis. The rotation angle is not known and will be generated by simulation of the robot system. However, its *initial* value, which defines the initial posture of the robot arm, is defined by the last part of the command. In the similar way *joint* 2 is defined with respect to the previous joint and so on until the *tcp* (tool central point) is defined. The *Initial* function specifies the initial values of the all joint displacements in the order as they appear in the command.

The *joint* statement generally can have the following form:

$$Joint\ n\ \text{Pre\_transform}\quad Joint\_transform\quad Post\_transform$$

where both the *pre-* and *post-transforms* are of the form

$$(shift\ x = a\ y = b\ z = c\ euler(angle1, angle2, angle3))$$

and in effect defines the homogeneous transformations. The joint transform is an elementary transform, which in case of the *revolute* joints represents rotation about the corresponding axis and for the *prismatic* ones the translation along the axis. Such complex transforms can be used for aligning the links axes with the joints axes. Note that the end of a command is denoted by a semicolon, and the command parts are delimited by spaces.

The *cylinder* and *cube* commands are used to define the corresponding simple objects, which are used to build the robot body and its links. For more complicated parts the *polyprism* commands can be used. The *Set* command serves to build a body as an assembly of the simpler objects, but also to connect the robot links objects to the corresponding joints, and hence to assemble the robot.

Finally the *render* command is used to create visualization pipelines for the robot body and their links, and connect it to the rendering engine. This command also allows assigning the colour using *RGB* (red-green-blue) colour scheme.

To apply a visualization script we need first to edit it as a textual file using suitable text editing tool such as *Notepad*. We save the file to an appropriate medium such as a memory stick, the fixed disk or other. To import it into the

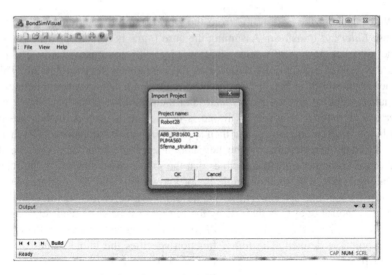

Fig. 9.78 Importing a project into the *VisualBondSim*

BondSimVisual program we open the program (Fig. 9.78), select the command *Import* in menu *File* and choose the *Project* subcommand. A Windows *Open file* dialogue opens, which we can use to browse the system to find the script file we wish to import. When we find it, we select it and press the *OK* button. A project import dialogue opens with the name of the file we have just selected (Fig. 9.78).

We can accept the file name as the name of the new project (or change it if we wish) and click *OK*; the file will be copied and stored into the project database under the accepted project name. This project name can be used to open the project file and render the robot to the screen. Thus, to open Puma 560 robot, we open first the *BondSimVisual*, click the *Open* toolbar button and from the projects combo box select the *Puma 560*. A new tabbed document opens showing the Puma 560 robot arm in its initial posture (Fig. 7.79).

The robot arm is shown as seen when viewing along the world z-axis. The BondSimVisual implements also the VTK's *interactor,* which allows rotating the complete scene by mouse clicking. This way we can change viewing direction at will (some other operations are supported as well).

The project file (script) of 3D Puma 560 model is relatively long because it is built of elementary components such as the cylinders, cubes, etc. It could be much shorter if the bodes are designed by suitable 3D CAD tool such as Catia, Solid Works, or others, and imported in the form of stl files.

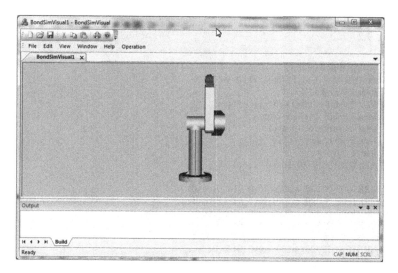

Fig. 9.79 Visualization of Puma 560 in initial posture and viewing along the world z-axis

To illustrate this point we consider 3D visual model of ABB IRB 1600 robot [30]. Its visualization script is shown in Box 9.2. It has a similar structure as the Puma 560 script in Box 9.1, but uses the *parts*, which represent the geometry of the robot. It is shorter than the Puma's script.

The 3D model was assembled from CAD drawings of the parts, which are downloaded from the robot manufacturer web page [31]. To implement the script it has been edited and saved as a text file and then imported similarly as we did with the Puma script. We assign to it the project name *ABB IRB1600_12*. In addition to the script we need to import the downloaded *parts* files as well. To do this, we apply the *Import* command, and then select the *CAD Part*....subcommand. In the dialogue window that opens we define the corresponding project name, to which the parts belong. In this case it is *ABB IRB1600_12*. Then, again the *Open* file dialogue appears, which we use to find and transfer the CAD part's stl files. These names are used as the parts names in the script in Box 9.2.

Box 9.2

```
!-------- ABB IRB1600_12 -------------------------------------------
Robot ABB_IRB1600_12 (euler -90.0 -90.0 0.0)
    Joint 1   revolute Z
    Joint 2   (shift x 150 z 486.5 ) revolute y
    Joint 3   (shift z 475) revolute y
    Joint 4 ( shift x 600 ) revolute x
    Joint 5  revolute y
    Joint 6  revolute x
    initial (0.0 0.0 0.0 0.0 0.0 0.0 0.0) ;


Part IRB1600_X-120_m2004_rev0_01-1_Body0;
Part IRB1600_X-120_m2004_rev0_01-7_Body1;
Part IRB1600_X-120_m2004_rev0_01-4_Body2;
Part IRB1600_X-120_m2004_rev0_01-5_Body3;
Part IRB1600_X-120_m2004_rev0_01-3_Body4;
Part IRB1600 X-120_m2004_rev0_01-2_Body5;
Part IRB1600_X-120_m2004_rev0_01-6_Body6;


Set ABB_IRB1600_12 add IRB1600_X-120_m2004_rev0_01-1_Body0 ;
Render ABB_IRB1600_12 color 0.89 0.423 0.039;


Set ABB_IRB1600_12#1 add IRB1600_X-120_m2004_rev0_01-7_Body1 ;
Render ABB_IRB1600_12#1 color 0.0 0.6 0.6 ;


Set ABB_IRB1600_12#2 add IRB1600_X-120_m2004_rev0_01-4_Body2
                        (shift x -150 z -486.5 );
Render ABB_IRB1600_12#2 color 0.0 0.2 0.8 ;


Set ABB_IRB1600_12#3 add IRB1600_X-120_m2004_rev0_01-5_Body3
                        (shift x -150 z -961.5 );
Render ABB_IRB1600_12#3 color 0.89 0.423 0.039 ;


Set ABB_IRB1600_12#4 add IRB1600_X-120_m2004_rev0_01-3_Body4
                        (shift x -750 z -961.5 );
Render ABB_IRB1600_12#4 color 0.1 0.1 1.0 ;


Set ABB_IRB1600_12#5 add IRB1600_X-120_m2004_rev0_01-2_Body5
                        (shift x -750 z -961.5 );
Render ABB_IRB1600_12#5 color 1.1 1.1 1.1 ;


Set ABB_IRB1600_12#6 add IRB1600_X-120_m2004_rev0_01-6_Body6
                        (shift x -750 z -961.5 );
Render ABB_IRB1600_12#6 color 0.1 0.1 0.1 ;


Probe Point1 ABB_IRB1600_12#6( shift x 65) refer ABB_IRB1600_12;

!------------ End --------------------------------------------------
```

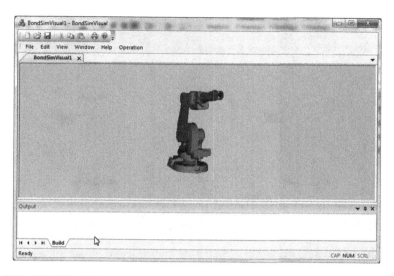

Fig. 9.80 ABB IRB 1600 robot in the initial posture (and slightly rotated to the right)

Figure 9.80 shows 3D visual model of robot **ABB IRB 1600** in the initial posture, slightly rotated to the right (by the mouse clicking).

9.7.3 Visualization of Robot Dynamics

Now we will join the simulation and visualization. We repeat simulation of Sect. 9.6.3 for simulation time of 1.0 s. We set the delay time 50 (ms). We will now view how the robot arm moves as the simulation advances.

To do this we need first to start the *BondSimVisual*, and open *Puma 560* project as already we did in Sect. 9.7.2. We obtain the robot arm in its initial posture as shown in Fig. 9.79. Now we will open the *Operation* menu and select the *Start IPC* command. This command creates a named pipe, creates a new thread for listening the client's messages, and calls the client to connect.

The next, we start *BondSim* application, and open project *Puma 560*. If we are using the both applications on the same computer, we can move BondSimVisual window to the left, the BondSim main window to the right so that visual scene is not covered by the window. Next we build the *Puma 560* project and in the *Simulation* menu select *Connect to IPC socket*. If the command is accepted the IPC communication is established and we may start the simulation. We do it in the usual way by selecting *Run* command. The simulation starts and we may observe in BondSimVisual screen how the Puma's arm moves during the simulation. The

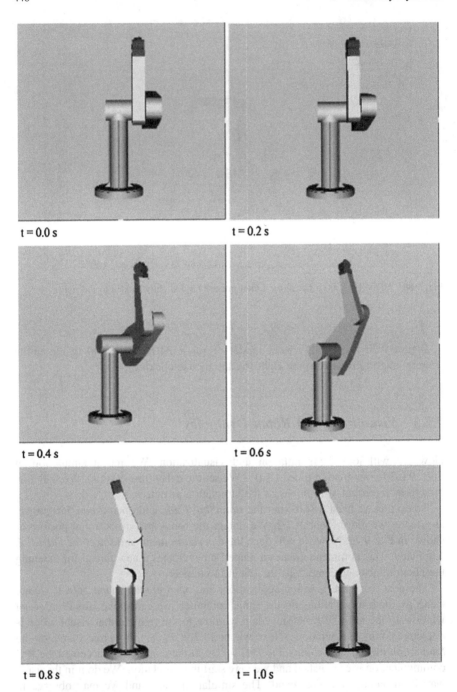

Fig. 9.81 The sequence of Puma 560 robot postures during the simulation

Fig. 9.82 The trajectories of Tcp by: (a) simulation (*thin line*), (b) visualization (*thick line*)

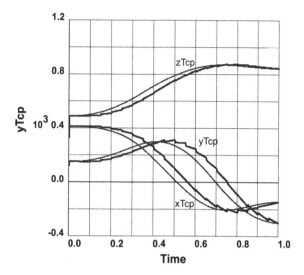

sequence of Puma 560 posturesthat were generated are shown in Fig. 9.81. The total elapsed time is 2.2 s.

Figure 9.82 shows the co-ordinates of the *Tcp* trajectory during the simulation. The thin lines represent the coordinates evaluated by the simulation, and the thick ones that returned by the IPC pipe from the visualization process.

We may see that the curves are shifted in time by the 50 ms delay we used for the generation of the visual scene. We may also note that the visually generated curves are not smooth ones because they are generated every 0.050 s. Otherwise they agree in the values.

References

1. Wittenburg J (1977) Dynamics of systems of rigid bodies. BG Teubner, Stuttgart
2. Haug EJ (1989) Computer-aided kinematics and dynamics of mechanical systems. Allyn and Bacon, Boston
3. Schiehlen W (1990) Multibody systems handbook. Springer, Berlin
4. von Schwerin R (1991) Multibody system simulation: numerical methods, algorithms, and software. Springer, Berlin
5. Shabana AA (1998) Dynamics of multibody systems, 2nd edn. Cambridge University Press, Cambridge
6. Rabier PJ, Rheinboldt WC (2000) Nonholonomic motion of rigid mechanical systems from a DAE viewpoint. SIAM, Philadelphia
7. Borri M, Trainelli L, Bottasso CL (2000) On representation and parameterizations of motions. Multibody Syst Dyn 4:129–193
8. Craig JJ (1986) Introduction to robotics: mechanics and control
9. Sciavicco L, Siciliano B (1996) Modeling and control of robot manipulators. McGraw-Hill, New York

10. Breedveld PC (1984) Physical systems theory in terms of bond graphs, PhD thesis, Technische Hochschool Twente, Entschede
11. Karnopp DC, Margolis DL, Rosenberg RC (2000) System dynamics: modeling and simulation of mechatronic systems, 3rd edn. Wiley, New York
12. Fahrenthold EP, Wargo JD (1994) Lagrangian bond graphs for solid continuum dynamics modeling. ASME J Dyn Syst Measur Control 116:178–192
13. Hairer E, Wanner G (1996) Solving ordinary differential equations II, stiff and differential-algebraic problems, 2nd Revisited edn. Springer, Berlin
14. Lioen WM, Swart JJB Test set for initial value problem solvers, Amsterdam. Available at http://www.cwi.nl/cwi/projects/IVPteseset/
15. Den Hartog JP (1956) Mechanical vibrations, 4th edn. McGraw-Hill, New York
16. Pan CH, Moskwa JJ (1996) An analysis of the effects of torque, engine geometry, and speed on choosing an engine inertia model to minimize prediction errors. ASME J Dynamic Syst Measur Control 118:181–187
17. Hesterman DC, Stone BJ (1994) A systematic approach to the torsional vibration of multi-cylinder reciprocating engines and pumps. Proc Inst Mech Eng 208:395–408
18. Damic V, Kesic P (1997) A system approach to modelling and simulation of multibody systems using acausal bond graphs. In: Marovic P, Soric J, Vrankovic N (eds) Proceedings of the 2nd congress of Croatian Society of Mechanics, Supetar, Croatia, pp 415–422
19. Damic V, Montgomery J, Koboevic N (1998) Application of automated modelling in design. In: Marjanovic P (ed) Proceedings of 5th international design conference, Dubrovnik, Croatia, pp 111–116
20. Damic V (1987) An approach to computer aided design of industrial robots using simulation package simulex. In: Vukobratovic M (ed) Proceedings of 5th Yugoslav symposium on applied robotics and flexible automatization, Bled, pp 28–39
21. Armstrong B, Khatib O, Burdick J (1986) The explicit dynamic model and inertial parameters of the PUMA 560 Arm. In: Proceedings of the IEEE international conference robotics and automation, vol 1, San Francisco, USA, pp 510–518
22. Corke PI, Armstrong B (1994) A search for consensus among model parameters reported for the PUMA 560 Robot. In: Proceedings of the IEEE international conference robotics and automation, San Diego, pp 1608–1613
23. Corke P (2011) Robotics, vision and control fundamental algorithms in matlab. Springer, Berlin
24. RobotStudio, http://new.abb.com/products/robotics/robotstudio. Accessed 09 May 2015
25. RoboGuide, http://www.fanucamerica.com/products/vision-software/ROBOGUIDE-simulation-software.aspx. Accessed 09 May 2015
26. KUKASim, http://www.kukarobotics.com/en/pressevents/productnews/NN_040630_KUKASim.htm. Accessed 09 May 2015
27. Grasp 10, www.bygsimulations.com. Accessed 27 May 2015
28. Vuskovic M (2014) Operating systems, inter-process communications. http://medusa.sdsu.edu/cs570/Lectures/Chapter9.pdf. Accessed Sept 2014
29. Schroeder W, Martin K, Lorensen B (1998) The Visualization toolkit, an object-oriented approach to 3D graphics, 2nd edn. Prentice Hall, Upper saddle river, NJ
30. Cohodar M (2012–2015), Private communications. University of Sarajevo, Faculty of Mechanical Engineering, Sarajevo, Bosnia and Herzegovina
31. http://new.abb.com/products/robotics/industrial-robots/irb-1600/irb-1600-cad. Accessed Feb 2015

Chapter 10
Continuous Systems

10.1 Introduction

Continuous systems are important in many engineering disciplines, such as structural mechanics, fluid mechanics, thermal systems, electrical field etc. They are important in mechatronics applications too, i.e. control of robotic manipulators taking account of flexibility of mechanical structure, sensor design, and micro-mechanics systems design. Solving such problems requires some form of discretisation. The methods that are generally used are finite difference and finite element. The both start with partial differential equations that describe the problem, but differ in the discretization. The first method uses suitable numerical approximation of the equation on a selected grid. Finite element methods, on the other hand, are based on discretization of the physical problem domain, dividing a continuous system or a continuous component into finite elements. Motion of elements typically is based on an assumed displacement field used to formulate the governing equations of element motion. The equations are formulated using different methods, such as the Lagrange equations, the D'Alembert-Lagrange principle, the Galerkin method, or others. Finite element methods nowadays dominate the scene in solid mechanics; many extremely powerful software packages based on these methods are available to solve various engineering problems. There are also related methods, such as finite-volume and boundary-element methods that are popular in fluid mechanics and the electrical engineering and field theory. We will not go into details of these methods here, but suggest that the interested reader consult suitable references. It should be noted that the available methods are not as powerful when applied to the mixed problems, e.g. solid body fluid interactions, complex electromechanical problems, continuous space—concentrated parameter problems, and similar. This is in particularly true if the interconnections are strong.

The question we would like to ask is where the place of the Bond graph methods in this area is. The strength of Bond graphs lies in its multidisciplinary paradigm and visual expressiveness. In many problems of mechatronics and micro-mechanics, the

© Springer-Verlag Berlin Heidelberg 2015
V. Damić and J. Montgomery, *Mechatronics by Bond Graphs*,
DOI 10.1007/978-3-662-49004-4_10

physical domain is not uniform. There are strong interactions between processes taking place in physically different fields. This is, for example, the case in computer-controlled drives of mechanical links that often are not too rigid, or in sensors where there are strong interactions of solid mechanics, fluid mechanics, electrical and thermal processes. Bond graphs are capable of helping to bridge the gap between these different fields.

In this final chapter we describe an approach for solving the problems dealing with continuous systems based on bond graphs. In the literature there are different approaches for solving continuous systems by bond graphs [1–3]. The approach described here is based on the component model philosophy developed in this book and implemented in BondSim. We start with a description of the general approach to modelling continuous systems based on component models. This will be applied first to the problem of modelling electric transmission lines. Next, a bond graph component model of a beam element, based on classic Euler-Lagrange theory, will be developed. We will end with two practical problems: package vibration testing and Coriolis mass-flow meters.

10.2   Spatial Discretisation of Continuous Systems

Continuous systems possess infinitely many degrees of freedom and, as such, are not directly amenable to numerical solution. Their solution requires some sort of discretisation. Various approaches are possible. Here we consider an approach that is compatible with the methods used in this book, i.e. representing the continuous system as an assemblage of bond graph components, then developing the mathematical model as a set of differential-algebraic equations that are solved directly. This kind of approach belongs to a class of methods known as the *Method of Lines* (MOL) [4]. To represent continuous systems—such as robotic links or transmission lines—by a bond graph component model, it is necessary to discretises them spatially. One of most powerful ways to do this is by the finite-element method. Other approaches also can be used, notably finite-difference approximations and even ad hoc lumping. We are not speaking in favour of any of these, but wish simply to explain how they can be implemented in a Bond graph setting.

A continuous system can be discretized by dividing it into a number of finite parts (elements). For example, a beam can be divided into number of small parts (Fig. 10.1). The parts are called finite to distinguish them from the differential elements used in mathematical analysis of the problem. The parts can also be of different forms. Thus, in finite element analysis, triangular or quadrilateral elements often are used in plane problems; tetrahedral and hexahedral (solid) elements for space problems; plate elements for analysis of plates, etc. In finite difference approximations one-, two- or three-dimensional grids are used.

The parts are assumed to be joined only at distinct nodes. Thus, discretisation is based on physical variables defined only at the distinct nodes. Values of variables inside the finite parts (elements) are described in terms of nodal variables.

Fig. 10.1 Continuous beam as an assemblage of finite parts

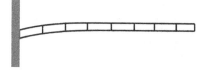

In the finite element method one group of variables, such as displacements in deformable body mechanics, fluid velocities in fluid mechanics, or field potential in electric fields, are represented inside the elements by polynomial interpolating functions. The approximation is smooth over an element, but there can be discontinuities in the derivatives of variables with respect to space coordinates at the transition from one element to the other. The expression for associated variables—e.g., forces, stresses, and currents—are found by applying a suitable method, such as Lagrange equations, the Principle of Virtual Work, and the like.

An element of the continuous domain can be represented by a bond graph component (Fig. 10.2a). Depending on the number of external nodes of the element that are used to join neighbouring elements, this component has a suitable number of power ports. These generally are compounded and, hence, can represent efforts and flows at the ports as vector or tensor quantities (Fig. 10.2b). Interconnection to other components is made by components consisting of effort or flow junctions, depending on the type of variables at the nodes. In solid continuum domains these typically are displacements. Displacements are not directly supported as port variables in bond graphs; thus, velocities will be used that correspond to the common flows at the effort junction nodes. Bond graph representation also shows assumed power flow thought the domain. The component model of underlying processes can be developed using different approaches, as already noted previously.

The finite-element method—as well as the finite-volume and boundary-element methods—uses pre-processors and post-processors that are, to a large extent, responsible for their successful application in engineering design and research.

Fig. 10.2 Finite element as bond graph component. **a** Four-node finite element. **b** Bond graph component connection diagram

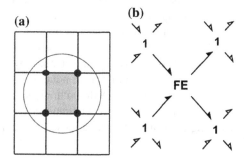

Fig. 10.3 A super element consisting of several finite elements. **a** Component representation. **b** Structure of the component

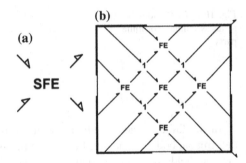

They are responsible for element mesh generation and displaying results in a user-friendly way. Such special purpose devices are not used in *BondSim*; the hierarchical component model approach we have developed can be used to simplify discretisation of continuous components by bond graph finite-element components.

To that end, a component can be defined that is represented by some number of bond graph finite-element components (Fig. 10.3). Such a component plays the role of a "super element". In the same way, a lower-level super-element component can be defined that is composed of such higher-level super elements. This way, a complex model consisting of a large number of finite elements can be constructed easily by using the component copy and insertion operations supported by the program. This technique is similar to finite-element sub-structuring [5].

In the next sections we apply this technique to solve some practical continuous system problems.

10.3   Model of Electric Transmission Line

Electrical transmission lines transmit electromagnetic energy between terminals. They can be of different forms, such as coaxial cables, parallel wires, strip lines. They find natural application in telecommunications and electric power engineering, and also as high-speed busses in modern digital computers. There are many references that treat modelling and analysis of electrical transmission lines [6]. We will not go into details here, but use some of those results to show how such lines can be modelled and analysed by bond graphs with *BondSim*.

A transmission line is characterised by a series resistance R' in ohm/m and an inductance L' in H/m of both conductors, and by a shunt capacitance C' in C/m and conductance G' in S/m. A differential element of the transmission line of length dx is described by the equivalent circuit of Fig. 10.4. This is used as a starting point for deriving the differential equations of the transmission line.

Fig. 10.4 Equivalent model of a transmission line of a length dx

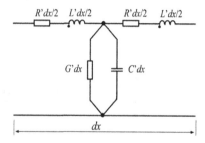

To develop a discrete model of the line, we approximate a differential line element by a line section of finite length. Such a section is represented by the component in Fig. 10.5a, which is created by dragging the *TSect* icon under the *Electrical Component, Transmittion Line*. Its model is shown in Fig. 10.5b.

The parameters of the model are

1. Section resistances $R = R' \cdot Len/2$
2. Section inductance $L = L' \cdot Len/2$
3. Shunt capacitance $C = C' \cdot Len$
4. Shunt resistance $Rs = 1/(G' \cdot Len)$ or, alternatively, $G = G' \cdot Len$

where *Len* is the length of the section.

To simplify generation of a transmission line model consisting of a large number of the sections, we define a component TLine, consisting of five line sections (Fig. 10.6).

We proceed in the same manner by defining a new line component containing, for example, two or more previously defined TLine components. In this way, it is not difficult to create a line component that, in a hierarchical fashion, contains a large number of the sections. As the number of sections increases and, hence, their length decreases, the model better approximates the line.

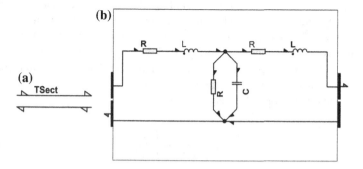

Fig. 10.5 The component model of a line section. **a** Representation. **b** Model

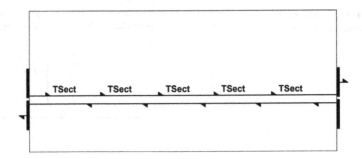

Fig. 10.6 TLine component composed of five line sections

To apply this modelling approach to transmission lines, we analyse the telephone line of [7] and compare the results given therein with those generated by *BondSim*. The parameters of the line are

1. *Length* 322 km(200 mi)
2. $R' = 0.006304$ ohm/m
3. $L' = 2.441 \times 10^{-6}$ H/m
4. $C' = 4.950 \times 10^{-12}$ F/m
5. $G' = 0.1801 \times 10^{-9}$ S/m

The line is divided into ten sections of length 32.2 km (20 mi) each. The parameters of the line segments are

1. $R = 101.5$ ohm
2. $L = 0.0393$ H
3. $C = 0.159 \times 10^{-6}$ F
4. $Rs = 0.172 \times 10^{6}$ ohm

The line is driven by a sinusoidal voltage source of 1 V amplitude, circular frequency of $\omega = 5000$ rad/s, and is loaded by characteristic impedance Z_0.

The model of the *Electric Line*, including the source and load, is shown in Fig. 10.7. The line is represented by the TL component, consisting of two TLine components of Fig. 10.6. Each of the line segments of the 32.2 km length is represented by TSec components of Fig. 10.5 with parameters as given above. We thus have $2 \times 5 = 10$ components. We next evaluate the characteristic impedance of the line, which is represented by component Z0.

The characteristic impedance is given by

$$Z_0 = \sqrt{\frac{R' + j\omega L'}{G' + j\omega C'}} \tag{10.1}$$

Fig. 10.7 Model of *Electric line*

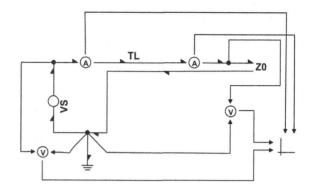

The modulus of the impedance is

$$|Z_0| = \left(\frac{R'^2 + \omega^2 L'^2}{G'^2 + \omega^2 C'^2}\right)^{1/4} = 744.991$$

and its argument is

$$\arg(Z_0) = 0.5 \cdot (a\tan(\omega L'/R') - a\tan(\omega C'/G')) = -0.234746 \text{ rad}$$

Hence, the characteristic impedance is

$$Z_0 = 744.991 \cdot e^{-0.2347455} = 724.558 - \text{j}173.282 \tag{10.2}$$

This way, the characteristic impedance can be represented by a resistor and a capacitor (Fig. 10.8) with resistance and capacitance, respectively, of

$$R0 = 724.558 \text{ ohm}$$
$$C0 = 1/(173.282 \times 5000) = 1.15419 \times 10^{-6} \text{ F}$$

In the model of Fig. 10.7, an ammeter and a voltmeter are inserted at both the sending and receiving ends of the line to provide output of the current and voltage, respectively. Outputs are fed to an output (display) component at the right. Before proceeding with the simulation, we evaluate another important characteristic of the line, the propagation parameter.

The propagation parameter is given by

$$\gamma = \alpha + j\beta = \sqrt{(R' + j\omega L') \cdot (G' + j\omega C')} \tag{10.3}$$

The modulus of the propagation factor is

$$|\gamma| = ((R'G' - \omega^2 L'C')^2 + \omega^2 (R'C' + G'L')^2)^{1/4} = 1.8439 \times 10^{-5} \text{ 1/m}$$

and its argument is

Fig. 10.8 Representation of
the characteristic impedance
Z0

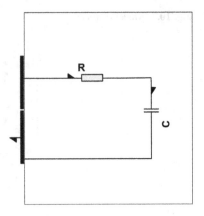

$$\arg(\gamma) = 0.5 \cdot a\tan\left(\frac{\omega(R'C' + G'L')}{R'G' - \omega^2 C'L'}\right) = 1.32954$$

Hence,

$$\gamma = 1.8439 \times 10^{-5} e^{j1.32954} = 4.40544 \times 10^{-6} + j1.7905 \times 10^{-5}\,1/m$$

This way we get the attenuation factor of the line $\alpha = 4.40544 \times 10^{-6}$ 1/m and
the phase constant $\beta = 1.7905 \times 10^{-5}$ 1/m.

In a transmission line loaded by the characteristic impedance, there are no
reflected waves, so voltage amplitudes along the line are given by

$$V(x) = V_s \cdot e^{-\gamma x} \tag{10.4}$$

The amplitude at the receiving end of line $x = Len$ is

$$V_r = V_s \cdot e^{-\alpha Len} = 0.2421\ V$$

The current at receiving end is

$$I_r = V_r / |Z_0| = 3.249 \times 10^{-4}\ A$$

A similar relationship holds for the current amplitudes, i.e.

$$I_r = I_s \cdot e^{-\alpha Len} \tag{10.5}$$

Hence,

$$I_s = I_r \cdot e^{\alpha Len} = 1.342 \times 10^{-3} \text{ A}$$

Finally, from (10.4), we find the time delay for voltage and current waves to reach the other end. This is given by

$$\Delta t = \beta \cdot Len/\omega = 0.001153 \text{ s}$$

Values found above correspond to steady-state sinusoidal wave transmission along the line. The transmission line problem in [7] is based on such steady-state frequency analysis.

The simulation based on the model developed above, on other hand, predicts the complete transient behaviour of the line and thus gives a more complete picture of the processes involved. This is, of course, approximate, as a finite number of the lumps are used; but the accuracy can be increased by increasing the number of sections.

We simulate processes in the line using a simulation interval of 0.05 s. The period of the supply voltage is $2\pi/\omega = 0.00126$ s. Thus, we use an output interval of 0.00001 s. Results are shown in Figs. 10.9, 10.10 and 10.11.

There is an initial period of about 0.01 s during which there are transients in the currents and the voltages, before they settle to a steady state. Voltage and current amplitudes at the receiving end are 0.2312 V and 0.3104 mA, respectively. At the sending end, the amplitude of the current is 1.392 mA, as in [7].

Figures 10.9 and 10.10 show a time delay before the waves reach the other end of the line. It is difficult to determine accurately when current and voltage start to increase from zero. Therefore, the delay is found by the difference in times when

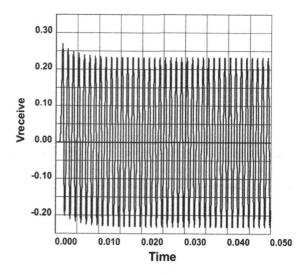

Fig. 10.9 The voltage at receiving end of the line

Fig. 10.10 The current at
receiving end of the line

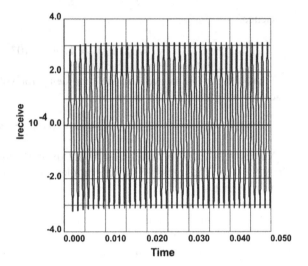

Fig. 10.11 The current at the
sending end of the line

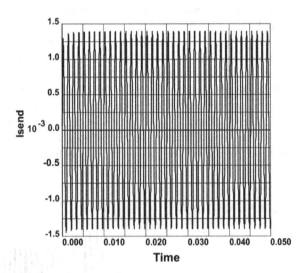

the voltage crosses the time axis for the first time. Thus, the delay is 0.001190 s.
The values found by simulations are close to those found analytically. Better
accuracy can be achieved by increasing the number of sections in which the line is
divided. The complete simulation uses 0.86 s of CPU time.

10.4 Bond Graph Model of a Beam

In this section we analyse a simple two-dimensional beam. This is a simple example of a more general deformable body and often is used to model elastic links in manipulators and other multibody systems. Interested readers are advised to consult any of the numerous references that deal with beams and flexible body dynamics. We point out here [8], which is more or less a standard reference on multibody system dynamics; and [9], because of its elegance and relevance to the component modelling approach used in this book.

We develop a bond graph model of two-dimensional beams using finite-element discretization. The beam will be described using Euler-Bernoulli theory—also known as thin-beam theory—in which shear deformation and rotary inertia are neglected. In spite of its simplicity, it is quite useful for solving various practical problems in flexible body dynamics. We will use it to solve problems of packaging systems vibration testing in Sect. 10.5. Another bond graph approach to Euler-Bernoulli beams is described in [3].

We begin by analysing an element of a beam that moves in a plane with respect to a body frame Oxz, which for this analysis is assumed to be fixed in a base frame (Fig. 10.12a). Figure 10.12b shows an element of the beam of length L with an attached element frame $O_e x_e z_e$. It will be assumed that the undeformed beam element is straight, having a uniform cross section and parallel to the body frame. Displacement of the element with respect to the body frame is described by two vertical displacements of its ends—w_1 and w_2—and two end slopes, θ_1 and θ_2.

A beam element can be represented by a BFE component with two ports, corresponding to the left and right ends of the element (Fig. 10.13).

The port flows are represented by vectors composed of linear and angular velocities of element ends

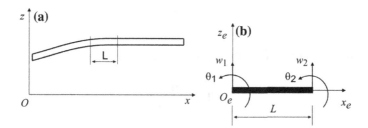

Fig. 10.12 A beam in planar motion. **a** An element of the beam. **b** The co-ordinates

$$\multimap \text{BFE} \multimap$$

Fig. 10.13 Component model of a beam element

$$\mathbf{f}_1 = \begin{pmatrix} \dot{w}_1 \\ \dot{\theta}_1 \end{pmatrix} \tag{10.6}$$

and

$$\mathbf{f}_2 = \begin{pmatrix} \dot{w}_2 \\ \dot{\theta}_2 \end{pmatrix} \tag{10.7}$$

The effort at the left port consists of force $F1$ in the transverse direction and moment $M1$ of action of the left part of the beam on the element

$$\mathbf{e}_1 = \begin{pmatrix} F_1 \\ M_1 \end{pmatrix} \tag{10.8}$$

Similarly, the effort at the right port is composed of a force $F2$ and a moment $M2$ of the action of the beam element on the beam part at the right of the element

$$\mathbf{e}_2 = \begin{pmatrix} F_2 \\ M_2 \end{pmatrix} \tag{10.9}$$

This way, at the right element end there is a reaction force and moment, and power flows through the element from the left to the right port (Fig. 10.13).

In a similar way, we introduce generalised displacements of the beam element ends

$$\mathbf{q}_1 = \begin{pmatrix} w_1 \\ \theta_1 \end{pmatrix} \tag{10.10}$$

and

$$\mathbf{q}_2 = \begin{pmatrix} w_2 \\ \theta_2 \end{pmatrix} \tag{10.11}$$

To simplify model development, we introduce end displacements

$$\mathbf{q} = \begin{pmatrix} \mathbf{q}_1 \\ \mathbf{q}_2 \end{pmatrix} \tag{10.12}$$

as generalised coordinates of the element. Comparing with (10.6) and (10.7), we have

$$\dot{\mathbf{q}} = \begin{pmatrix} \mathbf{f}_1 \\ \mathbf{f}_2 \end{pmatrix} \tag{10.13}$$

Transverse displacement of a point on the beam axis is described by [5]

$$w = \mathbf{S}\mathbf{q} \tag{10.14}$$

where $\mathbf{S}$ is the matrix of shape functions

$$\mathbf{S} = (S_1 \quad S_2 \quad S_3 \quad S_4) \tag{10.15}$$

With

$$\left.\begin{array}{l} S_1 = 1 - 3\xi^2 + 2\xi^3 \\ S_2 = L(\xi - 2\xi^2 + \xi^3) \\ S_3 = 3\xi - 2\xi^3 \\ S_4 = L(-\xi^2 + \xi^3) \end{array}\right\} \tag{10.16}$$

and $\xi = x_e/L$. Differentiating (10.14) with respect to time, we get the velocity of the beam transverse displacement

$$v = \mathbf{S}\begin{pmatrix} \mathbf{f}_1 \\ \mathbf{f}_2 \end{pmatrix} \tag{10.17}$$

Next, we develop the dynamic equation of element motion. A convenient way to do this is by using the Lagrange equations

$$\frac{d}{dt}\left(\frac{\partial T}{\partial \dot{\mathbf{q}}}\right) - \left(\frac{\partial T}{\partial \mathbf{q}}\right) = \mathbf{Q} \tag{10.18}$$

where T is the kinetic energy of the beam element and $\mathbf{Q}$ is a vector of generalised forces. The generalised forces include effects of element elastic forces, as well as those of forces at the element boundary, i.e. at the ports.

The kinetic energy of a beam element is given by

$$T = \frac{1}{2}\int_0^L \rho A v^2 dx \tag{10.19}$$

where ρ is the mass density of the beam, and A is the element's cross-sectional area. Using (10.17), we get

$$v^2 = (\mathbf{f}_1^T \quad \mathbf{f}_2^T)\mathbf{S}^T\mathbf{S}\begin{pmatrix} \mathbf{f}_1 \\ \mathbf{f}_2 \end{pmatrix} \tag{10.20}$$

By substituting in (10.19), the kinetic energy function can be written as

$$T = \frac{1}{2} \begin{pmatrix} \mathbf{f}_1^T & \mathbf{f}_2^T \end{pmatrix} \mathbf{M} \begin{pmatrix} \mathbf{f}_1 \\ \mathbf{f}_2 \end{pmatrix} \tag{10.21}$$

where $\mathbf{M}$ is the consistent mass matrix of the beam element. It is given by

$$\mathbf{M} = m \int_0^1 \mathbf{S}^T \mathbf{S} \, d\xi \tag{10.22}$$

where $m = \rho A L$ is the element mass. Using the shape functions of (10.16) and evaluating integrals in (10.22), the mass matrix can be written as [5]

$$\mathbf{M} = \frac{m}{420} \begin{pmatrix} 156 & 22L & 54 & -13L \\ 22L & 4L^2 & 13L & -3L^2 \\ 54 & 13L & 156 & -22L \\ -13L & -3L^2 & -22L & 4L^2 \end{pmatrix} \tag{10.23}$$

We also can represent the mass matrix (10.23) in the block form

$$\mathbf{M} = \begin{pmatrix} \mathbf{M}_{11} & \mathbf{M}_{12} \\ \mathbf{M}_{21} & \mathbf{M}_{22} \end{pmatrix} \tag{10.24}$$

where 2×2 blocks are found easily by inspection of (10.23)
   If we denote the momentum of the beam element by

$$\mathbf{p} = \partial T / \partial \dot{\mathbf{q}} \tag{10.25}$$

we get, from (10.13) and (10.21),

$$\mathbf{p} = \mathbf{M} \begin{pmatrix} \mathbf{f}_1 \\ \mathbf{f}_2 \end{pmatrix} \tag{10.26}$$

Thus,

$$\frac{d}{dt} \left(\frac{\partial T}{\partial \dot{\mathbf{q}}} \right) = \dot{\mathbf{p}} \tag{10.27}$$

In addition, $\partial T / \partial \mathbf{q} = 0$, because the kinetic energy depends only on the velocities.
   We now turn to the right side of (10.18). The part of the generalised forces due to elastic forces can be found from the strain energy V of the element. The other part is given by efforts at the element ends (ports). Thus, we have

$$Q = -\frac{\partial V}{\partial q} + \begin{pmatrix} e_1 \\ -e_2 \end{pmatrix} \tag{10.28}$$

Using simple beam theory, the beam element strain energy is given by

$$V = \frac{1}{2} \int_0^L EI \left(\frac{d^2 w}{dx^2} \right)^2 dx \tag{10.29}$$

where E is the Young modulus of elasticity and I is the second moment of the beam cross-section. Substituting from (10.14), we find

$$V = \frac{1}{2} q^T K q \tag{10.30}$$

where K is the stiffness matrix, given by

$$K = \frac{EI}{L^3} \int_0^1 \frac{d^2 S^T}{d\xi^2} \frac{d^2 S}{d\xi^2} d\xi \tag{10.31}$$

Using shape function of Eq. (10.16) and evaluating integrals, we get [5]

$$K = \frac{EI}{L^3} \begin{pmatrix} 12 & 6L & -12 & 6L \\ 6L & 4L^2 & -6L & 2L^2 \\ -12 & -6L & 12 & -6L \\ 6L & 2L^2 & -6L & 4L^2 \end{pmatrix} \tag{10.32}$$

This matrix also can be represented in 2×2 block form

$$K = \begin{pmatrix} K_{11} & K_{12} \\ K_{21} & K_{22} \end{pmatrix} \tag{10.33}$$

Now, from (10.30)

$$\frac{\partial V}{\partial q} = Kq \tag{10.34}$$

After substituting in Eq. (10.28), we have

$$Q = -K \begin{pmatrix} q_1 \\ q_2 \end{pmatrix} + \begin{pmatrix} e_1 \\ -e_2 \end{pmatrix} \tag{10.35}$$

Finally, substituting from (10.27) and (10.35) into (10.18), we get the governing equation of beam element dynamics

$$\frac{d\mathbf{p}}{dt} + \mathbf{K}\begin{pmatrix} \mathbf{q}_1 \\ \mathbf{q}_2 \end{pmatrix} = \begin{pmatrix} \mathbf{e}_1 \\ -\mathbf{e}_2 \end{pmatrix} \tag{10.36}$$

We further modify this equation by introducing the beam damping

$$\frac{d\mathbf{p}}{dt} + \mathbf{R}\begin{pmatrix} \mathbf{f}_1 \\ \mathbf{f}_2 \end{pmatrix} + \mathbf{K}\begin{pmatrix} \mathbf{q}_1 \\ \mathbf{q}_2 \end{pmatrix} = \begin{pmatrix} \mathbf{e}_1 \\ -\mathbf{e}_2 \end{pmatrix} \tag{10.37}$$

The damping matrix $\mathbf{R}$ is defined by the relation

$$\mathbf{R} = \alpha\mathbf{M} + \beta\mathbf{K} \tag{10.38}$$

where α and β are suitable constants. This type of damping is known as Rayleigh damping [5].

The equation of element motion, as given by (10.26) and (10.37), can be readily represented by a bond graph model of the BFE component shown in Fig. 10.14.

Note that the first row of (10.37) corresponds to the left port variables, and the second row to those of the right port. The variable at one effort junction at the left component port is the linear velocity; at the other, it is the angular velocity. A similar situation holds for the right effort junctions. The inertia of the beam element is represented by a four-port inertial element I, corresponding to (10.26) and the first term in (10.37). Elasticity of the element is represented by a four-port capacitive element C, corresponding to the third term of (10.37) and (10.13). Finally, a four-port resistive element R describes Rayleigh type damping in the beam, as given by the second terms of (10.37) and of (10.38).

To complete the model, it is necessary to define model parameters and element constitutive relations. Model parameters can be defined conveniently at the level of beam element document. We define the physical parameters first. These include the element length, mass, modulus of elasticity. We then define expressions for mass, stiffness, and damping matrix elements. The constitutive relations of I, C, and R elements are defined at corresponding ports as given by (10.26) and (10.37), respectively.

Fig. 10.14 Bond graph beam element BFE

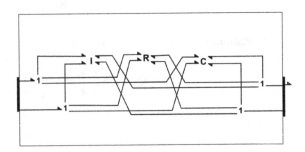

The model of a beam is created using component models that correspond to the elements into which the beam is divided. To simplify generating models that consist of a large number of elements, a technique similar to that used in Sect. 10.3 can be applied. Thus, we define a component consisting of five element component models; then another consisting of the five previously defined components; and so forth. This way, using hierarchical decomposition of the beam, it is relatively easy to generate a beam model consisting of a large number of finite element component models by copying and inserting. We apply this technique in the next section.

10.5 A Packaging System Analysis

10.5.1 Description of the Problem

A product-package system typically consists of an outer container, the cushion, the product, and a critical element. The critical element is the most fragile component of the product (e.g. electric boards). It is the part that is most easily damaged by a mechanical shock or by vibrations. The goal in distribution packaging (transport of packaging) is to provide a correct design for packaging so that its contents arrive safely at its destination. Vibration is associated with all transportation modes, although each mode has its own characteristic frequencies and amplitudes. The most troublesome frequencies are below 30 Hz because they are most prevalent in vehicles and it is difficult to isolate products from them [10].

In thesis [10], a simple packaging system is designed consisting of a container body with a cantilever beam as the critical element carrying a device at its end (Fig. 10.15). The package container plate was fixed on the table of a testing machine used for vibration testing of the package. One of the thesis goals was to predict the system's behaviour under vibration testing by simulation based on a model of the package system. The model of the package was developed using the bond graph methodology described in this book together with BondSim.

This section is partially based on this work. The goal was to show how the bond graph model of a typical vibration testing set-up can be developed in a systematic way within the BondSim programming environment. Using basic data from [10], a

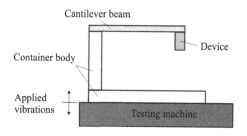

Fig. 10.15 The packaging system subjected to vibration testing

frequency response of the system was generated to simulate a typical laboratory vibration testing procedure.

10.5.2 Bond Graph Model Development

The package system model was developed using an approach similar to the floating frame approach of [8]. The coordinate frames used for the description of system motion are shown in Fig. 10.16. The critical part of the system is a beam clamped to a container body that is parallel to the container base plate.

The motion of the complete system is described with respect to the reference coordinate frame $O_r x_r z_r$ fixed to the ground (the testing machine body). Next, the container body frame $O_b x_b z_b$ is defined with axes parallel to that of the base frame, and that translates vertically with respect to the former.

The beam is modelled as an assemblage of the Eulcr-Bernoulli beam elements of Sect. 10.4. Figure 10.16 shows a typical beam element with the attached element frame. In the initial (undeformed) position, the element frames are parallel to the container body frame. There is also a device body frame $C_d x' z'$ attached to the body of the device at its mass centre and which moves with it.

The motion of the body frame is described by coordinate z_0, given as a function of time $z_0 = z_0(t)$. This describes the motion of the package during vibration testing. We will consider two common forms, the impulse and sinusoidal. Other forms could be used, as well.

Fig. 10.16 The co-ordinate systems of the package

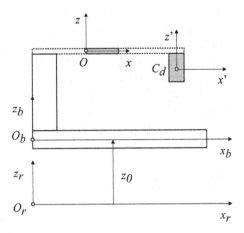

10.5.2.1 Model of Beam Elements

Beam elements will be represented using the component model of Sect. 3.4. In this problem the elements frames are not fixed with respect to the global reference frame, but translate jointly with the container body frame. Thus, in addition to the generalised displacements of the element with respect to the container frame, the matrix of generalised displacements also will contain the z_0 co-ordinate of the container body frame. The generalised displacement matrix of a typical element thus reads

$$\mathbf{q} = \begin{pmatrix} z_0 \\ \mathbf{q}_1 \\ \mathbf{q}_2 \end{pmatrix} \tag{10.39}$$

where $\mathbf{q}_1$ and $\mathbf{q}_2$ are defined by (10.10) and (10.11), respectively.

Transverse displacement of the beam axis with respect to the base frame is given by

$$z = z_0 + \mathbf{S} \begin{pmatrix} \mathbf{q}_1 \\ \mathbf{q}_2 \end{pmatrix} \tag{10.40}$$

where $\mathbf{S}$ is the shaping function matrix of (10.15) and (10.16). The corresponding velocities are

$$v = v_0 + \mathbf{S}\mathbf{f} \tag{10.41}$$

where $\mathbf{f}$ is vector of flows (see (10.13)) and

$$v_0 = \frac{dz_0}{dt} \tag{10.42}$$

To develop the equation for the beam element motion, Lagrange's equations were used, as in Sect. 10.4. These read

$$\frac{d}{dt}\left(\frac{\partial T}{\partial \dot{\mathbf{q}}}\right) - \left(\frac{\partial T}{\partial \mathbf{q}}\right) = \mathbf{Q} \tag{10.43}$$

where T is the kinetic energy of the element and $\mathbf{Q}$ is vector of generalised forces.

The generalised forces now read (see (10.33) and (10.35))

$$\mathbf{Q} = -\begin{pmatrix} 0 & \mathbf{0} & \mathbf{0} \\ 0 & \mathbf{K}_{11} & \mathbf{K}_{12} \\ 0 & \mathbf{K}_{21} & \mathbf{K}_{22} \end{pmatrix} \begin{pmatrix} z_0 \\ \mathbf{q}_1 \\ \mathbf{q}_2 \end{pmatrix} + \begin{pmatrix} F_0 \\ \mathbf{e}_1 \\ -\mathbf{e}_2 \end{pmatrix} \tag{10.44}$$

where $\mathbf{K}$ is the element stiffness matrix and F_0 is a force corresponding to the effect of container body translation on the element.

The kinetic energy of a beam element is given by

$$T = \frac{1}{2} \int_0^L \rho A v^2 dx \qquad (10.45)$$

where ρ is mass density of the beam element, and A is element's cross-sectional area. Using (10.41) we get

$$v^2 = (v_0 + \mathbf{f}^T \mathbf{S}^T)(v_0 + \mathbf{S}\mathbf{f}) = v_0^2 + 2v_0 \mathbf{S}\mathbf{f} + \mathbf{f}^T \mathbf{S}^T \mathbf{S}\mathbf{f} \qquad (10.46)$$

where superscript T denotes the transposition operation. By substituting (10.46) into (10.45), the kinetic energy expression can be written as

$$T = \frac{1}{2} m v_0^2 + v_0 \mathbf{N} \begin{pmatrix} \mathbf{f}_1 \\ \mathbf{f}_2 \end{pmatrix} + \frac{1}{2} \begin{pmatrix} \mathbf{f}_1^T & \mathbf{f}_2^T \end{pmatrix} \mathbf{M} \begin{pmatrix} \mathbf{f}_1 \\ \mathbf{f}_2 \end{pmatrix} \qquad (10.47)$$

where m is the element mass, $\mathbf{M}$ is element consistent mass matrix given by (10.22), and

$$\mathbf{N} = m \int_0^1 \mathbf{S} d\xi \qquad (10.48)$$

Using the shape functions of (10.15) and (10.16), we get

$$\mathbf{N} = \frac{m}{12} \begin{pmatrix} 6 & L & 6 & -L \end{pmatrix} \qquad (10.49)$$

This matrix can be written as 1×2 block matrix

$$\mathbf{N} = \begin{pmatrix} \mathbf{N}_1 & \mathbf{N}_2 \end{pmatrix} \qquad (10.50)$$

Next, we define momentum of the beam element by

$$\mathbf{p} = \partial T / \partial \dot{\mathbf{q}} \qquad (10.51)$$

From (10.47) we get

$$\mathbf{p} = \begin{pmatrix} m & \mathbf{N}_1 & \mathbf{N}_2 \\ \mathbf{N}_1^T & \mathbf{M}_{11} & \mathbf{M}_{12} \\ \mathbf{N}_2^T & \mathbf{M}_{21} & \mathbf{M}_{22} \end{pmatrix} \begin{pmatrix} v_0 \\ \mathbf{f}_1 \\ \mathbf{f}_2 \end{pmatrix} \qquad (10.52)$$

Fig. 10.17 Model of the
package beam element

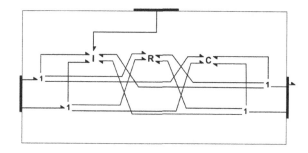

Thus, by substituting from (10.51), and (10.44) into (10.43), the equation of
beam element motion can be written as

$$\frac{d\mathbf{p}}{dt} + \begin{pmatrix} 0 & 0 & 0 \\ 0 & \mathbf{K}_{11} & \mathbf{K}_{12} \\ 0 & \mathbf{K}_{21} & \mathbf{K}_{22} \end{pmatrix} \begin{pmatrix} z_0 \\ \mathbf{q}_1 \\ \mathbf{q}_2 \end{pmatrix} = \begin{pmatrix} F_0 \\ \mathbf{e}_1 \\ -\mathbf{e}_2 \end{pmatrix} \qquad (10.53)$$

We modify this equation further by introducing beam damping

$$\frac{d\mathbf{p}}{dt} + \begin{pmatrix} 0 & 0 & 0 \\ 0 & \mathbf{R}_{11} & \mathbf{R}_{12} \\ 0 & \mathbf{R}_{21} & \mathbf{R}_{22} \end{pmatrix} \begin{pmatrix} v_0 \\ \mathbf{f}_1 \\ \mathbf{f}_2 \end{pmatrix} + \begin{pmatrix} 0 & 0 & 0 \\ 0 & \mathbf{K}_{11} & \mathbf{K}_{12} \\ 0 & \mathbf{K}_{21} & \mathbf{K}_{22} \end{pmatrix} \begin{pmatrix} z_0 \\ \mathbf{q}_1 \\ \mathbf{q}_2 \end{pmatrix} = \begin{pmatrix} F_0 \\ \mathbf{e}_1 \\ -\mathbf{e}_2 \end{pmatrix}$$

$$(10.54)$$

where $\mathbf{R}$ is defined by (10.38)

Comparing (10.54) and (10.52) to (10.37) and (10.26), we conclude that the
same component model as in Fig. 10.14 can be used with the added port corresponding to the body frame translation effect (Fig. 10.17). However, the constitutive relation at the inertial elements ports should be changed to the form given by
(10.52). This additional port shows that there is transfer of power, because of the
motion of the container body frame.

10.5.2.2   Motion of the Device

To model motion of the device, the rigid body model of Sect. 9.2 is used. There is
some difference, however, the motion of the device is referred to the container body
frame, not to the base (inertial) frame. Because the container body translates with
respect to the base, there is an additional (inertial) force acting at the device mass
centre, as was found above for the beam elements. Thus, in addition to the port
corresponding to joining the end of the cantilever beam, there is a port that corresponds to the effects of translation of the container body frame (Fig. 10.18). The
effort and flow of the container port are inertial force—owing to the container frame
acceleration—and velocity of the container frame translation, respectively. The

Fig. 10.18 Representation of device body in a translating frame

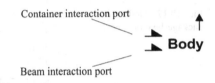

Container interaction port

Body

Beam interaction port

Fig. 10.19 Component model of the device body

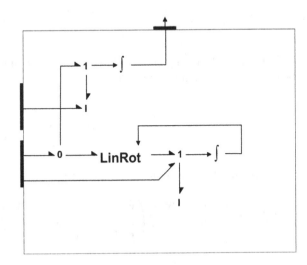

efforts and flows at the beam ports are of the form given by (10.9) and (10.7), respectively. It was assumed that the beam axis undergoes z-axis transverse displacement only, ignoring the beam x-axis strains. There is also an output port to supply information on the relative transverse displacement of the device mass centre with respect to the container body frame.

The model of the device body created using the plane body model of Fig. 9.3 is shown in Fig. 10.19. The effort junction at the top-left of the document window corresponds to the transverse velocity component of the device mass centre with respect to the base frame.

This velocity can be expressed as

$$v_d = v_0 + v_d^b \qquad (10.55)$$

where the term with the superscript b denotes the velocity component with respect to the container body frame, and v_0 is its translation velocity. Thus, the kinetic energy of the device translation is given by

$$T_{Cd} = m_d(v_0 + v_d^b)^2/2 \qquad (10.56)$$

Analogous to the approach used in the beam element model, we define the flow vector corresponding to the mass centre motion as

$$\mathbf{f}_d = \begin{pmatrix} v_0 \\ v_d^b \end{pmatrix} \tag{10.57}$$

The translation momentum of the device, defined by

$$\mathbf{p}_d = \partial T_{Cd}/\partial \mathbf{f}_d \tag{10.58}$$

now reads

$$\mathbf{p}_d = \begin{pmatrix} m_d & m_d \\ m_d & m_d \end{pmatrix} \mathbf{f}_d \tag{10.59}$$

This way, the device translation dynamics is given by

$$\dot{\mathbf{p}}_d = \mathbf{e}_d \tag{10.60}$$

where effort $\mathbf{e}_d$ consists of forces at the component ports

$$\mathbf{e}_d = \begin{pmatrix} F_0 \\ F_d \end{pmatrix} \tag{10.61}$$

Following (10.59) and (10.60), the device translation dynamic is represented in Fig. 10.19 with a two-port inertial element, where one port corresponds to interaction with the moving container body and the other to that with the beam end to which it is connected. Because the device is clamped to the cantilever, its angular velocity is equal to the angular velocity of beam cross section.

10.5.2.3 Model of the System

We give a short description of the model of package system vibration testing. The complete model can be accessed in the BondSim program library under the project name *Package Vibration Testing*. The system level model is given in Fig. 10.20.

It consists of two main components: Package and Testing machine. There is also an output component used to display time and frequency plots of the package container and device body positions. The Testing machine is represented simply by a flow source generating vertical motion of the machine table. We will use two types of test inputs. Similarly, as in Sect. 6.2.3, to generate a frequency response plot of device motion, we apply to the package a velocity pulse of short duration. The form of the pulse is shown in Fig. 10.21a. The corresponding package container position is shown in Fig. 10.21b.

Fig. 10.20 System model of
package vibration testing

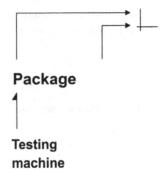

Package

**Testing
machine**

Fig. 10.21 The form of
pulses generated. **a** Velocity.
b Position

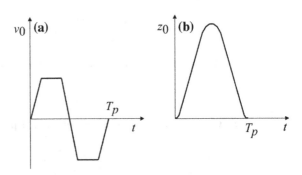

Another type input is a sinusoidal function of frequency f

$$v_0 = v_{p0} \sin(2\pi f t) \tag{10.62}$$

Amplitude v_{p0} is chosen such that the peak acceleration is constant, i.e.

$$v_{p0} = \frac{a_{\max}}{2\pi f} \tag{10.63}$$

The package consists of a container wall and base plate represented by the `Wall`
component (Fig. 10.22), fixed to the vibration machine table and to which is
connected a `Cantilever` component that carries the component `Device`.

Interconnection of the cantilever beam to the container is shown in Fig. 10.23.
The effort junction is the transverse displacement node of the container. The
junction port, together with the `Clamp` port, is connected to the `Wall` output port.
The port—a compounded one—has efforts and flows at the container wall represented by

$$\mathbf{e}_{con} = \begin{pmatrix} F_0 \\ \mathbf{e}_1 \end{pmatrix} \tag{10.64}$$

Fig. 10.22 The basic structure of the package system

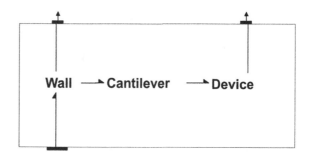

Fig. 10.23 Interactions at the container wall

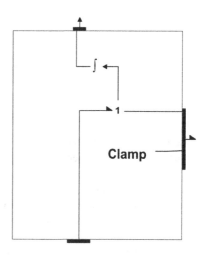

and

$$\mathbf{f}_{con} = \begin{pmatrix} v_0 \\ \mathbf{f}_1 \end{pmatrix} \qquad (10.65)$$

Here, v_0 and F_0 are, respectively, its velocity and the total force that the testing machine supplies to the container. Effort component $\mathbf{e}_1$ consists of the transverse force and moment at the clamping of the cantilever beam to the container. The component in the beam axis direction was neglected. Similarly, $\mathbf{f}_1$ consists of the linear velocity and the angular velocity of the beam end with respect to the container.

The end of the cantilever is clamped to the container wall, which is represented by the Clamp component. This component consists of two zero-flow sources that force the linear velocity and the angular velocity of the beam end to be zero.

The cantilever beam is divided into four subsections (Fig. 10.24), each of which is represented by a SectB component consisting of five BFE components (Fig. 10.25). The bond graph representation of this finite-element beam

Fig. 10.24 Cantilever beam
represented by four beam
sections

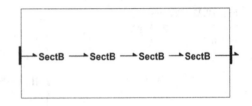

Fig. 10.25 Beam section as a
super element of five BFEs

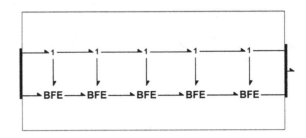

discretization is depicted in Fig. 10.17. Overall, the cantilever is divided into 20
finite element components.

The effort junctions appearing above each BFE component are the
finite-element's translation velocity nodes. The effort variables at the ports connected to the BFE components are forces corresponding to the distributed inertia of
the element translation. All such forces, including that of the device, give the total
force generated by the Test machine. This way, the mechanical effect of the
machine on the beam elements is clearly visible.

Finally, returning to Fig. 10.22, there is a component Device that models the
device at the cantilever end. This consists of the Body component of Figs. 10.18
and 10.19. It was not possible to use this component directly, due to the way in
which the ports were compounded.

10.5.3   Evaluation of Vibration Test Characteristics

Now we can simulate the vibration characteristics of the package. The parameters
used for simulation are given in Table 10.1. The damping parameters α and β of the
beam model (see (10.38)) were selected such that they correspond approximately to
5 % of the critical damping ratio in the range of natural frequencies of interest 20–
50 Hz, e.g. $\alpha = 8$ and $\beta = 4 \times 10^{-4}$.

We first generate a frequency response diagram for the device body transverse
motion. Hence, the package is subjected to a velocity pulse generated by the Test
machine component according to Fig. 10.21. The strength of the pulse is 0.5 m/s
and its duration is $Tp = 0.001$ s. The system settle-time is about 0.5 s. As noted in

Table 10.1 Parameters of the package

Property		Value
Cantilever	Length	0.100 m
	Width	0.0258 m
	Height	0.0055 m
Material		Acrylic (PMMA)
Modulus of elasticity		3.1 GPa
Density		1200 kg/m^3
Device block	Mass	0.0362 kg
	Moment of inertia	5.168×10^{-6} kg m^2
	x	−0.0128 m
	z	0.0126 m

Note The device body data include the part of the beam clamped to it. x and y are co-ordinates of the mid-point of the cantilever beam end section with respect to the device centroidal body frame

Fig. 10.26 The device position response to the velocity pulse input

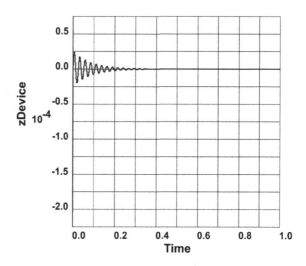

Sect. 6.2.3 (*Response to Harmonic Excitation*), we append more zeros to the response to create a better resolution of the frequency response. Thus, a simulation interval of 1 s is taken along with a fairly small output interval of 0.0001 s. The rather tight error tolerances of 10^{-8} were selected. The transient of the device body position (with respect to the container) is shown in Fig. 10.26. At the beginning, the device experiences a relatively large initial jump −0.0002042026 in the negative direction, and then settles down to zero after about 0.3 s.

The data gathered during the simulation are used to generate a frequency response plot by applying the Fast Fourier Transform (FFT). As explained in Sect. 6.2.3, this is done by right-clicking the time plot and selecting *Continuous*

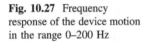

Fig. 10.27 Frequency response of the device motion in the range 0–200 Hz

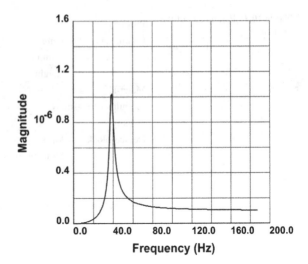

Fourier Transform from the drop-down menu. We enlarge the part of the plot in the range of 0–100 Hz by right-clicking on the frequency plot, selecting the *Expand* command, and drag the right edge of the frequency window to about 100 Hz. The resulting diagram is shown in Fig. 10.27.

The diagram resembles the familiar response diagrams of single-degree-of-freedom system vibrations induced by the motion of the base [11]. The maximum amplitude occurs in the vicinity of 39.0 Hz.

In a similar way, we can find the *Continuous Fourier Transform* of the input (Fig. 10.28). The amplitude of the container displacement is practically constant over the frequency range of interest at about 1.0×10^{-7} m and, hence, it approximates the ideal impulse fairly well. Comparing this with Fig. 10.27, it can be seen that the maximum displacement transmissibility ratio of the device to container displacement is a little above 10.

We next apply a sinusoidal vibration to the container according to (10.62) and (10.63) with a frequency of 39.0 Hz and a peak acceleration of $a_{max} = 0.5$ g. This requires changing the source effort relation in the `Testing machine` (Fig. 10.20) component, then re-building the model.

The response curve is shown in Fig. 10.29. It was generated with same simulation parameters as in the previous run. The amplitudes at beginning steadily grow until they reach a steady value of 0.0008375 m. The amplitude of the container vibration is 8.170×10^{-5} m; thus, we have the transmissibility ratio of 10.25, as found previously.

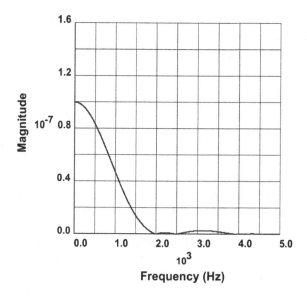

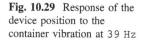

Fig. 10.28 Amplitude-frequency diagram of impulsive displacement of the container

Fig. 10.29 Response of the
device position to the
container vibration at 3 9 Hz

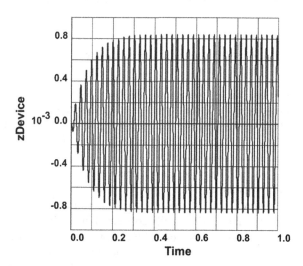

10.6   Coriolis Mass Flowmeters

10.6.1   Problem Statement

The subject of this subsection is modelling and simulation of Coriolis Mass
Flowmeters. *Coriolis mass flowmeters* (CMF) are widely used for direct flow mass
metering in the petrol, food, pharmaceutical, and chemical industries. The accuracy

of typical commercial CMFs can be as low as ±0.10 % of the reading, practically over the complete measurement range [12]. They can be used for gases, water, petrol products, milk, honey, and even for unstable fluids. CMFs are highly insensitive to temperature and pressure variations. They are really very accurate and reliable instruments.

The modern CMFs, however, are pretty complex devices consisting of high quality transducers connected to Digital Signal Processing (DSP) units, which enables such outstanding performances. In the first edition of this book CMF were analyzed mostly from the transducer point of view not going into the details of control and operation. Now, we will try to develop a complete model of the device. The work in this section was conducted in close cooperation with Jörg Gebhardt and Frank Kassubek of ABB Corporate research. The geometrical device model, as well as control, that will be developed, however, do not correspond to any concrete CMF (produced by ABB or other manufacturers), but is based on a hypothetical device. The goal was to develop a physical model of the device based on Bond graphs in *BondSim* environment and show by simulation a typical behavior of such devices under appropriate control.

It should be noted that the study of this section requires some knowledge on theory of elasticity, vibrations and communications technology. The readers not interested in CMF technology can skip this subsection.

10.6.2   Principle of Operations

Coriolis mass flowmeters (CMF) are very accurate and reliable instruments. The range of fluid densities that can be measured are as low as 1 kg/m^3 or less to over 3000 kg/m^3; they can be used for gases, water, petrol products, milk, honey and others. CMFs are highly insensitive to temperature and pressure variations. The principle of their operation is rather simple. We will explain it with an example of a straight parallel tube CMF. There are other configurations as well, but the basic principles are the same.

Figure 10.30 shows sketch of a Coriolis mass flowmeter consisting of two parallel tubes hydraulically connected in parallel. The incoming flow is divided into two separate and practically identical tubes; the flows again combine at their exits.

The tubes' ends are clamped to the meter body. They are excited by a drive—located at their middle—and vibrates in their first vibration mode. On the left and the right of the driver are two sensors that detect velocity of displacements between the tubes. The actuator and sensors are usually simple voice coil devices. These are used for the control of the meter and the measurement of the phase displacement between the oscillating tubes, which is used to determine the mass flow rate of the fluid through the meter. As a rule, CMFs use two tubes, the better to minimize the effects of vibration associated with the environment in which the meter is installed for operation.

Fig. 10.30 Scheme of
Coriolis mass flowmeter with
two straight parallel tubes

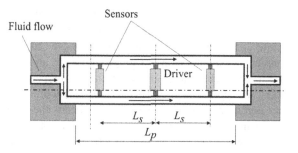

   To explain CMF operation, Fig. 10.31 shows the exaggerated displacement of
one of the tubes moving up with fluid flowing through it. Two small elements of the
tube are shown at symmetrical positions to the left and the right of the driver.

   During the vibration, the tube elements rotate with angular velocity ω as shown
and, because of the relative velocity V_f of the fluid flowing through the tube, there is a
Coriolis acceleration given by $a_{cor} = 2V_f\omega$. As a result, there is an apparent (inertial)
Coriolis force on the tube element of the opposite sense. The Coriolis force acting on
a tube element is given by $F_{cor} = dm \cdot a_{cor}$, where dm is the mass of the fluid
contained in the tube element. If dL is the length of the tube element, the fluid mass
contained in the element is equal $dm = \rho A_s dL$, where ρ is fluid density and A_s is the
area of the wetted cross-section of the tube. Thus, the Coriolis force on the element is
$F_{cor} = 2\rho A_s V_f \omega dL$. Hence, because the mass flow rate through the tube is $Q_m = \rho A_s V_f$,
the Coriolis force is $F_{cor} = 2 Q_m \omega dL$. That is, the Coriolis force is proportional to the
mass flow rate. Because it appears as a load distributed along the tube, the tube
elastically deforms as shown in Fig. 10.31. Note that because the Coriolis forces on
the left and right parts of the tube are of opposite directions, the left and the right tube
parts deform differently. Thus, the tube sections in the planes of the sensors don't
cross the zero positions at the same instants of time; rather, there is a time delay. As
the analysis of [13] shows, the time delay is proportional to the mass flow rate.
Hence, by evaluating the time delay, accurate information on the mass flow rate can
be generated. This is not a technically easy problem: The time delays are of the order
of microseconds, and their accurate measurement is critical. CMF manufacturers are
careful not to expose many of the details of how this is really done. However, we will
propose a possible alternative solution to this problem.

   Mathematical modelling of Coriolis mass flowmeters has attracted the attention
of both academic and industry researchers. The common objective is to explain
more precisely the operation of the meters, and to improve their design, testing, and

Fig. 10.31 Principle of the
flowmeter operation

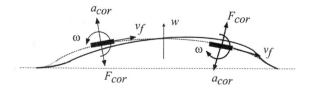

application. The main focus has been the evaluation of CMF sensitivity factors, vibration control of the tube with the specified amplitude and operating frequency close to fundamental natural frequency of the tubes, the effects of meter geometry and environment conditions on performance, and others. Different approaches have been used. Thus, in [14], rather simple lumped-parameter models were used. In [13], the tubes are modelled as Euler–Bernoulli beams and problems were solved by perturbation theory. Timoshenko beam theory was applied in [15], and the problem was analyzed using a finite-element program. CMF analysis based on curved beam theory was used in [16]. We will show that problems can be readily solved using the bond graph component method. A CMF transducer that uses curved tubes will be analyzed using Timoshenko beam theory. The complete model will be developed including its digital controls. The behavior of such CMF system will be investigated by simulation.

10.6.3   Dynamics of Curved CMF Tubes

Figure 10.32 shows a sketch of a curved tube through which fluid flows. The Coriolis mass flowmeters typically consist of two such tubes. They are fixed at their ends and under action of an actuator vibrates orthogonal to their planes. The tubes vibrate in opposite directions. Their relative transverse displacements are measured by two sensors positioned symmetrically with respect to their mid sections.

The CMF tubes can have quite different configuration depending on the manufacturers of the CMF. We will analyze a meter whose tubes have the form depicted in the figure and which consists of circular and straight parts. Such tube parts allow analytic treatment of the eigenmodes as shown in Irie et al. [17] and Howson and Jemah [18].

10.6.3.1   Out-of-Plane Vibrations of Circular Tubes

The equations of out-of-plane free vibrations of a circular tube based on Timoshenko theory is presented below following Irie et al. [17] and Howson and

Fig. 10.32 Vibration of curved Coriolis tube

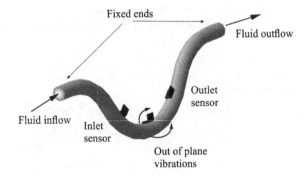

Fig. 10.33 Forces and displacements of a circular element

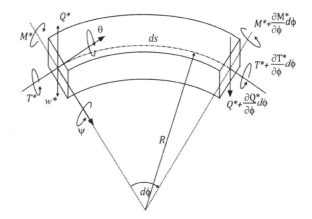

Jemah [18]. The tube has a constant radius of the curvature and the uniform cross-section and material properties and is filled with a still fluid.

An infinitesimal element of the tube with subtended angle $d\varphi$ about the center of a circle of radius R is shown in Fig. 10.33. The local element coordinate frame is shown in the center of the left cross-sections and consist of the radial ψ (axis 1), transversal w (axis 2), and tangential θ (axis 3) directions. The corresponding variables are rotation angle due to pure bending ψ, the transverse displacement due to shear w, and the torsion angle θ. Along these axes act the bending moment M^*, shearing force Q^*, and torsional moment T^*. The equations of dynamic equilibrium of shear, bending and torsion are given by

$$\frac{1}{R}\frac{\partial Q^*}{\partial \phi} = \rho A \frac{\partial^2 w^*}{\partial t^2}$$

$$\frac{1}{R}\frac{\partial M^*}{\partial \phi} + \frac{1}{R}T^* - Q^* = \rho I_1 \frac{\partial^2 \psi}{\partial t^2} \qquad (10.66)$$

$$\frac{1}{R}\frac{\partial T^*}{\partial \phi} - \frac{1}{R}M^* = \rho I_p \frac{\partial^2 \theta}{\partial t^2}$$

Quantity ρA represents the mass of tube including the contained fluid per unit length. Similarly ρI_1 is moment of inertia of the tube including fluid about rotation axis 1 per unit length, and ρI_p is moment of inertia of tube about tangential axis per unit length. Equivalent mass per tube length has the form $\rho A = \rho_t A_t + \rho_f A_f$, where subscripts t denotes quantities referred to the tube cross section, and f to inside cross section containing the fluid. The similar relation holds for flexure, i.e. $\rho I_1 = \rho_t I_{1t} + \rho_f I_{1f}$. In torsion, however the effect of fluid is neglected and hence $\rho I_p = \rho_t I_{pt}$. For the circular cross-sections we have $I_{pt} = 2I_{1t}$.

The stress-strain and strain-displacement equations are given by

$$M^* = \frac{EI_{1t}}{R}\left(\frac{\partial \psi}{\partial \phi} + \theta\right)$$

$$T^* = \frac{GJ_t}{R}\left(\frac{\partial \theta}{\partial \phi} - \psi\right)$$

$$(10.67)$$

$$Q^* = k'A_tG\left(\frac{1}{R}\frac{\partial w^*}{\partial \phi} + \psi\right)$$

where EI_{1t}, GJ_t and k' are the flexural rigidity, torsional rigidity and shape factor of the tube, respectively; J_t is St. Venant's torsional constant of the tube.

We consider free vibration of tubes in which all variables can be expressed as

$$\chi(\phi, t) = \bar{X}(\phi)e^{j\omega t} \qquad (10.68)$$

where

$$\chi \equiv \psi, w, \theta, M^*, Q^*, T^*$$
$$\bar{X} \equiv \bar{\Psi}, \bar{W}, \bar{\Theta}, \bar{M}, \bar{Q}, \bar{T}$$

Substituting into (10.66) and (10.67) we obtain

$$\frac{1}{R}\frac{d\bar{Q}}{d\phi} + \rho A\omega^2 \bar{W} = 0$$

$$\frac{1}{R}\frac{d\bar{M}}{d\phi} + \frac{1}{R}\bar{T} - \bar{Q} + \rho I_1\omega^2 \bar{\Psi} = 0 \qquad (10.69)$$

$$\frac{1}{R}\frac{d\bar{T}}{d\phi} - \frac{1}{R}\bar{M} + \rho I_p\omega^2 \bar{\Theta} = 0$$

and

$$\bar{M} = \frac{EI_{1t}}{R}\left(\frac{d\bar{\Psi}}{d\phi} + \bar{\Theta}\right)$$

$$\bar{T} = \frac{GJ_t}{R}\left(\frac{d\bar{\Theta}}{d\phi} - \bar{\Psi}\right)$$

$$(10.70)$$

$$\bar{Q} = k'A_tG\left(\frac{1}{R}\frac{d\bar{W}}{d\phi} + \bar{\Psi}\right)$$

Following Irie et al. [17] the following dimensionless variables and parameters are introduced:

$$\Psi = \bar{\Psi}, W = \frac{\bar{W}}{R}, \Theta = \bar{\Theta}, (M,T) = \frac{R}{EI_{1t}}(\bar{M},\bar{T}), Q = \frac{R^2}{EI_{1t}}\bar{Q},$$

$$s_1^2 = \frac{\rho A R^2}{\rho I_1}, s_{1t}^2 = \frac{A_t R^2}{I_{1t}}, \eta = \frac{\rho I_p}{\rho I_1}, \mu = \frac{GJ_t}{EI_{1t}}, k_q = k'\frac{G}{E}s_{1t}^2, \tag{10.71}$$

$$\gamma^2 = \frac{\rho I_1 R^2 \omega^2}{EI_{1t}}, \lambda^2 = \frac{\rho A R^4 \omega^2}{EI_{1t}} = s_1^2 \gamma^2$$

Equations (10.69) and (10.67) now read

$$\frac{dQ}{d\phi} + s_1^2 \gamma^2 W = 0$$

$$\frac{dM}{d\phi} + T - Q + \gamma^2 \Psi = 0 \tag{10.72}$$

$$\frac{dT}{d\phi} - M + \eta \gamma^2 \Theta = 0$$

and

$$M = \frac{d\Psi}{\partial \phi} + \Theta$$

$$T = \mu\left(\frac{d\Theta}{d\phi} - \Psi\right) \tag{10.73}$$

$$Q = k_q\left(\frac{dW}{d\phi} + \Psi\right)$$

Substituting (10.73) into (10.72) one obtains

$$k_q\frac{d\Psi}{d\phi} + k_q\frac{d^2 W}{d\phi^2} + s_1^2 \gamma^2 W = 0$$

$$\frac{d^2\Psi}{d\phi^2} - \left(\mu + k_q - \gamma^2\right)\Psi - k_q\frac{dW}{d\phi} + (1+\mu)\frac{d\Theta}{d\phi} = 0 \tag{10.74}$$

$$-(1+\mu)\frac{d\Psi}{d\phi} + \mu\frac{d^2\Theta}{d\phi^2} + \left(\eta\gamma^2 - 1\right)\Theta = 0$$

These are linear differential equations of second order. Thus, we may assume their solutions in the form

$$\Psi = Ce^{p\phi}$$

$$W = qCe^{p\phi} \tag{10.75}$$

$$\Theta = rCe^{p\phi}$$

Substituting into (10.74) we obtain the following matrix equation

$$
\begin{pmatrix}
k_q p & k_q p^2 + s_1^2 \gamma^2 & 0 \\
p^2 - (\mu + k_q - \gamma^2) & -k_q p & (1+\mu)p \\
-(1+\mu)p & 0 & \mu p^2 + (\eta \gamma^2 - 1)
\end{pmatrix}
\begin{pmatrix} 1 \\ q \\ r \end{pmatrix} = 0 \qquad (10.76)
$$

This is a homogenous linear equation for q and r. It has solutions if the determinant of the system matrix is equal to zero. Expanding this determinant we obtain a bicubic equation

$$
a_6 p^6 + a_4 p^4 + a_2 p^2 + a_0 = 0 \qquad (10.77)
$$

where the coefficients are given by

$$
\begin{aligned}
a_6 &= -\mu k_q \\
a_4 &= -\left(s_1^2 \mu + k_q(\mu+\eta)\right)\gamma^2 - 2\mu k_q \\
a_2 &= -\left(s_1^2(\mu+\eta) + k_q \eta\right)\gamma^4 - \left(2\mu s_1^2 - k_q\left(1+\mu(\eta+s_1^2)\right)\right)\gamma^2 - \mu k_q \\
a_0 &= -s_1^2 \gamma^2 \left[\eta \gamma^4 - \left(1 + (\mu + k_q)\eta\right)\gamma^2 + (\mu + k_q)\right]
\end{aligned} \qquad (10.78)
$$

This is the characteristic equation for the problem, whose solutions are the eigenvalues of out-of-plane tube vibrations. These eigenvalues can be found using e.g. Cardano's formulas. The corresponding eigenfunctions constants q and r can be easily found from (10.76):

$$
q = -\frac{k_q p}{k_q p^2 + s_1^2 \gamma^2} \qquad (10.79)
$$

$$
r = \frac{(1+\mu)p}{\mu p^2 + \eta \gamma^2 - 1} \qquad (10.80)
$$

Because there are six eigenvalues we may express the beam bending, torsion and transverse displacements by

$$
\begin{aligned}
\Psi &= \sum_{i=1}^{6} C_i e^{p_i \phi} \\
W &= \sum_{i=1}^{6} q_i C_i e^{p_i \phi} \\
\Theta &= \sum_{i=1}^{6} r_i C_i e^{p_i \phi}
\end{aligned} \qquad (10.81)
$$

The constants C_i can be found by applying appropriate boundary conditions to the circular element.

The equations for the straight tube parts can be found from the above equations by allowing the radius of curvature to tend to infinity. Their solutions have a similar form as above.

10.6.3.2 Vibration of the Curved Tube

An example for geometry of a CMF is shown in Fig. 10.34. It may be decomposed into two short straight tube parts at the start and end of the tube, two small circular arcs and the main one in the middle. The tube is clamped at the nodes 0 and 8. Nodes 1 and 7 are the connection nodes between the straight and curved tube parts at the beginning and end, and nodes 2 and 6 are the connection nodes between the small and main circular arcs. The main arc is further divided onto sub-arcs by nodes 3, 4 and 5. At these nodes the sensors' and the actuator's coils or magnets and their bodies (supports) are connected, which we represent by concentrated masses. Their effects will be taken into account later. We consider now only the tubes filled with still fluid.

The figure also shows the coordinate systems used. The origin of the global coordinate frame is placed at the midpoint between left and right tube ends. Axis z is directed to the right, y axis upwards and x axis into the undeformed tube centerline plane. The local tube coordinates $\psi_i w_i \theta_i$ are defined for each of the circular parts in the same way as we did earlier (Fig. 10.33). For the straight parts the coordinates are simple, just the coordinate s along the parts centerlines.

Irie et al. in [17] analyzed structures consisting of only a single arc. It can be shown that their approach based on transfer matrices can be extended to multi arch structures such as in Fig. 10.34. The transfer matrices approach of Irie et al. [17] is interesting because it is compatible with the Bond graph component approach. There is also another possible better known approach based on stiffness matrices, proposed by Howden and Jemah [18]. It is compatible with the classical finite element displacement based approach.

For our calculations, we start at the left end by using equations of the form (10.81) and applying the boundary conditions for that part. The left straight part is

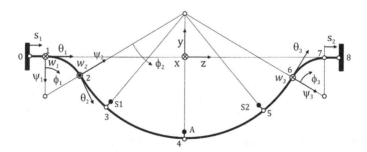

Fig. 10.34 Model of the tube

clamped at the left node 0, and on the right at node 1 is connected to the small circular tube. At the connection points they have the same displacements and slopes. We continue along the small arc and again apply the boundary conditions at its left and right nodes. We continue along the main arc, and so on until the last straight part at the right end of the tube. These lead to a frequency equation from which we can find the natural frequencies of tube vibrations. For each of these frequencies we can find the values of the corresponding constants in (10.81) for the tube parts, and thus the forms of the vibration shapes. The complete procedure was implemented in a program written in C++, which generates the natural frequencies of the tube and the mode shapes coefficients. We give the final results only. In Fig. 10.35 are shown the first three natural frequencies and mode shapes for a specific Coriolis tube filled with water. It shows all three vibration displacements—bending, torsion and transverse. For lower modes the bending dominates. For higher modes (not shown) the torsion is more influential. The tube sensors measure the transverse displacements. In the fundamental mode (1) the transverse displacements are symmetric with respect to the mid of the tube. In the second mode it is antisymmetric. Note also that the tube bending always has the opposite symmetry with respect to transverse displacement and torsion.

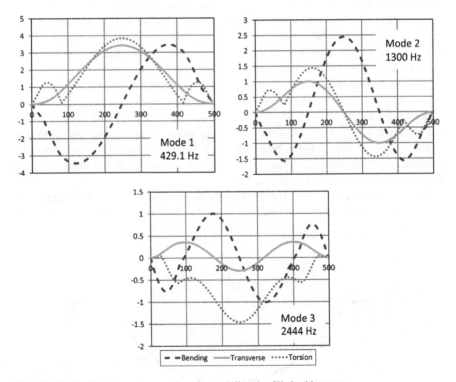

Fig. 10.35 The first three eigenmodes of a coriolis tube filled with water

10.6.3.3 Dynamics of the Perturbed Motion of CMF Tube

Now we will consider the dynamical model of CMF tube including masses of sensor and actuator parts (the coils, magnets and holders), but also the effects of fluid flow—the Coriolis and centrifugal inertial forces. We consider the main circular parts, which play the fundamental role. The dynamics of small arcs and short straight parts are much simpler and their role is minor. The equations of motion have the following form

$$\frac{\partial M^*}{\partial s^*} + \frac{1}{R}T^* - Q^* = \left(\rho I_1 + J_{s1}\delta(s^* - s_{s1}^*) + J_{a1}\delta(s^* - s_a^*) + J_{s1}\delta(s^* - s_{s2}^*)\right)\frac{\partial^2 \psi}{\partial t^{*2}}$$

$$\frac{\partial Q^*}{\partial s^*} + F_a^*(t^*)\delta(s^* - s_a^*) = \left(\rho A + m_s\delta(s^* - s_{s1}^*) + m_a\delta(s^* - s_a^*) + m_s\delta(s^* - s_{s2}^*)\right)\frac{\partial^2 w^*}{\partial t^{*2}}$$

$$+ 2\rho_f A_f V_f^* \frac{\partial^2 w^*}{\partial s^* \partial t^*} + \rho_f A_f V_f^{*2}\frac{\partial^2 w^*}{\partial s^{*2}}$$

$$\frac{\partial T^*}{\partial s^*} - \frac{1}{R}M^* = \left(\rho I_p + J_{s3}\delta(s^* - s_{s1}^*) + J_{a3}\delta(s^* - s_a^*) + J_{s3}\delta(s^* - s_{s2}^*)\right)\frac{\partial^2 \theta}{\partial t^{*2}}$$

$$(10.82)$$

where $ds^* = Rd\phi$ is differential of the arc length along the tube centerline. Comparing with (10.66) there are two main differences. One is the effect of fluid flow with mean velocity V_f^* on the tube motion. It is described by the second and third terms on the right hand side of the second (10.82). They represent the Coriolis inertial force and centrifugal force, respectively. The other term describes the interaction of the measuring tube with the attached sensors' and actuator's bodies containing their coils or magnets. The sensor and actuator parts attached to the tube for out-of-plane motion are described respectively by mass moments of inertia J_{s1} and J_{a1} with respect to the radial axes, J_{s3} and J_{a3} with respect to the tangential axes, and their total mass m_s and m_a. Because these are the concentrated parameters they are taken into account in distributed model (10.82) by use of *Dirac* $\delta(s^*)$ functions. Their positions along the arc are denoted by the arc lengths $s_{s1}^*, s_a^*,$ and s_{s2}^*, respectively. The actuator also delivers the forces to the measuring tubes. It is represented by the concentrated force $F_a^*\delta(s^* - s_a^*)$ acting at the position of actuator coils or magnets in the transverse direction as can be seen in (10.82).

Substituting stress-strain and strain-displacement Eqs. (10.67), normalizing by (10.71) we obtain the dynamic equations of CMF tube written compactly as

$$L(\mathbf{w}(\phi, t)) + F_a(t)\delta(\phi - \phi_a)\mathbf{e}_2 = M(\mathbf{w}(\phi, t)) + M_a(\mathbf{w}(\phi, t))$$

$$+ \left(2\lambda_{1f}^2 V_f \frac{\partial^2 w}{\partial \phi \partial t} + \lambda_{1f}^2 V_f^2 \frac{\partial^2 w}{\partial \phi^2}\right)\mathbf{e}_2 \qquad (10.83)$$

where we introduced the state vector

$$\mathbf{w}(\phi,t) = (\psi \quad w \quad \theta)^T \tag{10.84}$$

Following Han et al. [19] we also introduced linear matrix operators defined as

$$L(\mathbf{w}(\phi,t)) = \begin{pmatrix} \frac{\partial^2}{\partial\phi^2} - (\mu+k_q) & -k_q\frac{\partial}{\partial\phi} & (1+\mu)\frac{\partial}{\partial\phi} \\ k_q\frac{\partial}{\partial\phi} & k_q\frac{\partial^2}{\partial\phi^2} & 0 \\ -(1+\mu)\frac{\partial}{\partial\phi} & 0 & \mu\frac{\partial^2}{\partial\phi^2} - 1 \end{pmatrix} \mathbf{w}(\phi,t) \tag{10.85}$$

and

$$M(\mathbf{w}(\phi,t)) = \begin{pmatrix} \gamma_1^2\frac{\partial^2}{\partial t^2} & 0 & 0 \\ 0 & \lambda_1^2\frac{\partial^2}{\partial t^2} & 0 \\ 0 & 0 & \eta\gamma_1^2\frac{\partial^2}{\partial t^2} \end{pmatrix} \mathbf{w}(\phi,t) \tag{10.86}$$

Note that γ_1 and λ_1 are parameters γ and λ defined in (10.71) that correspond to the fundamental natural frequency of the unperturbed tube ω_1^*. Time is also normalized by the same frequency, i.e. $t = t * \omega_1^*$.

Operator M_a takes into account inertias of the sensors and actuator coils or magnets and is given bellow:

$$M_a(\mathbf{w}(\phi,t)) = diag\begin{pmatrix} \gamma_{s1}^2\delta(\phi-\phi_{s1}) + \gamma_{a1}^2\delta(\phi-\phi_a) + \gamma_{s1}^2\delta(\phi-\phi_{s2}), \\ \lambda_s^2\delta(\phi-\phi_{s1}) + \lambda_a^2\delta(\phi-\phi_a) + \lambda_s^2\delta(\phi-\phi_{s2}), \\ \gamma_{s3}^2\delta(\phi-\phi_{s1}) + \gamma_{a3}^2\delta(\phi-\phi_a) + \gamma_{s3}^2\delta(\phi-\phi_{s2}) \end{pmatrix} \frac{\partial^2\mathbf{w}(\phi,t)}{\partial t^2} \tag{10.87}$$

We define constants

$$\gamma_{s1}^2 = \frac{J_{s1}R\omega_1^{*2}}{EI_{1t}}, \quad \gamma_{a1}^2 = \frac{J_{a1}R\omega_1^{*2}}{EI_{1t}}, \quad \lambda_s^2 = \frac{m_sR^3\omega_1^{*2}}{EI_{1t}}, \quad \lambda_a^2 = \frac{m_aR^3\omega_1^{*2}}{EI_{1t}}, \quad \gamma_{s3}^2 = \frac{J_{s3}R\omega_1^{*2}}{EI_{1t}},$$

$$\gamma_{a3}^2 = \frac{J_{a3}R\omega_1^{*2}}{EI_{1t}}, \quad \lambda_{1f}^2 = \frac{\rho_fA_fR^4\omega_1^{*2}}{EI_{1t}}, \quad F_a = \frac{F_a^*R^2}{EI_{1t}}, \quad V_f = \frac{V_f^*}{R\omega_1^*} \tag{10.88}$$

Finally $e_2 = (0,1,0)^T$ in (10.83) is unit vector of the second (transverse) axis.

Similar equations hold for small circular arcs and the straight part at the front and end of the measuring tube, but are simpler because there are no the sensors and actuator.

The motion of the tube can be treated as a perturbation of the pure tube motion, which was considered in the previous subsection. To find its motion we will apply the method of eigenfunction expansion (Han et al. [19]). According to which the motion of the perturbed tube can be written in the form

$$\mathbf{w}(u,t) = \sum_{n=1}^{\infty} q_n(t)\mathbf{W}_n(u) \tag{10.89}$$

where $\mathbf{W}_n(u)$ are eigenfunctions of the unperturbed tube found previously and $q_n(t)$ are the generalized coordinates. The eigenfunctions have different forms on different parts of the tube, and u denotes the corresponding coordinates along the tube, which on the straight part is simply the tube length, and on the curved the corresponding central angle (Fig. 10.34).

We apply this equation to the dynamical equation of the tube, multiply the resulting equations by the eigenfunctions of the unperturbed tube and integrate along the tube. The procedure leads to a linear system of equations which agrees with the perturbed motion in the first n eigenmodes.

We approximate the dynamics of the bent tube by retaining the first three eigenmodes. We expect that this is a reasonable accurate model for the study of Coriolis mass flow meters under the control. We obtain a model of the perturbed tube having three generalized coordinates. The state of the system is defined by vector

$$\mathbf{q} = (q_1 \quad q_2 \quad q_3)^T \tag{10.90}$$

The model of the system can be put in the form

$$\mathbf{M}\ddot{\mathbf{q}} + \mathbf{B}\dot{\mathbf{q}} + \mathbf{K}\mathbf{q} = \mathbf{F} \tag{10.91}$$

where

$$\mathbf{M} = \begin{pmatrix} \mu_1 + m_{11} & m_{12} & m_{13} \\ m_{21} & \mu_2 + m_{22} & m_{23} \\ m_{31} & m_{32} & \mu_3 + m_{33} \end{pmatrix},$$

$$\mathbf{B} = \begin{pmatrix} 2\zeta_1\mu_1\omega_1 + 2\lambda_{1f}^2 V_f\alpha_{11} & 2\lambda_{1f}^2 V_f\alpha_{12} & 2\lambda_{1f}^2 V_f\alpha_{13} \\ 2\lambda_{1f}^2 V_f\alpha_{21} & 2\zeta_2\mu_2\omega_2 + 2\lambda_{1f}^2 V_f\alpha_{22} & 2\lambda_{1f}^2 V_f\alpha_{23} \\ 2\lambda_{1f}^2 V_f\alpha_{31} & 2\lambda_{1f}^2 V_f\alpha_{32} & 2\zeta_3\mu_3\omega_3 + 2\lambda_{1f}^2 V_f\alpha_{33} \end{pmatrix},$$

$$\mathbf{K} = \begin{pmatrix} \mu_1\omega_1^2 + \lambda_{1f}^2 V_f^2\beta_{11} & \lambda_{1f}^2 V_f^2\beta_{12} & \lambda_{1f}^2 V_f^2\beta_{13} \\ \lambda_{1f}^2 V_f^2\beta_{21} & \mu_2\omega_2^2 + \lambda_{1f}^2 V_f^2\beta_{22} & \lambda_{1f}^2 V_f^2\beta_{23}q_3 \\ \lambda_{1f}^2 V_f^2\beta_{31} & \lambda_{1f}^2 V_f^2\beta_{32}q_2 & \mu_3\omega_3^2 + \lambda_{1f}^2 V_f^2\beta_{33} \end{pmatrix},$$

$$\mathbf{F} = F_a(t)\begin{pmatrix} W_1(\phi_a) \\ W_2(\phi_a) \\ W_3(\phi_a) \end{pmatrix}$$

$$\tag{10.92}$$

Note that mass matrix $\mathbf{M}$ is composed of the modal masses μ_i of the unperturbed tube on its diagonal, and inertial parameters m_{ij}, which describe the inertial effect of the sensors and actuator bodies, and their coils and magnets. Due to parity of the eigenmodes these parameters satisfy: $m_{12} = m_{21} = m_{23} = m_{32} = 0$, and $m_{13} = m_{31}$.

The damping matrix $\mathbf{B}$ contains the linear damping terms $2\zeta_i\mu_i\omega_i$ on the diagonal, and the Coriolis force terms $2\lambda_{1f}^2 V_f\alpha_{ij}$. Due to parity of the eigenmodes αij satisfy: $\alpha_{11} = \alpha_{13} = \alpha_{22} = \alpha_{31} = \alpha_{33} = 0$, $\alpha_{12} = -\alpha_{21}$, and $\alpha_{23} = -\alpha_{32}$. Only the linear damping term is responsible for the dissipation of the tube mechanical energy. The Coriolis forces are in effect gyroscopic and are energetically neutral.

The third matrix $\mathbf{K}$ describes the tube stiffness terms of the form $\mu_1\omega_1^2$ on the diagonal and centrifugal terms of the form $\lambda_{1f}^2 V_f^2\beta_{ij}$. The stiffness terms are described as product of the tube modal masses and the square of the natural frequencies of the unperturbed tube. The centrifugal stiffness parameters βij has similar properties as the inertial parameters: $\beta_{12} = \beta_{21} = \beta_{23} = \beta_{32} = 0$, and $\beta_{13} = \beta_{31}$. Thus the centrifugal forces reduce the tube eigenfrequencies at higher fluid flows (they are of the order of the squared mean fluid velocity). The values of the parameters are not specified here because they depend on characteristics of the measurement tubes, the actuator and sensors employed. One set of these parameters are used in the simulation project discussed later.

10.6.4   Bond Graph Model of CMF Transducer

Using the model developed in the previous section it is a relatively simple problem to develop the corresponding Bond Graph model. We give a short description of it.

We introduce the following vectors

$$
\mathbf{q} = \begin{pmatrix} q_1 \\ q_2 \\ q_3 \end{pmatrix}, \quad
\mathbf{v}^* = \begin{pmatrix} v_1^* \\ v_2^* \\ v_3^* \end{pmatrix}, \quad
\mathbf{p}^* = \begin{pmatrix} p_1^* \\ p_2^* \\ p_3^* \end{pmatrix}, \quad
\mathbf{F} = \begin{pmatrix} F_1 \\ F_2 \\ F_3 \end{pmatrix} \tag{10.93}
$$

where $\mathbf{q}$ is vector of generalized coordinates and $\mathbf{v}^*$ is the corresponding vector of generalized velocities

$$
\mathbf{v}^* = \frac{d\mathbf{q}}{dt^*} \tag{10.94}
$$

The generalized momentum vector $\mathbf{p}^*$ is defined as

$$
\mathbf{p}^* = \mathbf{M}\mathbf{v}^* \tag{10.95}
$$

where $\mathbf{M}$ is the mass matrix given in (10.92). Because of the form of the first three eigenfunctions (see Fig. 10.35) some terms are zero. Thus, we have

$$\mathbf{M} = \begin{pmatrix} \mu_1 + m_{11} & 0 & m_{13} \\ 0 & \mu_2 + m_{22} & 0 \\ m_{31} & 0 & \mu_3 + m_{33} \end{pmatrix} \tag{10.96}$$

We can write (10.91) in un-normalized form as

$$\frac{d\mathbf{p}^*}{dt^*} + \mathbf{B}^*\mathbf{v}^* + \mathbf{K}^*\mathbf{q} = \mathbf{F}^*(t^*) \tag{10.97}$$

where

$$\mathbf{B}^* = \begin{pmatrix} 2\zeta_1\mu_1\omega_1^* & 2\lambda_{1f}^2\alpha_{12}\left(\frac{V_f^*}{R}\right) & 0 \\ 2\lambda_{1f}^2\alpha_{21}\left(\frac{V_f^*}{R}\right) & 2\zeta_2\mu_2\omega_2^* & 2\lambda_{1f}^2\alpha_{23}\left(\frac{V_f^*}{R}\right) \\ 0 & 2\lambda_{1f}^2\alpha_{32}\left(\frac{V_f^*}{R}\right) & 2\zeta_3\mu_3\omega_3^* \end{pmatrix} \tag{10.98}$$

Comparing with matrix $\mathbf{B}$ in (10.92) the eigenfrequencies are the un-normalized values in rad/s. The diagonal terms contain only the damping terms; the off diagonal terms are antisymmetric.

The stiffness matrix is given by

$$\mathbf{K}^* = \begin{pmatrix} \mu_1\omega_1^{*2} + \lambda_{1f}^2\beta_{11}\left(\frac{V_f^*}{R}\right)^2 & 0 & \lambda_{1f}^2\beta_{13}\left(\frac{V_f^*}{R}\right)^2 \\ 0 & \mu_2\omega_2^{*2} + \lambda_{1f}^2\beta_{22}\left(\frac{V_f^*}{R}\right)^2 & 0 \\ \lambda_{1f}^2\beta_{31}\left(\frac{V_f^*}{R}\right)^2 & 0 & \mu_3\omega_3^{*2} + \lambda_{1f}^2\beta_{33}\left(\frac{V_f^*}{R}\right)^2 \end{pmatrix} \tag{10.99}$$

It can be shown that the coefficients $\beta_{ii} < 0$, $(i = 1, 2, 3)$, and thus the centrifugal inertial forces effectively lower the tube stiffness, a well-known effect (Stack et al. [15]).

Finally, the actuator force is given by (see 10.92 and 10.71)

$$\mathbf{F}^*(t^*) = \begin{pmatrix} \frac{\omega_1^{*2}R^2}{EI_{1t}}W_1(\phi_a) \\ \frac{\omega_1^{*2}R^2}{EI_{1t}}W_2(\phi_a) \\ \frac{\omega_1^{*2}R^2}{EI_{1t}}W_3(\phi_a) \end{pmatrix} F_a^*(t^*) \tag{10.100}$$

To find the velocity measured by the sensors from (10.89) we may write for the main arc

$$\mathbf{w}(\phi_2, t) = \sum_{n=1}^{3} q_n(t)\mathbf{W}_n(\phi_2)$$

Thus un-normalized transverse displacement is given by

$$w^*(\phi_2, t) = R\sum_{n=1}^{3} q_n(t)W_n(\phi_2)$$

The transverse tube velocity is given by

$$V^*(\phi_2, t) = R\sum_{n=1}^{3} \frac{dq_n}{dt^*}W_n(\phi_2)$$

Hence, the tube velocities measured by the sensors are given by

$$V_i^* = R\left[v_1^* W_1(\phi_{si}) + v_2^* W_2(\phi_{si}) + v_3^* W_3(\phi_{si})\right], \quad i = 1, 2 \tag{10.101}$$

Similarly, the velocity of the point of application of the actuator force is given by

$$V_a^* = R\left[v_1^* W_1(\phi_a) + v_2^* W_2(\phi_a) + v_3^* W_3(\phi_a)\right] \tag{10.102}$$

We represent the model of the CMF transducer by the Bond graph component model shown in Fig. 10.36. Because this component will be used for modelling the behavior of the CMF under control it has only signal ports. Going from the left the first port serves for input of control signal generated by digital control circuits. Next two deliver voltage signals generated by the sensors. Finally, the last port gives information on the voltage across a shunt resistor in the actuator driver circuit, which gives information on the current in the actuator coil.

The model of transducer is shown in Fig. 10.37. It represents CMF transducer consisting of two parallel Measuring Pipes, driven by an Actuator. Two function components model the sensors. Thus the output of the left sensor is given by

$$Va = 0.5 * (Vs1\_up - Vs1\_dn) * Ksen * 1000 \tag{10.103}$$

Fig. 10.36 Coriolis mass
flowmeter component model

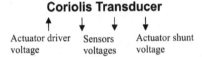

Coriolis Transducer

Actuator driver   Sensors   Actuator shunt
voltage   voltages   voltage

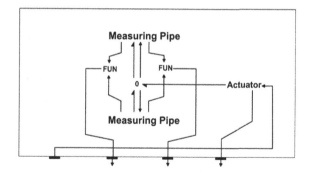

Fig. 10.37 Model of Coriolis Transducer component

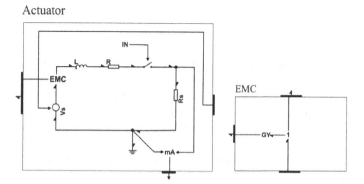

Fig. 10.38 Model of Actuator

where *Ksen* is the sensor gain in V/(m/s), and is expressed in mV. *Vs1_up* and *Vs1_dn* are the velocities of the pipe at position of the sensor on upper and downward measuring pipe. The analogous statement holds for the right sensor.

A model of Actuator is shown in Fig. 10.38. The L and R represent the actuator coil inductance and resistance, and EMC models electro-mechanical conversion between the electrical and mechanical actuator ports. As shown in Fig. 10.38 right it is represented simply by a Gyrator, similarly as in other electrical motors. The gyrator ratio is *Kemc* expressed in N/A or V/(m/s). The *Rs* is a shunt used to get voltage in mV, which gives information on the current flowing through the actuator coils.

A model of the Measuring Pipe is shown in Fig. 10.39. The I, R, and C components represent three ports inertial, resistive and capacitive components describing the corresponding terms in (10.95) and (10.97). The 1 represents three 1-junctions, which describes vector summation implied by (10.97). Similarly, the SE component consists of three controlled SE (source effort) elementary components, which represent the actuator force vector given by (10.100). The control signal

Fig. 10.39 Model of
Measuring Pipe

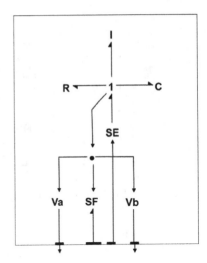

is the force signal, generated by the `Actuator`, which is picked at 0-junction in the `Coriolis Transducer` component model (Fig. 10.37).

In the same vein `Va` and `Vb` components consist of three functions and a summator, which implements (10.101). The input signals are the velocity components picked at 1-junctions in component 1. The dot denotes the component, which represent branching out the velocities signals.

Finally there is a `SF` component (compare with the central part of Fig. 10.37). It has similar structure as `Va` and `Vb` components and evaluates the velocity of the point where the actuator is connected as given by (10.102). In spite that we do not explicitly use this quantity for the control it is necessary to evaluate this in order to correctly interconnect the I-R-C model of the pipe and the `Actuator`. Note that generalized coordinates q in (10.93) are not the physical quantities, but convenient dimensionless quantities.

10.6.5 Control of CMF Transducer

The control of CMF transducer is a mysterious field. Practically there is not much information in the literature on this subject. Here we will develop a control algorithm borrowing some ideas from *Phase Locked Loops* (PLL) field (see e.g. Best [21]).

A PLL is a circuit synchronizing an output signal (generated by an oscillator) with an input signal both in frequency and phase. In the synchronized state, usually called the *locked* state, the phase error between the oscillator output and reference input is zero or remains constant. PLL consists of three basic blocks: a voltage-controlled oscillator (VCO), a phase detector (PD), and a loop filter (LF).

Fig. 10.40 Scheme of a PLL

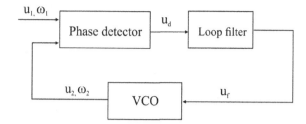

There are various kinds of PLL. We consider here so called all digital PLLs (ADPLLs) following Best [21]. They have similar structures as the other kinds of PLLs, but are realized entirely digitally. A typical scheme of PLL is shown in Fig. 10.40. We left out the optional divide-by-N counter for frequency down-scaling. There are different kinds of PD, which are based on different principles. It generates at its output a signal proportional to the phase difference between the reference input and VCO generated signals. The LF is really a controller, typically of PI type.

The radian frequency ω_2 generated by VCO is proportional to output u_f of the LF,

$$\omega_2 = \omega_0 + K_0 u_f \tag{10.104}$$

where ω_0 is so the center frequency of VCO, and K_0 is VCO gain in (rad/s)/V.

We now return to the problem of CMF control. It has basically two goals:

- To drive system from some start up frequency into close vicinity of the fundamental natural frequency of the measuring tubes and to hold it there.
- To drive system in such way that amplitude of voltages measured by sensor has a predefined value.

To gain more information on behavior of the CMF transducer in the vicinity of fundamental frequency we approximate its steady state frequency response by that of a simple harmonic oscillator. Note that we are not interested in the transverse displacements, but in their velocities, which leads the first by $\pi/2$. The corresponding frequency response is shown in Fig. 10.41 for a high value of Q factor of 5000 (corresponding to the damping coefficient of 0.0001). At the resonance the velocity is in phase with driving force, i.e. the phase difference is zero. In vicinity of the resonance the phase-frequency curve is very steep. It can be shown that tangent to the curve at the resonance has a gain of $-2Q$. A half of the bandwidth (normalized by the fundamental frequency) is $1/(2Q) = 0.0001$, and phase difference at the end of the bandwidth zone is $\pm\pi/4$. Therefore, the linear region around resonance is very short. The amplitude starts rising when the frequency of driving signal become close to the fundamental natural frequency.

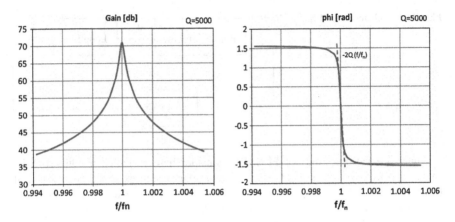

Fig. 10.41 Amplitude and phase frequency response in vicinity of resonance

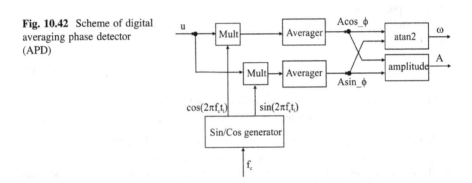

Fig. 10.42 Scheme of digital averaging phase detector (APD)

It should be noted that this behavior is generally masked by transients, which exist at the start of the motion. Due to very low damping in high quality CMFs the transients die very slowly (e.g. for about 5–10 s).

To find phase of sensor signals, as well their amplitudes, we will apply a digital averaging phase detector described in Best [21, Fig. 11.5]. As a DCO (Digital Controlled Oscillator) we use a Sin/Cos generator based on the recursive discrete-time sinusoidal oscillator of Turner [20]. The scheme of the detector is shown in Fig. 10.42.

Sin/Cos generates on its outputs in each sampling instant t_i the quadrature signals $\cos(2\pi f_c t_i)$ and $\sin(2\pi f_c t_i)$, where f_c is the input frequency in Hz. The input signal is multiplied by these signals and the products are collected in a buffer and averaged over a suitable time interval. As the averaging interval one full period of the quadrature signals is selected. If f_s is the sampling frequency of the corresponding analog to digital converter (ADC), then the number of sampling instances N over which the accumulated products are averaged is equal to the smallest integer not less than f_s/f_c, i.e. *ceiling(f_s/f_c)*.

Denoting by n the current value of the time the averaged values are calculated simply as

$$A \cos \_\varphi = \frac{1}{N} \sum_{i=0}^{N-1} u_{n+i} \cos(2\pi f_c t_{n+i})$$

$$A \sin \_\varphi = \frac{1}{N} \sum_{i=0}^{N-1} u_{n+i} \sin(2\pi f_c t_{n+i})$$

(10.105)

It is also possible to extend the averaging process over several periods of the generated quadrature functions.

The amplitude of the input signal is calculated as

$$A = \sqrt{(A \cos \_\phi)^2 + (A \sin \_\phi)^2}$$

(10.106)

and the phase by

$$\phi = \text{atan2}(A \sin \_\phi, A \cos \_\phi,)$$

(10.107)

Because atan2 function is not implemented in *BondSim*, the following formula based on tangent of a half angle is used

$$\text{atan2}(y, x) = 2 \, arctg \frac{\sqrt{x^2 + y^2} - x}{y}, \quad y \neq 0$$

(10.108)

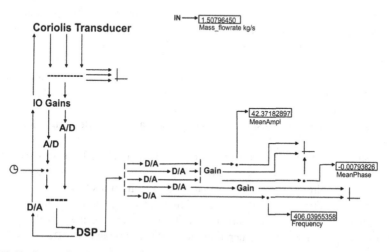

Fig. 10.43 System level model of CMF

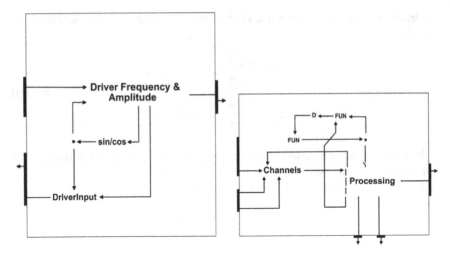

Fig. 10.44 DSP component (*left*) and `Driver Frequency & Amplitude` component (*right*)

We will now describe the proposed model of CMF control system. Figure 10.43 shows BondSim model of the complete CMF system. Starting from the top of the figure we have `Coriolis Transducer` component, which we have discussed in the previous section. There are three output signals—from the left and right sensors and shunt voltage we have already discussed. They are connected to a component from which they branch to the display. The sensors signals are further amplified in `IO Gains` component and then converted into the digital form by `A/D` converters. The clock component defines the sampling rate of the converters, and defines starting ending of the digital processing. The signals are further preceded to the `DSP` (Digital Signal Processing) component where main processing was done and which is the heart of the complete system. The `DSP` generates a voltage signal which is after `D/A` conversion and scaling, returned back to the `Coriolis Transducer`. It is used to drive the tubes actuator. The other signals are output as well, such as *Mean Amplitude, Mean Phase, Signals Phase Difference, Operating Frequency*, etc. These are first converted into analog domain by `D/A` conversion components, scaled and displayed. The phases are converted into degrees for display.

The structure of `DSP` component is shown in Fig. 10.44 (left). It consists of three main components. The `Driver Frequency & Amplitude` generates the frequency and amplitude for driving actuator of Coriolis transducer. The `sin/cos` generates the values of the quadrature signals at every sampling instance. It is implemented as described in Turner [20]. These values are used in the previous component and also for generating the driver voltage in `DriverInput` component.

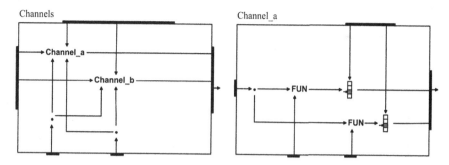

Fig. 10.45 Accumulation of the products

The generated signal has the form

$$U_c(t_i) = A_c \cos(2\pi f_c t_i) \qquad (10.109)$$

where amplitude A_c and the frequency f_c are generated by the Driver Frequency & Amplitude. The structure of Driver Frequency & Amplitude component is shown in Fig. 10.44 (right).

The component consists of component Channels, which calculates and stores the terms used for averaging operation in (10.105), and the component Processing where all processing is really done. The structure of the Channels is shown in Fig. 10.45 left. It consists of two Channel_a and Channel_b components, which process the voltage signals generated by the sensors and received at the DSP input port (Fig. 10.43).

The structure of e.g. Channel_a is shown in Fig. 10.45 right. The sensor voltage branches to the function components, which multiply the current value of the signal by the corresponding $\cos(2\pi f_c t_i)$ and $\sin(2\pi f_c t_i)$ values obtained from the bottom ports. The results are pushed into the buffers in form of the arrays discussed in Sect. 4.6.4. The buffers are reset at in the beginning of the current evaluation cycle by the top signals generated by the Processing (Fig. 10.44 right).

The components at the top of Fig. 10.44 right serve to trigger the processing. As (10.105) shows, it is necessary to calculate N product terms. The N depends on the current value of the operating frequency f_c and is evaluated in the Processing. The initial value of the counter (the memory component D) is set to $-N$. At every sampling interval the product of the velocity signals and sin/cos terms are evaluated and stored into the buffers as explained, and the counter is increase by one. The Processing has a trigger port, which activates the processing in the component when the counter becomes positive. The structure of the Processing is shown in Fig. 10.46.

The Channel_a and Channel_b evaluate the amplitude and phase of sensors signals as has been explained before (see (10.105)–(10.108)). Their structure is shown in Fig. 10.47.

Processing

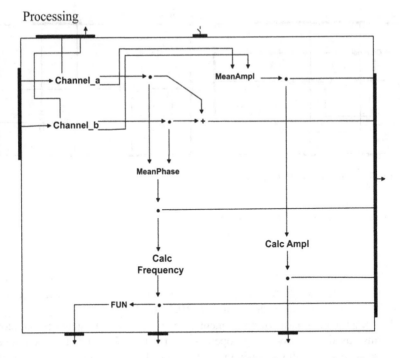

Fig. 10.46 Processing of the signals and generation of driving frequency and amplitude

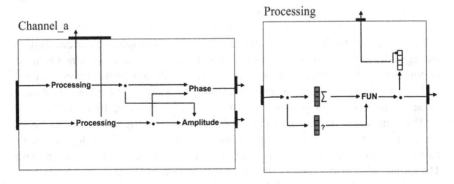

Fig. 10.47 Averaging and calculation of the signal amplitude and phase

As Fig. 10.47 right shows, to apply the averaging according to (10.105) three special digital components described in Sect. 4.6.4 are used—the buffer summation, buffer size and the reset operation. The first two serves to evaluate the sum of the products stored in the buffers, and the number of the products stored. Dividing these quantities the averages values of $A \cos \_\varphi$ and $A \sin \_\varphi$ are calculated. After these

Fig. 10.48 Change of mean amplitude and phase with time

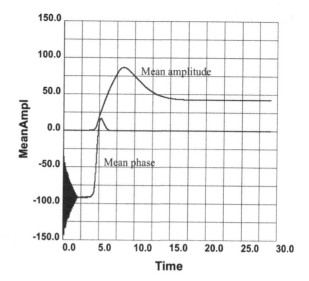

values are evaluated the reset signal is sent to the buffer to reset the buffer array indexes to zero (the array operations are zero based). The `Amplitudes` and `Phases` in Fig. 10.47 left are evaluated according to (10.106)–(10.108).

The function `MeanAmpl` in Fig. 10.46 evaluates the arithmetic mean of the sensors amplitudes. It is used as the input to `CalcAmpl` component, which represent a common *PID* controller, similar to that discussed in Chap. 8. This controller generates the amplitude A_c of the control signal (10.41), which drives the CMF actuator. Similarly, the `MeanPhase` evaluates the arithmetic mean of the sensors phases. It is used as the input to `CalcFreq` component, which represent a conventional *PID* controller. It generates the *operation frequency f_c*, which is used both in `sin/cos` generator (Fig. 10.44), and in the actuator driver voltage (10.109). The summator ('+') evaluates differences between phases of signal b and that of signal a. The generated signals are also sent to the right output port, and thus out of the `DSP` component (Fig. 10.43), where they are after post-processing sent to the displays as described before.

10.6.6 Simulation of CMF Control Loop

To show typical behavior of CMF under control a project *CMF system* is created following the approach described above. Because sampling frequency of A/D converters was about 52 kHz, the maximum step size was set to 9.5 μs, and the output interval to 0.0019 s. The simulation time was set to 40 s. The simulation was stopped after 30 s because all variables already have settled to their steady-state

Fig. 10.49 Search for natural
frequency

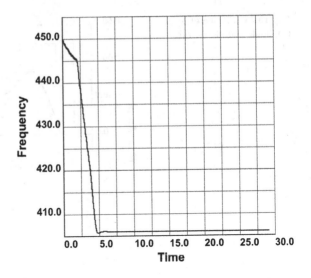

Fig. 10.50 Plot of the
sensors signal phase
difference in degrees

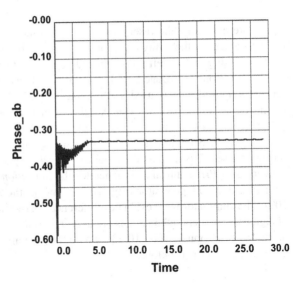

values. The complete simulation lasts about 2400 s of wall clock time.[1] The results
are shown in Fig. 10.48, 10.49, 10.50 and 10.51.

The value of the operating frequency was set to 450 Hz, and set point of the
mean amplitude to $30\sqrt{2} \approx 42.4264$ mV. Figures 10.48 and 10.49 show that
operating frequency becomes close to the fundamental natural frequency after 5 s,
and after about 6 s the mean signal phase drops practically to zero.

---

[1]The simulation was run on a laptop with i7 quad core processor under Windows 7.1.

Fig. 10.51 Change of the
1.8 ohm shunt voltage in mV

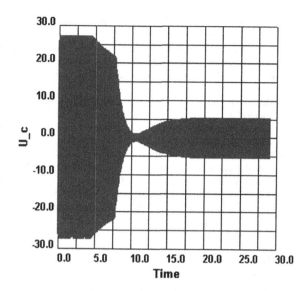

The mean amplitude starts rising only when the frequency approaches the fundamental natural frequency and settles down to value very close to the required set point after about 20 s. This is due to very slow transient of the tubes. The operating frequency settled to about 406.039 Hz, which agrees well with the calculated value of 406.2 Hz.

From Fig. 10.50 it can be seen that the sensors phase difference very quickly settles to a constant value of −0.3261°. Because the mass flowrate is 1.508 kg/s this means that the sensitivity of the transducer is 0.216°/(kg/s). Finally Fig. 10.51 shows that maximum current drawn by the actuator was 15 mA, and in the steady-state it drops down to only a few milliamps.

It could be concluded that the averaged phase detector works very well. The resulting curves are pretty smooth and regular. By applying a suitable quick search algorithm it is possible to shorten the starting time for operating frequency to reach the natural frequency, i.e. the first 4–5 s. However, the transients dominate the amplitude response.

References

1. Fahrenthold EP, Wargo JD (1994) Lagrangian bond graphs for solid continuum dynamics modeling. ASME J Dyn Syst Meas Control 116:178–192
2. Fahrenthold EP, Venkataraman M (1996) Eulerian bond graphs for fluid continuum dynamics modeling. ASME J Dyn Syst Meas Control 118:48–57
3. Karnopp DC, Margolis DL, Rosenberg RC (2000) System dynamics: modeling and simulation of mechatronic systems, 3rd edn. Wiley, New York
4. Schiesser WE (1991) The numerical method of lines. Academic Press, San Diego

5. Cook RD, Malkus DS, Plesha ME (1989) Concept and applications of finite element analysis, 3rd edn. Wiley, New York
6. Johansson J, Lundgren U (1997) EMC of telecommunication lines. A Master thesis from the Fieldbusters. http://jota.sm.luth.se/~d92-uln/master/Theory/4
7. Keown J (2001) OrCAD PSpice and circuit analysis, 4th edn. Prentice Hall, Upper Saddle River
8. Shabana AA (1998) Dynamics of multibody systems, 2nd edn. Cambridge University Press, Cambridge
9. Borri M, Trainelli L, Bottasso CL (2000) On representation and parameterizations of motions. Multibody Syst Dyn 4:129–193
10. Jaram V (2001) Evaluation of bond graph based object oriented approach to determination of natural frequencies of packaging system elements. MS thesis, Department of Packaging Science of Rochester Institute of Technology, Rochester, New York
11. Rao SS (1995) Mechanical vibrations, 3rd edn. Addison-Wesley, Reading
12. Kesic P, Damic V, Ljustina AM (2000) The coriolis flowmeter for measurement of petroleum and its products. Nafta, Zagreb 51:103–111
13. Raszillier H, Durst F (1991) Coriolis-effect in mass flow metering. Arch Appl Mech 61:192–214
14. Plache KO (1979) Coriolis gyroscopic flow meter. Mechanical Engineering, March 1979, pp 36–41
15. Stack CP, Garnett RB, Pawlas GE (1993) A finite element for the vibration analysis of a fluid-conveying timoshenko beam. Am Inst Aeronaut Astronaut AIAA-93-1552-CP
16. Sultan G, Hemp J (1989) Modelling of a coriolis mass flowmeter. J Sound Vib 132:473–489
17. Irie T, Yamada G, Tanaka K (1982) Natural frequencies of out-of-plane vibration of arcs. Trans ASME 49:910–913
18. Howson WP, Jemah AK (1999) Exact out-of-plane natural frequencies of curved timoshenko beams. J Eng Mech 125:19–25
19. Han SM, Benaroya Haym, Wei T (1999) Dynamics of transversely vibrating beams using four engineering theories. J Sound Vib 225(5):935–988
20. Clay ST (2003) Recursive discrete-time sinusoidal oscillators. IEEE Signal Process Mag 101–111
21. Best RE (2007) Phase locked loops, 6th edn. McGraw-Hill Co., New York

Appendix

A.1 Installation of BondSim

The book uses the program *BondSim 2014* which can be downloaded from the program web page given below. The currently there are two versions of the software: BondSim 2014 Basic and Professional.

The Basic version is a limited version of *BondSim* program, which permits building the models of up to 1500 modelling components. This version is really designed to be used with the book. This is often enough for many problems. The readers can use methods explained in the book for developing their own mid-size modelling projects. For comparison Puma 560 project of Chap. 9 contains about 1200 components.

The professional version, which is not free, has no restrictions on the size of the modelling projects. In addition it includes *BondSimVisual* program, which jointly with *BondSim* supports visualization of motion of mechatronics system in 3D space, as discussed in Chap. 9.

The programs can be run on PC computers under Windows 7, 8 or 10 operating systems. Prior to the installation they require the .NET Framework version 4.5. If it is not already contained as a part of the operating system, the framework can be freely downloaded from the Microsoft Windows Update website.

The complete instructions on download and installation of BondSim program can be found on the web page given below.

A.2 Launching and Using BondSim

BondSim can be launched as any other Windows application, e.g. by double-clicking the *BondSim 2014* icon on the desktop, or by using *Start button*, select *All Programs*, and then choosing *BondSim 2014*. Because the program internally changes the computer at the start the operating system usually asks the user if she or he want to allow such changes. If the user clicks the yes the program opens and the user can continue with the work.

© Springer-Verlag Berlin Heidelberg 2015
V. Damić and J. Montgomery, *Mechatronics by Bond Graphs*,
DOI 10.1007/978-3-662-49004-4

The book explains how to use the program. Chap. 4, for example, describes the visual environment, the main program commands, editing tools, and how to develop the Bond Graph models. Modelling and simulation of practical mechatronic systems using *BondSim* is described in the application part of the book, starting with Chap. 6. The first section of these chapters describes the procedure on relatively simple problems in a step-by-step manner. The later sections treat more demanding problems.

Interested readers can order a full version of the program by visiting the program web page. We would also be very happy for any feedback, criticism, advice or support for further work. Suggestions on other mechatronic or other relevant problems are truly welcome.

A.3   Contact Addresses

The authors can be reached at

vdamic@unidu.hr, info@bondsimulation.com
haedickemontgome@compuserve.de

The readers are welcome to visit *BondSim* program web page

http://www.bondsimulation.com.

Index

© Springer-Verlag Berlin Heidelberg 2015
V. Damić and J. Montgomery, *Mechatronics by Bond Graphs*,
DOI 10.1007/978-3-662-49004-4

Printed in the United States
By Bookmasters